MANAGING A MATERIAL WORLD

ENVIRONMENT & POLICY

VOLUME 13

The titles published in this series are listed at the end of this volume.

A C.I.P. Catalogue record for this book is available from the Library of Congress.

ISBN 0-7923-5153-3 (HB)
ISBN 0-7923-5206-8 (PB)

Published by Kluwer Academic Publishers,
P.O. Box 17, 3300 AA Dordrecht, The Netherlands.

Sold and distributed in North, Central and South America
by Kluwer Academic Publishers,
101 Philip Drive, Norwell, MA 02061, U.S.A.

In all other countries, sold and distributed
by Kluwer Academic Publishers,
P.O. Box 322, 3300 AH Dordrecht, The Netherlands.

Printed on acid-free paper

Printed in the Netherlands.

Managing a Material World

Perspectives in Industrial Ecology

An edited collection of papers based upon the International Conference on the Occasion of the 25th Anniversary of the Institute for Environmental Studies of the Free University Amsterdam, The Netherlands

Edited by

Pier Vellinga

Institute for Environmental Studies,
Free University Amsterdam,
Amsterdam, The Netherlands

Frans Berkhout

Science Policy Research Unit,
University of Sussex,
Brighton, UK

and

Joyeeta Gupta

Institute for Environmental Studies,
Free University Amsterdam,
Amsterdam, The Netherlands

KLUWER ACADEMIC PUBLISHERS

DORDRECHT / BOSTON / LONDON

Contents

Preface v

List of Abbreviations ix

1. Rationale for a physical account of economic activities 1
   ROBERT U. AYRES

2. Substance flows through environment and society 21
   MICHAEL CHADWICK

3. Towards the end of waste 31
   ROBERT A. FROSCH

4. Dematerialisation 45
   ERNST ULRICH VON WEIZSÄCKER

5. Analytical tools for chain management 55
   HELIAS UDO DE HAES, GJALT HUPPES & GEERT DE SNOO

6. Material flow accounts: definitions and data 85
   IDDO WERNICK

7. Environmental research and modelling 97
   HARMEN VERBRUGGEN

8. Software for material flow analysis 111
JOS BOELENS AND XANDER OLSTHOORN

9. Integrating life cycle assessment and economic evaluation 127
JANE POWELL, AMELIA CRAIGHILL AND DAVID PEARCE

10. Dematerialisation and rematerialisation 147
SANDER DE BRUYN

11. Aggregate resource efficiency 165
FRANS BERKHOUT

12. The industrial ecology of lead and electric vehicles 191
ROBERT H. SOCOLOW AND VALERIE M. THOMAS

13. Dissipative emissions 217
ROB VAN DER VEEREN

14. Recycling of materials 229
PIETER VAN BEUKERING AND RANDALL CURLEE

15. Philips Sound and Vision 239
JACQUELINE CRAMER

16. Unilever 251
CHRIS DUTILH

17. AT&T 259
BRADEN R. ALLENBY

18. Dow Europe 267
CLAUDE FUSSLER

19. The control of waste materials in Germany 275
ULF D. JAECKEL

20. Dematerialisation and innovation policy 285
J.L.A. JANSEN

21. The substance flow approach 297
MARIUS ENTHOVEN

22. The EU eco-label — 307
MARCO LOPRIENO

23. Towards industrial transformation — 321
PIER VELLINGA, JOYEETA GUPTA AND FRANS BERKHOUT

Index — 345

Preface

As we reach the end of the 20th century, the question of how to meet human needs and preferences while safeguarding the global environment is a major concern for many of us. If present trends are extrapolated into the future, they raise serious questions. Will our children, grandchildren and the generations thereafter be able to enjoy the quality of life we now strive for? Can the demands for food of 10 billion or more people be met while preserving biodiversity? Will all people have access to not only adequate food, potable water, clothing but also to housing, transport, electronics and recreation opportunities now common to the elite of the industrialised countries? Will the life support systems be stable and will nature continue to be the source of inspiration for all kinds of human activities?

Extrapolation of present trends in resource use and in resource efficiency increases may not be the best way of imagining future economic and ecological problems. History indicates that human ingenuity has helped to adjust resource use to environmental conditions. But not in all cases. There is also evidence that some cultures have collapsed as a result of the over-exploitation of natural resources.

One thing is for certain, a pre-condition for survival is a proper understanding of signals of change and vulnerability. Since the late sixties a series of signals has made society aware of the potential problems of over-exploitation and pollution of the earth's resources. In the nineties several concerns have come together to provide the basis for international efforts and new approaches to safeguard the global environment.

Concern about depletion of resources (Malthus revisited) first emerged in the late sixties and early seventies at the end of a period of high economic growth in the world economy. With technological change, the discovery of new reserves and slower growth rates, this concern diminished in the eighties.

In the nineties the focus has shifted from limits on the input side of the economy to limits on the assimilative capacity and resilience of global life support systems such as the ozone layer, the climate system, biodiversity, water quality and land use.

A second concern has been with pollution. Visible and noxious industrial emissions have been managed progressively through end-of-pipe and emergency measures. Over time this approach evolved into a reconsideration of the production process. In the 1990s management has moved beyond the process to include entire chains of material and substance flows. New heuristics whose policy implications are only now being understood have been developed to assess the resource and environmental burdens of production and consumption in a systematic way.

A third concern is about fulfilling the needs of a growing world population. In the event that the developing world follows a development path similar to that of the industrialised world, the volume of goods and services would grow by many orders of magnitude.

Theoretical analysis suggests that technological change and related changes in our economic incentive structure may be able to cope with the constraints of a finite set of resources at both the input and the output side of the economy. But how can we be sure that the changes required in production and consumption processes will come about? This calls for foresight, insight, creativity and most importantly, the political will to bring about the necessary changes.

How can humans ensure that the right choices are made in the trade off between present use and safeguarding options for future generations? How can we ensure that the quality of life of future generations is not degraded by present over-exploitation?

The major question at the threshold of the 21st century is whether the global market economy can deal with this challenge. Is our political system capable of including future needs in present day decisions?

Research institutes world-wide can help to clarify the issues and the choices. Moreover they can help to develop the necessary concepts, tools and instruments that help producers, consumers, their interest groups and governments to adjust their policies and practices. It is for this reason that the Institute for Environmental Studies of the Vrije Universiteit of Amsterdam has published this book on the occasion of its 25th anniversary.

This book is about the management of energy and material substance flows through environment and society. It starts with a series of chapters illustrating the need for a physical account of the economy which matches the more usual monetary account. These chapters show that the availability of environmental resources and the stability of the global life support systems are much too often taken for granted in our money economy. We need economic instruments and indicators that reflect the crucial role of the earth's natural capital. These chapters also illustrate that nature is full of paradoxes. For instance catastrophes are typical in nature, while this is what humanity will be trying to avoid.

The next series of chapters illustrate research issues and tools. These chapters deal with the question of value in environmental studies, and whether economic valuation can be reconciled with other systems of value. Economic valuation approaches are applied in life cycle assessment, which has previously avoided this form of assessment. The challenges and tensions between economic and ecological analysis and modelling are also discussed. They illustrate how value judgement based on social and time preferences will always be an important element in the development of rules and incentives.

These chapters are followed by a set of chapters that demonstrate the value of a systems approach in environmental analysis. A series of widely-held beliefs are tested in the form of testing hypothesis. Of particular interest is the finding that the shift from an industrial economy to a services economy does not automatically lead to a decrease of the materials intensity in the economy. It also illustrates that dematerialisation, as an overall guiding principle, can lead to controversial and not necessarily environment friendly results. Finally, it shows that recycling of materials on an international scale may be both economically and environmentally beneficial.

The book continues with case studies about progress in the development of resource efficiency policies by the private sector and governments. Case studies written by the representatives of Dow Europe, Unilever, Philips Sound and Vision and AT & T reflect the state of the art in industry. The case studies illustrate how environment has become a strategic element in the design of processes, products and marketing. The case studies of government initiatives deal with closed cycle management in Germany, product innovation processes in the Netherlands and life cycle based product policies in the European Union.

The final part of the book is about policy developments in the national and international context. The concluding chapter sums up the key issues covered in the book and argues in favour of comprehensive research activities covering the relevant aspects of industrial transformation. We need a better understanding of how the macro-incentive structure (including fiscal, trade and liability aspects) effects production and consumption systems. The physical throughputs associated with industrial activities need to be considered through the framework of industrial ecology. Industrial Transformation research should particularly consider technological innovation and industrial ecology. A much better understanding of the patterns and dynamics of consumption is also needed, including research into consumers needs and preferences and how they are shaped by available infrastructure, logistics, marketing and advertisement.

This book has been written by a team that includes the most forward thinking scientists on the issue. It is a handbook for all those involved in decision making about product design and ecoefficiency, about environmental and resource use policies. Moreover it can serve as a handbook for all those who are studying with the aim to become involved in these issues. We take this opportunity to thank all the authors who devoted so much time and energy to write chapters for this book. The team of people at IVM who helped to work the raw chapters into finished products, are particularly recognised for their commitment and contributions. They are Joke Daamen, Michiel van Drunen, Fons Groot, Els Hunfeld, Dorine Lambrichts, Sylvana Rooseman and Hasse Goosen. Finally the financial support of the Vrije Universiteit USF-Programme and of the Netherlands Research Foundation is gratefully acknowledged.

The editors,

Pier Vellinga, Frans Berkhout and Joyeeta Gupta

List of Abbreviations

BAT	Best Available Techniques
BCSD	Business Council for Sustainable Development
B-mfa	Bulk-Material Flow Analysis
CGE	Computable General Equilibrium
CLTM	Dutch Committee for Long-term Environmental Policy
CML	Centre for Environmental Studies, (Leiden University, NL),
CSERGE	Centre for Social and Economic Research on the Global Environment
CSG	Council of State Governments (USA)
CST	Culture, Structure and Technology
DBMS	Database Management System
DFE	Design for Environment
DMI	Direct Materials Input
DRI	Direct Resource Input
DSD	Duales System Deutschland
ECC	Environmental Competence Centre
EFTEC	Economics for the Environment Consultancy
ERM	Environmental Resources Management
EST	Efficient State of Technology
EU	European Union
EU-EMAS	(European Union) - Environmental Management and Audit Scheme
EV	Economic Valuation
EVA	Economic Valuation Analysis
GDP	Gross Domestic Product
GNP	Gross National Product
GWP	Global Warming Potential

HMSO	Her Majesty's Stationery Office
ICCET	Imperial College Centre for Environmental Technology
IEM	Instituut voor Europees Milieubeleid
IPPC	Integrated Pollution Prevention and Control
ISO	International Organisation for Standardisation
ISO-EMS	Environmental Management System
IVA	Royal Swedish Academy of Engineering Sciences
IVL	Swedish Environmental Research Institute, Stockholm.
LCA	Life Cycle Analysis
LCANET	Life Cycle Assessment Network
LCI	Life Cycle Inventory
MCP	Material Composition of Products
MFA	Materials Flow Analysis
MIPS	Material Intensity Per Service
MPC	Material-Product-Chain
MRF	Materials Recycling Facility
MS	Microsoft
NGO	Non-Governmental Organisations
NHANES II	National Health and Nutrition Examination Surveys
OBIA	Overall Business Impact Assessment
ODP	Ozone Depleting Potential
OECD	Organisation of Economic Cooperation and Development
PC	Personal Computer
PCD	Philips Corporate Design
PCI	Product Composition of Income
PEM	Proton-Exchange-Membrane
PIA	Product Improvement Analysis
PNGV	Partnership for a New Generation of Vehicles (USA)
R&D	Research and Development
RCRA	Resource Conservation and Recovery Act (USA)
RIVM	The National Institute of Public Health and Environmental Protection (NL)
SETAC	Society for Environmental Toxicology and Chemistry
SFA	Substance Flow Analysis
SLI	Starting Lighting Ignition
SNDF	South Norfolk District Council
STD	Sustainable Technology Development
STRETCH	**S**election of s**TR**ategic **E**nvironmen**T**al **Ch**allenges
TMR	Total Materials Requirement
UK	United Kingdom
UNCED	United Nation Conference on Environment and Development
UNEP	United Nations Environment Programme

UR	Unilever Research
US	United States
WBCSD	World Business Council for Sustainable Development
WRR	Scientific Council for Government Policies (the Netherlands)
WTA	Willingness To Accept
WTO	World Trade Organisation
WTP	Willingness To Pay
WWW	World Wide Web
ZEV	Zero-Emission Vehicle

Chemicals and Chemical Substances

Ca	Calcium
Cd	Cadmium
CFCs	Chlorofluorocarbons
CO_2	Carbon dioxide
Cu	Copper
DDT	Dichlorodiphenyltrichloroethane
Hg	Mercury
K	Potassium
N	Nitrogen
NH_3	Ammonia
NH_4^+	Ammonium
NO_3^-	Nitrate
NOx	Nitrogen oxide(s)
P	Phosphorus
PCBs	Polychlorinated Biphenyls
PE	Polyethylene
PVC	Polyvinylchloride
SO_2	Sulphur dioxide
Zn	Zinc

Chapter 1

Rationale for a physical account of economic activities

ROBERT U. AYRES
Sandoz Professor of Management and the Environment, The European Institute of Business Administration, Fontainebleau Cedex

Key words: mass flows, mass balances, physical-economic models

Abstract: This article argues that quantification, selection and simplification of information helps to make analysis of problematic situations easy and to forecast with some degree of reliability. Neo-classical economics has attempted to quantify and simplify the real world, but some of its simplifications have severe limitations. For example, current growth theory does not recognise that the factors of production can change over time. Environmental services tend to be excluded. These simplifications cast doubt about the relevance of economic models as a forecasting tool.

1. PREFACE: WHY QUANTIFY?

The logical starting point is a question: Why do we need to measure things? Why do we use numerical measures? The proper but pedantic answer is, roughly: To compare quantities more exactly than "big" or "small" or even "bigger than" or "smaller than". In the economic arena, we want to know what we have, what we have gained, what we have lost, where we are and where we are going. A realistic answer is cruder: in our world, that which can be measured precisely counts more — has more "weight" — than what is not measured. It has been said that "what gets measured gets done".

So far, so good. Accounting is the activity of compiling numerical measures of things or actions, in a form that makes comparison and analysis easy. Or, at least, possible. Accounting, in economic activities, is almost exclusively concerned with a single measure, namely money. Accountants keep track of money inflows, outflows, and balances. They perform this function, as a rule, for business and financial enterprises, governments and (increasingly) households. Accountants are essential to the operation of a modern economic system. They allow a businessman to know whether the business is profitable or not in a given period. They are the basis for determining tax liabilities. Comparing accounts over different periods the business man can see trends: he can determine whether the business is growing or shrinking, and how fast. Over longer periods it is possible to see changes in the trends. Is growth accelerating or decelerating? Are profits and /or inventories growing faster or slower than volume? Is the debt/equity ratio increasing or decreasing? This kind of knowledge, in turn, permits the businessman to make rational decisions about market strategy, investment, pricing, wage costs, borrowing, dividend payments and so forth.

The importance of monetary accounting for governments and families is comparable, although the specific questions and issues are different in detail. Governments worry about tax revenues, trade balances, debt, interest rates and long-term liabilities. Families worry about meeting current bills, credit card balances, mortgage payments, tax liabilities, college costs, and retirement. Small differences can be very important. As Dickens' Mr. Micawber said[1] "Annual income twenty pounds, annual expenditure nineteen pounds nineteen six, result happiness. Annual income twenty pounds, annual expenditure twenty pound ought six, result misery."

Monetary accounts are certainly important. If the number of accountants and the fees of accounting firms are a basis for judgement, the profession is growing in importance. But, as we know all too well, money is not the measure of everything. Some things are literally priceless. There is no way to put a price on life, or liberty, or on human happiness.[2] More relevant to this book, there is no good way to assign monetary values to most of the essential services provided by the environment, ranging from benign climate, breathable air and fresh water to biodiversity, nutrient recycling and waste assimilation.

Yet, many environmental features, can be quantified in terms of other, non-monetary measures. In the absence of reliable monetary measures (i.e. prices) such quantification is all the more necessary for making rational policy decisions. Just as the businessman needs to know whether the business that supports him is profitable and growing or unprofitable and declining, and how fast, we as a society need to know how the environment that sustains all life — and, especially, human life — is doing. Is it healthy and stable against perturbations? Is it prospering? Or is it sickening, perhaps to the verge of collapse?

These questions are difficult and deep. In the first place, they can only be answered when we understand very clearly and precisely what we mean by terms like "healthy" or "sick" as applied to the environment. Such an understanding requires scientific knowledge, which implies the development of a body of theory based on "hard" (i.e. quantifiable) observational data. Such analysis requires knowledge of the current and changing thermodynamic state of the air, the water, the soil, the weather and climate, and the biota, of course. The thermodynamic state encompasses temperatures, pressures, humidity, wind velocity, chemical and physical composition, and radiation spectrum, at all locations and times.

Yet direct measurement alone is not enough and never can be enough for purposes of forecasting the future. The environment is too complex and heterogeneous to understand in terms of quantification, without accompanying theory. In any case, there can never be enough instruments to make all those measurements. Even if there were, the very existence of so many measuring devices would interfere with and modify the system being measured. (This is a version of the famous Heisenberg "uncertainty principle".) Moreover, as information scientists have begun to realise, there can never be a digital computer large enough to record all the data, still less analyse the multiple statistical correlations to ascertain causal relationships and generate accurate forecasts. And if such an all-powerful computer could exist, its programs would be more complex than the world itself, it's operation would require more energy than the economy can spare, and it would be inherently incapable of processing data in "real time". For these reasons, "brute force" data analysis is not the final answer. Data must be gathered, to be sure. But it must be averaged and aggregated over various spatial and time scales, and it's significant information content must be enhanced in other ways.

The essence of real science, therefore, is selection and simplification. Vastly simplified world-models are needed, at first. The trick is to build deeper understanding step-by-step by starting with simple models that explain the most fundamental phenomenon, adding complexity (or making changes in the assumptions) only when the existing models are clearly inconsistent with the reality they are attempting to explain.

2. DE-SIMPLIFICATION IN ECONOMICS

Neoclassical economics is a prime example of this evolutionary process in action. It has proved fruitful in explaining certain fundamental social phenomena, namely the operation of exchange markets under various conditions. However one of the major simplifications in economic theory, as it developed in the 19th century, was that production and consumption constitute a closed circle, as illustrated in *Figure 1a.* This misleading circularity implies, among other things, that production and consumption are immaterial i.e. that the economic system produces *only* services, not products (e.g. Ayres & Kneese 1969). Production functions that are widely used in neoclassical economics assume unlimited substitutability between factors of production, whether capital and labour, or capital labour and resources.[3] The not-so-hidden implication is that capital can be produced by labour, or that resources can be produced by labour and/or capital. Mainstream neoclassical theorists economists explain, of course, that they merely mean that resource inputs at the margin can be reduced by judicious investment of capital or labour. So be it.

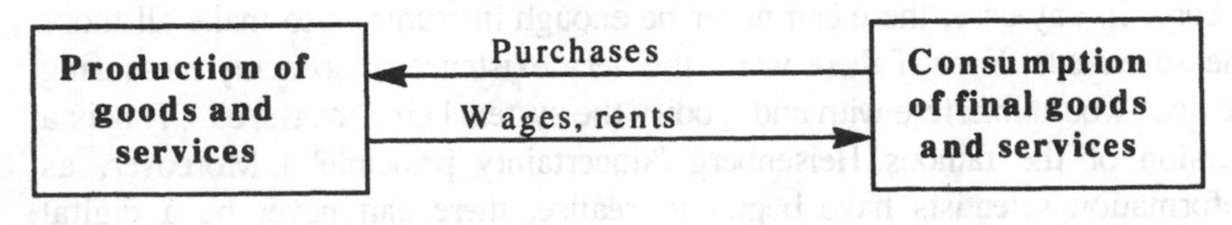

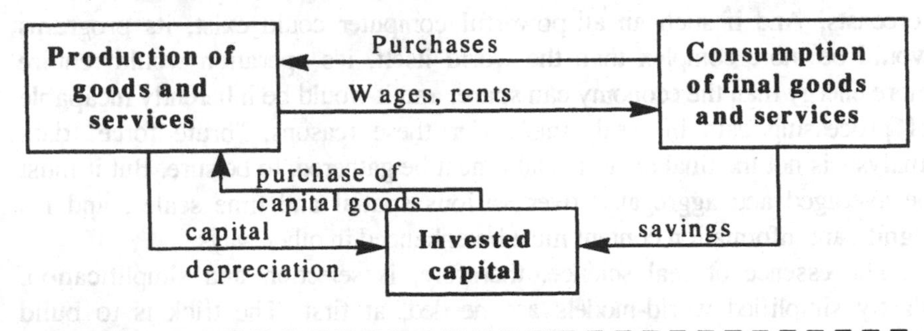

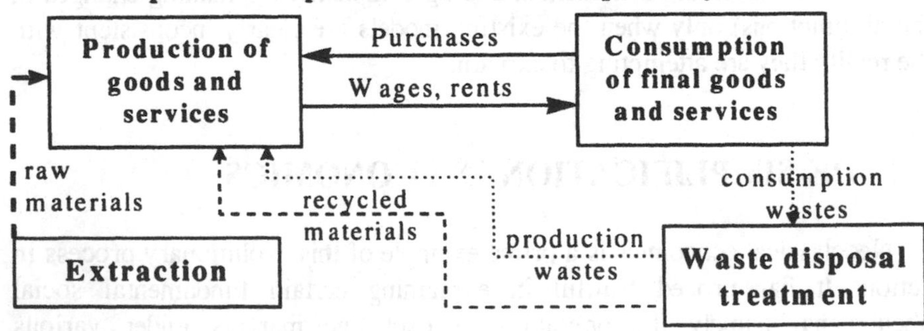

Figure 1. Closed and open production consumption system

Of course, the simple one-sector closed (static) model of *Figure 1a* has been extended in a variety of ways, by adding investment to make it dynamic,

Figure 1b and by "opening" it to include extraction and waste disposal services as environmental linkages (*Figure 1c*). The open version can, of course, also be dynamic.

However, despite the added complexity, it must be said that insights from this oversimplified theory about the production and consumption of pure services can tell us nothing important about the relationships between man and the natural environment, since the intermediate link between economic activity and material resources has been left out. The well-known "Limits to Growth" study, sponsored by the Club of Rome, attempted to use a different and slightly less simple model to demonstrate the vulnerability of the natural environment to human economic activity (Meadows *et al* 1972, 1974). The Meadows group neglected to start from the established and accepted neoclassical paradigm, however, or even to acknowledge it's relevance. The Meadows' conclusions, based on a different set of economic assumptions, were immediately challenged by a number of neoclassical theorists, primarily on the grounds that the potential for resource substitutability (by technology and capital) had been ignored.[4]

Yet, economics is mostly about goods and markets, after all. "Any good that cannot be bought or sold in exchange for money is out of the domain and interest of economics" (Nebbia 1981).[5] A new body of theory is needed to repair the omission. It must encompass the features of neoclassical economics, since economic activity has an important impact on the environment, and *vice versa.* Human impacts on the environment are, *ipso facto*, caused by land-use changes and by the economic activities of extraction, concentration, reduction, transformation, manufacturing, consumption and disposal of waste materials (and energy). But since man-nature interactions are not pure monetary transactions governed by market mechanisms, in general, they cannot be understood in terms of a pure market model of the world. Other elements must be added, also.

Actually, one branch of economics has always considered physical flows of commodities as a starting point. It began with Quesnay's simple "tableau" model of the French economy in the 18th century. Interindustry accounting was revived by the early Soviet central planners in the 1920s, and elaborated especially by Wassily Leontief (a Russian émigré) and his followers in the U.S. However, input-output economics (so called), has been limited to material and energy flows that were accompanied by economic transactions.

All materials extracted from the natural environment are potential wastes (Ayres & Kneese 1969, 1989). Most become actual wastes in a short time. Wastes have traditionally been regarded as economic zeros, and waste disposal has been treated as a "free good" in traditional neoclassical theory. This approximation was perhaps justifiable in the past, when population densities were low and industrial activities were small and localised. It is no longer justifiable, however. The new elements that must be added to the economic models are material stocks and flows, and physical (i.e. thermodynamic) and biological relationships, including relationships between pollution and productivity.

The step-by-step approach to theory development, starting from the circular flow model, would seem to involve the addition of subsidiary loops and feedbacks of a partially non-monetary nature. Something tangible must flow apart from money. During the 1970s there was also a significant — if short lived — movement, supported mainly by engineers and natural scientists, promoting "net energy analysis". It's more modest proponents thought of net energy analysis as a tool for screening energy supply investment projects, such as nuclear power plants, solar satellites, and the like in terms of the ratio of net energy output to energy inputs over a project life. It was, in effect, an application of what we now call Life Cycle Analysis, or LCA.[6]

Meanwhile, input output economists in the early 1970s worked intensively on developing materials and energy input matrices and waste/pollution output matrices linked to standard I-O models.[7] Some linear programming activity-type models were also developed, including engineering components, mainly for regional (e.g. river basin) assessments and energy supply analysis. There were some attempts to extend and elaborate the rather abstract "activity analysis scheme introduced by Koopmans (Koopmans 1951), to incorporate more realistic forms of process analysis (e.g. Manne & Markowitz 1961; Nordhaus 1973; Ayres & Cummings-Saxton 1975). Most mainstream economists will understandably tend to regard these modifications and add-ons as peripheral to the main body of theory.

Nevertheless, hybrid engineering-economic "optimal growth" models with linear programming cost-minimisation modules[8], made their appearance in the mid-1970s, ostensibly to help governments formulate long-term energy policy in response to the 1974 "energy crisis" e.g. (Cherniavsky 1974), (Hudson & Jorgenson 1974), (Haefele *et al* 1976), (Hoffman & Jorgenson 1977), ETA-MACRO developed by Alan Manne and co-workers (Manne 1977, 1979), MARKAL (Fishbone *et al* 1983) and so on.

The engineering-economic optimal growth modelling tradition took on new life as the problem of climate warming due to greenhouse gas emissions began to attract the attention of governments, resulting in the creation of the Intergovernmental Panel on Climate Change (IPCC). This body needed to assess alternative long-term emission scenarios in the context of various regulatory regimes. The obvious (albeit not necessarily appropriate) tools were at hand: second and third generation versions of the earlier engineering-economic models, e.g. EFOM (Van der Voort *et al* 1985), MESSAGE, (Messner & Strubegger 1987, 1995), ETA-MACRO (III) (Manne & Richels 1992), ETSAP (Kram 1993) and MARKAL-MACRO (Manne & Wene

1994) are currently used for purposes of assessing long-range costs of greenhouse gas regulation.[9] I will have more to say on this subject later.

For most readers of this book, the "missing link" is the explicit role of mass flows. The energy-economic models mentioned in the previous paragraph utilise mass flows only implicitly (inside the L-P module) except in one place. The least-cost combination of energy production technologies is selected for each level of demand, and the result is a GNP trajectory and an energy output trajectory. But, for the climate-warming problem, it is necessary to convert energy output into greenhouse-gas (GHG) output, which is what determines climate. This is done by attaching mass emissions coefficients to each of the energy production technologies. The climate change problem is not the only one that depends on mass flows. On the contrary, waste emissions (mass flows) are one of the two generic factors affecting the environment. (The other is land-use, which also involves physical and biological relationships). For instance, there is increasing concern over global acidification (resulting from excessive sulphur and nitrogen oxide emissions), global eutrophication from excessive use of nitrogenous and other fertilisers, and "toxification" of soils and groundwater from excessive mobilisation of chlorinated chemicals and heavy metals. All of these phenomena are related to mass flows.

These mass flows are generated by economic activity. So much is clear. In the tradition established by the energy-economic models one need only attach mass-coefficients to the appropriate activity levels, to forecast the associated mass flows. This part might be difficult and tricky in practice, perhaps, but it is straightforward in principle. But what about the impact of these mass flows on the economy and on welfare? How do they affect productivity? Does the economic system still approach some sort of quasi-Walrasian equilibrium in the presence of these non-economic transactions? If so, how? If not, what policy interventions would eliminate the externalities? And, since the system is presumably in or near a general equilibrium now (albeit the "wrong" one) what economic consequences would these policy interventions have? Would growth be accelerated or decelerated? Would unemployment increase or decrease? These are all key questions.

Some economists may assume that all of these questions can be addressed without specific knowledge of the physical flows and transformations. I doubt it very much. The core issue can be restated thus: how far (and in what specific directions) has the existing economy departed from the Pareto-optimal competitive equilibrium that would exist if all environmental resource/service flows were paid their appropriate shares of the social product? How much would each of those environmental services be worth if they were really exchanged (bought and sold) in a truly competitive free market? More succinctly, how big is the distortion arising from indivisibilities (public goods) and other externalities?

The answer that is often proposed boils down to "contingent valuation" (CV). This means, in practice, a more-or less sophisticated survey of consumers willingness-to-pay for a *gedanken* purchase of access to some desired environmental service or amenities and willingness to accept money for the *gedanken* sale (i.e. loss) of some such service. There are fundamental difficulties of interpretation with any survey approach, of course, which I need not review here. There is a further problem of extrapolating partial equilibrium data (even if it were absolutely reliable) to a general equilibrium situation. And there is a very fundamental problem that people cannot put a meaningful price on an environmental service to society as a whole, the personal impact of the absence of which they literally cannot imagine. People think they can imagine "climate warming" perhaps, and they might even be willing to pay a bit for warmer winters. But people far from the ocean will understandably have difficulty translating sea-level rise or *El Nino* frequency into personal terms. The nitrogen cycle is an even better illustration of this problem. Very few experts, if any, could put a meaningful price on the personal impact of massive eutrophication of coastal waters, for instance.

But suppose all these problems were overcome. Suppose it were possible by means of CV techniques to estimate the social cost of (for instance) a marginal ton of PCBs or organotins in the oceans, or a ton of arsenic or cadmium in the soil. It would nevertheless be necessary to also know the total quantities of these substances emitted by different activities. Otherwise we cannot know how to allocate the taxes or how to frame the regulations. Indeed, it is also important to know where and how these substances are used (and disposed of) for the same reasons. It is true that, if we knew the "right" prices for every commodity and waste, we could theoretically adjust the economy to maximise social welfare. But, as a practical matter, we cannot hope to determine the right prices by CV techniques alone.

What remains? I think that it is necessary to simulate a general equilibrium economy using not one but two kinds of information. CV is one of them. Mass flows and process data constitute the other. Shadow payments (shadow prices per unit times quantities) for an imaginary flow of *environmental* services must be matched (in equilibrium) by shadow costs of services purchased from the "rest of the economy" by the eco-service sector. The latter consists precisely of environmental protection (waste treatment and pollution abatement) and rehabilitation services. This equilibrium shadow economy may be using imaginary money, but the balancing condition for money (income equals outgo) must be satisfied even for imaginary currency. If the eco-service "sector" is to be paid for it's services, it must also pay for services rendered to it.

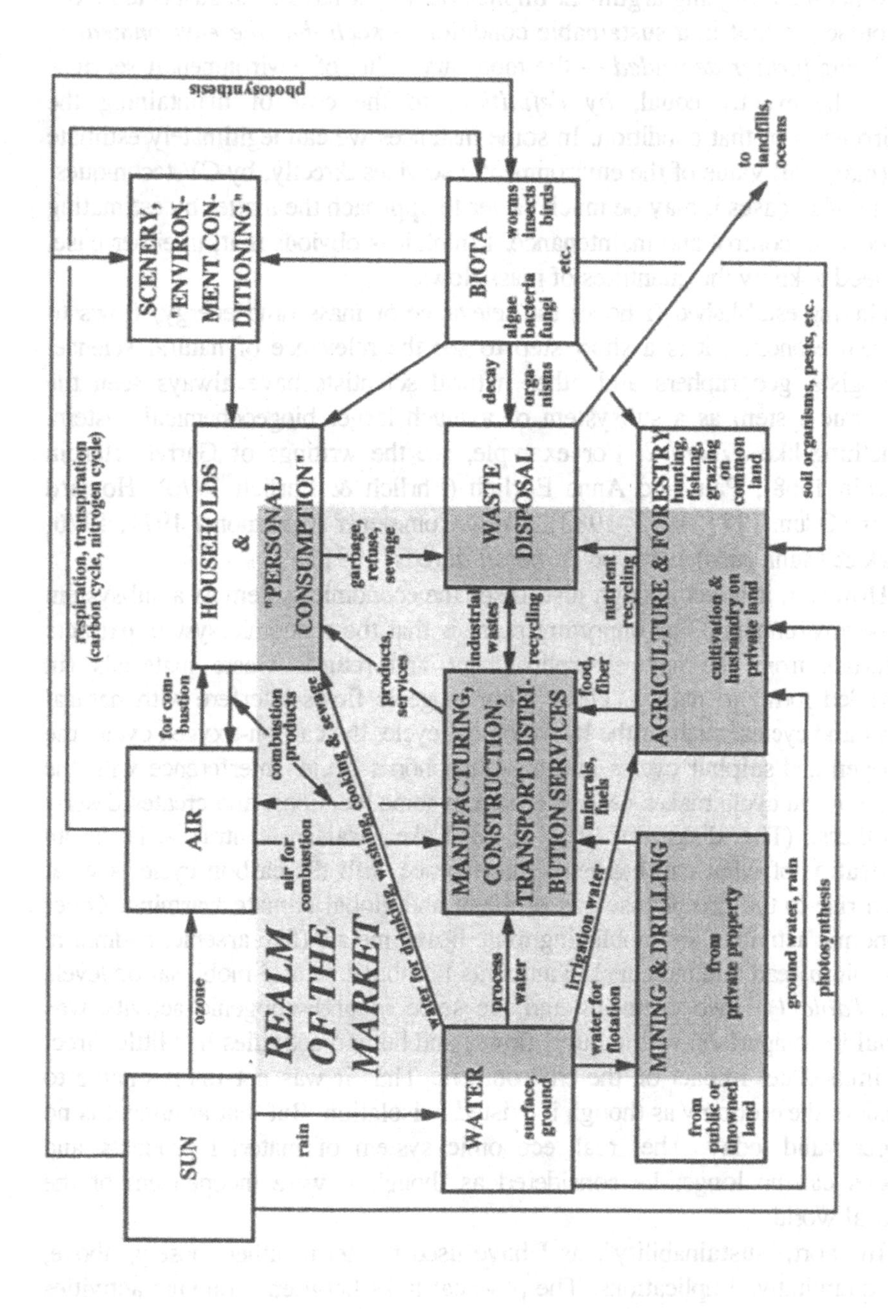

Figure 2. Realm of the market

What the foregoing argument implies (as might have been suspected from the outset) is that in a sustainable condition — *such that the environment is not being further degraded* — the monetary value of environmental services would be exactly equal, *by definition*, to the cost of maintaining the environment in that condition. In some instances we can legitimately estimate the (marginal) value of the environmental services directly, by CV techniques. But in other cases it may be much easier to approach the matter by estimating the costs of control and maintenance. I think it is obvious that, in either case, we need to know the quantities of mass flows.

Having established (I hope) the relevance of mass (and energy) flows to the real economy, it is a short step to see the relevance of natural science. Ecologists, geographers and other natural scientists have always seen the economic system as a subsystem of a much larger biogeochemical system, something like *Figure 2*. For example, see the writings of Garrett Hardin (Hardin 1968), Paul and Anne Ehrlich (Ehrlich & Ehrlich 1970), Howard Odum (Odum 1971, 1973, 1983), Barry Commoner (Commoner 1971, 1976; Clark & Munn (eds) 1986; or Turner *et al* (eds) 1991).

However, it is not enough just to see the economic system as a subsystem of the environment. The important point is that the economic system extracts materials from the natural environment and returns waste materials (in degraded form) to nature. These anthropogenic flows interfere with natural flows and cycles, such as the hydrological cycle, the carbon-oxygen cycle, the nitrogen and sulphur cycles and the phosphorus cycle. Interference with the hydrological cycle makes deserts bloom in some locations, and creates deserts in others. (The disastrous decline of Lake Aral in central Asia is an illustration of what can happen). Interference with the carbon cycle is what gives rise to the greenhouse gas problem and global climate warming. Other economic activities are mobilising toxic heavy metals (like arsenic, cadmium, chromium, lead and mercury) in amounts far above natural mobilisation levels (see *Table 1*). Two centuries ago the scale of anthropogenic activity was trivial in comparison with natural flows, and human activities had little direct (or immediate) impact on the environment. Thus it was not unreasonable to consider the economy as though it existed in isolation. But that argument is no longer valid today. The `real' economic system of material products and wastes can no longer be considered as though it were independent of the natural world.

In short, "sustainability", as I have used the term rather loosely, above, has quantitative implications. The physical links between economic activities and the environment can no longer be ignored. They include both resource extraction and waste disposal, on the one hand, and environmental impacts on economic activities (especially agriculture, of course) on the other. I have

Table 1. Indicators of unsustainability (ratios)

Metal	Anthropogenic flow to natural flow		Cumulative extraction to top soil inventory Azar et al. 1994
	Galloway et al. 1982	Azar et al. 1994	
Antimony (Sb)	38	6.0	-
Arsenic (As)	4	0.33	-
Cadmium (Cd)	20	3.9	3.0
Chromium (Cr)	2	4.6	2.6
Copper (Cu)	14	24.0	23.0
Lead (Pb)	333	12.0	19.0
Mercury (Hg)	-	6.5	17.0
Nickel (Ni)	4	4.8	2.0
Selenium (Se)	5	2.0	-
Vanadium (V)	3	0.32	-
Zinc (Z)	23	8.3	6.9

used the term "industrial metabolism" to convey the essence of the interaction (Ayres 1989; Ayres & Simonis 1994). Others have used the more or less equivalent term "industrial ecology" (Billen *et al* 1983; Socolow *et al* 1994; Graedel & Allenby 1995).

Perhaps the very first explicit acknowledgement of the importance of material flows at the highest level of government, was the creation in 1992 of a so-called Enquete Commission by the German Bundestag. Its mandate was

- To identify the most important problems involved in industrial substance cycles, including their historical background, and to propose possible solutions;
- To develop scientifically founded and socially acceptable assessment criteria for comparative life cycle analyses;
- To assess applications, larger substance groups, and finished products, both from the perspective of the manufacturing and processing industries, and from the perspective of the final consumer;
- To describe potential alternative development scenarios for the production, processing and disposal of substances (paths for the future), taking into consideration technical, economic, ecological and social parameters;
- To intensify the chemical and industrial policy dialogue in order to improve the conditions for reaching a consensus in society;
- To submit to the German Bundestag recommendations for legislative and political action.

The Enquete Commission submitted its report (available in English) in the form of a book, entitled *Responsibility for the Future: Options for Sustainable Management of Substance Chains and Material Flows* (Enquete Commission 1994).

3. MASS FLOWS AND MASS BALANCES

Having acknowledged the relevance of mass flows, we come to the problem of measurement. A rather obvious first step would be to construct a mass balance for the national economy, displaying material inputs, waste flows at various stages, production flows and material accumulation within the anthroposphere. It is interesting to note that this "simple" piece of analysis has never yet been carried out consistently for any country, although there has been significant recent progress. Allen Kneese and I thought to do it for the U.S. in the late 1960s, but we were forced to recognise that many of the important flows were not quantified and that we did not have nearly enough data to do a credible job (Ayres & Kneese 1968). I tried it again in 1989 (Ayres 1989), but the results at that time were not much better.[10] The German Enquete Commission did not even make the attempt.

A recent attempt for the U.K. — which also does not balance — nicely illustrates all of the problems.[11] Conventional statistics do not include waste flows from mines, quarries, agriculture or forestry. (In most countries they also do not count aggregates such as sand and gravel.) Nor do they include flows of "free" goods, such as air and water. There is a temptation to start from "finished" materials, such as metals, paper, rubber and plastic. But this neglects not only mine wastes from preceding steps in the process chain but also salable but low value by-products such as blast furnace slag. It is likely to double-count some inputs, such as petroleum and natural gas, which are the source of petrochemicals (from which synthetic rubber and plastics are obtained), unless great care is taken. (Taking care, in this case, requires the use of fairly detailed process descriptions). It is also necessary to be careful to count imports and exports of materials, including finished products.

To make matters worse, there are no reliable statistics on intermediate process losses of such "harmless" materials as calcium sulphate, ferrous sulphate, calcium chloride, sodium chloride and water vapour. Many chemical processes (such as acid neutralisation) generate such wastes, which must be counted in order to balance inputs and outputs. This can only be done by incorporating mass balances at the intermediate process level. Finally, there are no reliable statistics on either the total quantities of materials embodied in durable goods and structures or on the net additions to this total in a given year. These can only be estimated indirectly if all the other flows are accurately identified and accounted for. In short, to construct a national materials balance involves creating a detailed physical model of the national economy. As indicated above, such a model must include technological elements, especially details of many materials/energy conversion processes.

The data problem has been attacked frontally, for the first time, by a joint venture coordinated by the World Resources Institute involving four major countries, Germany, the Netherlands, the US and Japan (Adriaanse *et al* 1997). The purpose of the first effort was to gather consistent time series data on material inputs (requirements) — excluding air and water — for all four countries, using the same categories, for a twenty year period (1975-1994). The project involved collecting data on some material flows, such as erosion losses in agriculture, mining overburden moved, mine concentration wastes, and earth moved in road construction, that have never been systematically compiled before.

This effort is incomplete inasmuch as it deals only with the "front end" of the problem (inputs) but not the "back end" (outputs, recycling, production and consumption wastes, and emissions). There is obviously some overlap, inasmuch as erosion, overburden, concentration waste and earth moved in construction projects are both inputs and outputs at the same time. However, much remains to be done in terms of compiling data on "useful" materials, as they are transformed and converted into final products, or lost via consumptive or dissipative processes *en route*. Examples of such dissipative processes include fuel combustion, lubrication, surface treatment, cleaning, bleaching, acid/alkali neutralisation, fertilisers, pesticides and so on (e.g. Ayres 1978).

4. PHYSICAL-ECONOMIC MODELS

Let us suppose, however, that such a physical model were available for each country and were updated regularly. While we are at it, let us suppose that each firm and each industry also developed such models at the appropriate level of aggregation. What sorts of questions could be answered with their help? In the first place, a model of this sort would permit national or regional authorities to identify and quantify aggregate waste flows. This provides a method of cross checking and supplementing data currently collected largely by other means, such as direct measurements at point sources and in environmental media. Such data can be very unreliable as regards emissions of trace pollutants, such as dioxins, especially from dispersed sources.

Physical models of the economy are also essential for purposes of building physical models of environmental processes, such as processes of environmental transfer and transformation of toxics. This is very important for understanding the environmental fates of persistent organochlorines and toxic heavy metals such as lead, cadmium, mercury and chromium. The problem in all of these cases is that a model of these environmental transport and transformation processes cannot be constructed from concentration measurements alone, no matter how numerous and accurate they may be. It is also necessary to include data on quantity inputs to environmental media, in space and time. Unfortunately, these are data that industry is very reluctant to provide.

Another use of physical-economic models would be to identify specific opportunities for large-scale economic process optimisation (which could be realised by a combination of regulation, taxes or subsidies). By this means "industrial ecosystems", along the lines of the famous Danish town of Kalundborg, where wastes from one industry can be utilised as inputs by another (Frosch & Gallopoulos 1989), could be encouraged. As a straightforward example, calcium sulphate from flue gas desulphurisation can be used in place of natural gypsum in the manufacture of wallboard. There are a number of other possibilities for "mining" wastes, or using waste products as resource inputs to secondary production processes (Nemerow 1995; Ayres & Ayres 1996).

A further application of physical-economic models of resource flows is in the realm of resource policy. Models such as the one used by the Meadows group in their Report to the Club of Rome are clearly too simplistic. But the economic models that were used to discredit the "limits to growth" thesis are also far too simplistic for any purpose other than the one for which they were created. At some point in time certain non-renewable resources *will* become significantly scarcer than they are now, as high quality reserves are exhausted. (This is already a reality insofar as Europe and the U.S. are concerned.) At that point, prices will begin to rise in real terms and alternative resources (not excluding conservation) will have to be developed, or major infrastructure modifications will have to be undertaken, or (more likely) both.

Since very large amounts of capital are involved, long lead times are unavoidable. It may not be optimal to wait until market signals (prices) indicate the need for change. In fact, while economic theory suggests that the market will respond, there is no basis in economic theory for assuming that the market response to such a price signal will occur rapidly enough to avoid a dangerous crisis. The reaction time may be much too long. There are good arguments for anticipatory planning. For example, it may be optimal to accelerate the development of solar photovoltaic (PV) and hydrogen technologies so that they are some way down the learning curve before a crisis occurs.

A final, and very important, use of physical-economic models would be to permit improved estimates of the potential for emission reduction by large-scale system modification via regulation or tax policy. A case in point is the current controversy over the "costs" of reducing carbon dioxide emissions. There is one very questionable aspect of the current methodology, which is based on the use of so-called computable general equilibrium (CGE) economic models. It is the fact that the economy is *assumed* to be in competitive equilibrium and on an optimal growth path at all times. Thus, any external constraint on the system, imposed by government, must *ipso facto* reduce economic growth below it's optimum level. Many energy experts with technical backgrounds note that there are apparently major opportunities for reducing energy consumption (hence pollution) and reducing costs at the same time, thus constituting a "free lunch" or "double dividend". This possibility is generally denied or dismissed as unimportant by mainstream economists. This dispute is not likely to be settled any time soon, and in any case is beyond the scope of this chapter.[12]

The probable explanation of the discrepancy between the two disparate views is that the CGE approach does not allow for the fact that new technologies normally penetrate markets less rapidly than would be optimal because of time lags in the system, notably the spread of information on the availability and performance of new products and technologies. It would seem, then, that government intervention to accelerate the adoption/diffusion process would actually accelerate economic growth, rather than inhibiting it.

In any case, the development of more realistic physical models of the economy would constitute a useful complement to the current generation of CGE models and would permit more realistic simulation of system response to perturbations.

5. CONCLUDING THOUGHTS

There are undoubtedly more reasons than I have noted in the preceding pages for extending the neoclassical paradigm into the physical realm. I am strongly tempted to discuss the implications for production functions and growth theory, for instance. Growth is possibly the central theme of post-Keynesian economics, and the current economic theory of growth is clearly inadequate. I doubt that a new and better theory can be formulated without a more explicit recognition of the role of physical resource inputs and physical materials, and the physical nature of capital.

To state the problem very succinctly, current growth theory does not recognise the possibility (indeed strong likelihood) that "factors of production" can change over time. The limiting factors that were appropriate at the beginning of the Industrial Revolution would have been land, livestock and labour (as the French physiocrats understood). A century later, when Alfred Marshall was formulating the neoclassical world view, "capital" (including land and livestock in a larger category) and labour were appropriate, at least for industrialising Western Europe and the US. Resources, at that time, were not "scarce"; they could be regarded as products of capital and labour.

Since the 1970's some growth theorists have been willing to acknowledge that available energy and materials play an under-rated role. But already the so-called "endogenous growth" theorists are beginning to look for ways of incorporating knowledge, in some form, explicitly into the production function. The traditional concepts of capital and labour are eroding and evolving. What is obviously missing, however, is a proxy for environmental services. This may still be tolerable, for the moment, but a century from now it is clear to me, at least, that environmental services will then be the scarcest factor of production, if social product is properly defined. Meanwhile labour and capital are already "non-scarce" in the western world. A century from now they are unlikely to be scarce at all, at least in the sense that a factor of production is supposed to be.

Yet, we are cheerfully using models constructed on the basis of a 1950 conceptualisation to inform policy decisions that are supposed to be relevant a century from now. I do not believe that this is appropriate. In any case, I think a major reconsideration of growth theory is needed. However, an adequate discussion of the subject would necessarily be quite extensive. I must leave it for another occasion. Meanwhile no exposition is ever complete and certainly not this one. Consequently it makes sense for me to stop at this point and let other authors have their say.

REFERENCES

Adriaanse, Albert, Stefan Bringezu, Allen Hammond, Yuichi Moriguchi, Eric Rodenburg, Donald Rogich & Hemut Schütz, *Resource Flows: The Material Basis of Industrial Economies* (ISBN 1-56973-209-4), World Resources Institute, Washington DC, with Wuppertal Institute, Germany, National Ministry of Housing, Netherlands & National Institute for Environmental Studies, Japan, 1997.

Ayres, Robert U., Resources, Environment & Economics: Applications of the Materials/Energy Balance Principle (ISBN 0-471-02627-1), John Wiley & Sons, New York, 1978.

Ayres, Robert U., ``Industrial Metabolism", in: Ausubel, Jesse & Hedy E. Sladovich(eds), *Technology & Environment*, National Academy Press, Washington DC, 1989.

Ayres, Robert U., ``On Economic Disequilibrium & Free Lunch", *Environmental & Resource Economics* 4, 1994 :435-454.

Ayres, Robert U. & Allan V. Kneese, ``Production, Consumption & Externalities", *American Economic Review*, June 1969. (AERE `Publication of Enduring Quality' Award, 1990)

Ayres, Robert U. & Allen V. Kneese. ``Pollution & Environmental Quality", in: Perloff(ed), *The Quality of Urban Development*, Johns Hopkins University Press, Baltimore, 1968.

Ayres, Robert U. & James Cummings-Saxton. ``The Materials-Process-Product Model: Theory & Applications", in: Vogeley(ed), *Mineral Materials Modeling - A State-of-the-Art Review* :178-244, Johns Hopkins University Press, Baltimore, 1975. (Resources for the Future, Inc., Washington, D.C.)

Ayres, Robert U. & Leslie W. Ayres, *Industrial Ecology: Closing the Materials Cycle* (ISBN 1-85898-397-5), Edward Elgar, Aldershott, UK, 1996.

Ayres, Robert U. & Udo E. Simonis (eds), *Industrial Metabolism; Restructuring for Sustainable Development* (UNUP-841), United Nations University Press, Tokyo, 1994. (ISBN 92-808-0841-9)

Billen, G., ``The Phison River System: A Conceptual Model of C, N & P Transformations in the Aquatic Continuum from Land to Sea", in: Wollast, Roland, Fred T. MacKenzie & Lei Chou(eds), *Interactions of C, N, P & S Biogeochemical Cycles & Global Change* :141-162 (Series: NATO ASI I: Global Environmental Change)(ISBN 3-540-53126-2 Springer-Verlag, Berlin, 1993. (Proceedings of the NATO Advanced Research Workshop on Interactions of C, N, P & S Biogeochemical Cycles, Melreux, Belgium, March 4-9, 1991)

Brunner, Paul H., Hans Daxbeck & Peter Baccini. ``Industrial Metabolism at the Regional & Local Level: A Case-study on a Swiss Region", in Ayres, Robert U. & Udo E. Simonis(eds), *Industrial Metabolism; Restructuring for Sustainable Development*, Chapter 8 :163-193(ISBN 92-808-0841-9), United Nations University Press, Tokyo, 1994.

Christensen, Paul P. & Richard G. Fritz,``A Reconsideration of the Physical Laws for Transforming Environmental Resources in Post-Keynesian Production Theory", *Review of Regional Studies* 10, Fall 1981.

Clark, William & R. E. Munn, *Sustainable Development of the Biosphere*, Technical Report, International Institute for Applied Systems Analysis, Laxenburg, Austria, 1986. (Cambridge University Press)

Commoner, Barry, *The Closing Circle*, Alfred Knopf, New York, 1971.

Commoner, Barry, *The Poverty of Power*, Bantam Books, New York, 1976.

Costanza, Robert, ``Embodied Energy & Economic Valuation", *Science* 210, December 12, 1980 :1219-1224.

Daly, Herman E., ``The Economic Thought of Frederick Soddy", in: *History of Political Economy* :469-488 12(4), Duke University Press, 1980.

Daly, Herman E., ``The Circular Flow of Exchange Value & the Linear Throughput of Matter-Energy", *Review of Social Economics*, 1985.

Dasgupta, Partha & G. Heal. ``The Optimal Depletion of Exhaustible Resources", in: *Symposium on the Economics of Exhaustible Resources*, Review of Economic Studies, 1974.

Ehrlich, P. H. & A. H. Ehrlich, *Population, Resources, Environment: Issues in Human Ecology*, W. H. Freeman & Company, San Francisco, 1970.

Enquete Commission of the German Bundestag on the Protection of Humanity & the Environment, *Responsibility for the Future: Options for Sustainable Management of Substance Chains & Material Flows; Interim Report* (ISBN 3-87081-044-4), Economica Verlag, Bonn, Germany, 1994. (translated from the German)

Frosch, Robert A. & Nicholas E. Gallopoulos, ``Strategies for Manufacturing", *Scientific American* 261(3), September 1989 :94-102.

Georgescu-Roegen, Nicholas, ``Myths About Energy & Matter", *Growth & Change* 10(1), 1979.

Graedel, Thomas & Braden R. Allenby, *Industrial Ecology*, Prentice-Hall, Englewood NJ, 1995.

Hardin, G., ``The Tragedy of the Commons", *Science* 162, 1968 :1243-48.

Huettner, David A., ``Net Energy Analysis: An Economic Assessment", *Science*, April 9, 1976 :101-104.

Jones, Peter, ``Mass Balance & the UK Economy", *Environmental Excellence*, undated :22-27.

Koopmans, Tjalling C. (ed), *Activity Analysis of Production & Allocation* (Series: Cowles Commission Monograph)(13), John Wiley & Sons, New York, 1951.

Manne, Alan S., ``ETA-Macro", in: Hitch, C. J.(ed), *Modeling Energy-Economy Interactions: Five Approaches* (Series: Research Paper R5), Resources for the Future, Washington DC, 1977.

Manne, Alan S. & Harry M. Markowitz (eds), *Studies in Process Analysis: Economy-Wide Production Capabilities*, John Wiley & Sons, New York, April 24-26, 1961. (Conference Sponsored by The Cowles Foundation for Research in Economics at Yale University)

Manne, Alan S. & Clas-Otto Wene. ``MARKAL/MACRO: A Linked Model for Energy Economy Analysis", in: *Advances in Systems Analysis: Modelling Energy-Related Emissions on a National & Global Level*, Forschungszentrum Jülich GmbH, Jülich, Germany, 1994.

Meadows, Donella H, Dennis L. Meadows, Jorgen Randers & William W. Behrens III, *The Limits to Growth: A Report for the Club of Rome's Project on the Predicament of Mankind*, Universe Books, New York, 1972.

Meadows, Donella H, Dennis L. Meadows, Jorgen Randers & William W. Behrens III, *Dynamics of Growth in a Finite World*, Wright-Allen Press, Cambridge MA, 1974. (Distributed by Productivity Press)

Meadows, Donella H, Dennis L. Meadows & Jorgen Randers, *Beyond the Limits: Confronting Global Collapse, Envisioning a Sustainable Future*, Chelsea Green Publishing Company, Post Mills VT, 1992.

Nebbia, Giorgio, ``Warenkunde/Merceologia: The Unifying Ground for Economic & Natural Sciences", *Forum Ware* 9(1-2), 1981 :28-30.

Nemerow, Nelson L., *Zero Pollution for Industry: Waste Minimization Through Industrial Complexes* (ISBN 0-471-12164-9), John Wiley & Sons, New York, 1995.

Nordhaus, William D., ``World Dynamics: Measurement without Data", *Economic Journal*, December 1973 :1156-1183.

Odum, Howard T., *Environment, Power & Society*, Wiley, New York, 1971.

Odum, Howard T., ``Energy, Ecology & Economics", *Ambio* 2(6), 1973 :220-227.

Odum, Howard T., *Systems Ecology: An Introduction*, John Wiley & Sons, New York, 1983.

Socolow, Robert H., C. J. Andrews, F. G. Berkhout & V. M. Thomas (eds), *Industrial Ecology & Global Change*, Cambridge University Press, Cambridge, UK, 1994.

Solow, Robert M., ``The Economics of Resources or the Resources of Economics", *American Economic Review* 64, 1974.

Stiglitz, Joseph, ``Growth with Exhaustible Natural Resources. Efficient & Optimal Growth Paths", *Review of Economic Studies*, 1974.

Stiglitz, Joseph. ``A Neoclassical Analysis of the Economics of Natural Resources", in: Smith, V. Kerry(ed), *Scarcity & Growth Reconsidered*, Resources for the Future, Washington DC, 1979.

Turner, B. H. II, William C. Clark, Robert W. Cates, John F. Richards, Jessica T. Matthews & William B. Meyer, *The Earth as Transformed by Human Action*, Cambridge University Press, Cambridge, UK, 1991.

---

[1] Charles Dickens *David Copperfield*.

[2] This is a ticklish subject, to be sure. In practice, society makes many political choices that amount to assigning a monetary value to saving a life. This is not done consistently, of course. Society will spend enormous sums to save the life of an astronaut (for instance), or to reduce the cancer risk from some obscure chemical (such as dioxin) while neglecting far more cost efficient ways of saving lives, such as providing better prenatal

care, enforcing speed limits, or prohibiting the sale of cigarettes to minors. Individual choices of careers also imply such valuations. Racing car drivers, motorcyclists, ski racers, boxers, lion tamers and high-wire circus performers take far greater risks than ordinary people; to a lesser extent, so do fire fighters, policemen, divers, coal miners and certain construction workers.

[3] For detailed critiques along these lines, see (Ayres 1978; Christensen & Fritz 1981; Daly 1985).

[4] See, especially, (Solow 1974; Stiglitz 1974; Dasgupta & Heal 1974). A number of papers were collected in a special issue of the *Review of Economic Studies* (1974). The controversy continued, however. In particular, Georgescu-Roegen harshly attacked the "Solow-Stiglitz" production-function approach as a "conjurer's trick" in his commentary on a paper by Stiglitz prepared for a 1979 conference held by Resources for the Future, entitled "Scarcity and Growth Reconsidered" (Georgescu-Roegen 1979); (Stiglitz 1979). This attack has recently been revived by Herman Daly, and a new round of comments on the issue appeared in a special issue of *Ecological Economics* Vol 22 (1997). Meanwhile, the Meadows group has not retreated. In fact, the principal authors of the original Report to the Club of Rome have argued that their 1972 predictions have been substantially confirmed by observed trends in the following two decades (Meadows *et al* 1992). This is more than can be claimed for neoclassical growth theory, incidentally.

[5] Nebbia dedicated his paper to Oskar Lange, who was one of the few economists of the twentieth century who emphasized the importance of material flows.

[6] However, some of the more sanguine proponents of this idea (following earlier suggestions by the Nobel Laureate chemist, Frederick Soddy) saw energy content as a measure of value (Daly 1980).. In any case, economists generally react negatively to any suggestion of an energy theory of value (e.g. Huettner 1976). The idea is not altogether dead, however (Costanza 1980).

[7] The main groups involved in this activity in the late 1960s and early 1970s were John Cumberland, Clopper Almon and their co-workers at the University of Maryland, Walter Isard and co-workers at Cornell University and the University of Pennsylvania, Allen V. Kneese, Ronald G. Ridker, C. Russell and co-workers at Resources for the Future Inc. (together with the present author), and Wassily Leontief and co-workers at Harvard and New York University.

[8] For those not familiar with L-P models, as applied to the energy scenario problem, they utilize energy cost functions for each "type" of energy (e.g. electric power from a coal-burning steam turbo-electric generator) together with capacity constraints and capacity addition costs and constraints. Energy demand is assumed to be a function of GNP (forecast by a macro-economic module), and is allocated among the supply options to minimize costs, subject to the capacity constraints. The resulting energy supply function (which requires labor and capital) is then fed back into the aggregate production function to satisfy the equilibrium condition (supply equals demand). The above description gives the flavor. However, these models are very complex in practice.

[9] All of these models utilize macroeconomic "drivers" which are essentially computable general equilibrium models. They all share a common feature: they assume that the economy grows in equilibrium, which implies that growth is not endogenous, but is determined by exogenous forces expressed as a smoothly increasing total factor productivity trend. Since growth is already optimal (by assumption) these models share another common implication: namely, that any government intervention (e.g. to regulate greenhouse gas emissions) must reduce growth. It is important to recognize that in reality economic growth must be endogenous, and that the most likely growth-inducing

---

mechanisms (Schumpeterian radical innovations) are unlikely to be consistent with either optimization or equilibrium.

[10] The most comprehensive attempt to date has been done for a small valley in Switzerland (Brunner *et al* 1994).

[11] Jones deliberately neglects atmospheric oxygen for combustion (approximately 439 million metric tons), by counting only the carbon content of atmospheric emissions. He apparently estimates the latter by assuming that all fuel carbon is converted to carbon dioxide. However, he apparently ignores the contribution from cement and lime production, as well as brewing. He also neglects a number of other inputs, such as grazing fodder consumed by livestock, ammonia (from natural gas and atmospheric nitrogen), phosphates, gypsum (the source of plaster), sulfur, sodium carbonate and salt. On the other hand he vastly overestimates chlorine (made from salt) while ignoring its co-product, caustic soda. On the output side he neglects water vapor produced by combustion (about 328 million metric tons, of which the hydrogen content C from fuel C amounts to 41 million metric tons) as well as respiration products of livestock. Some of these omissions are easily corrected, but others are not.

[12] For a review of the arguments and the literature, see (Ayres 1994).

Chapter 2

Substance flows through environment and society

A natural scientist's perspective

MICHAEL CHADWICK

Director, Lead-Europe, Leadership pour L'Environnement et le Développement, Geneva.

Key words: **dynamic ecosystems, nutrients, materials cycling, system fatigue**

Abstract: **The cycling of materials in natural and semi-natural ecosystems is complex and should only be taken as a model of materials use in human societies with recognition of these complexities. Whilst overall common features can be discerned, in detail there is large variability. This variability is in respect of the relative size of nutrient compartments from ecosystem to ecosystem, the rate of flow between compartments, differences from one element to another and between species of a single element, the degree of overall loss from systems (leakiness under equilibrium and disturbance conditions) and the equilibrium - time interaction of the ecosystem. Analysis of such differences has equal validity in drawing lessons for materials use in human societies as the recognition of the broader generalisations**

1. INTRODUCTION: PARALLELS AND PITFALLS

The cycling of materials within ecosystems (along with the flow of energy through them) is a basic feature of ecosystems. These cycles and flows are supported by structures of organisms. Energy *flows*, materials *cycle*. This has led those who are interested in more sustainable patterns of human living, lifestyles that are less intrusive or disruptive of natural processes, to draw parallels with natural and semi-natural systems. They advocate that the goal of a truly sustainable human society should be to operate in a way akin to natural systems. This finds its most complete expression, perhaps, in the Gaia hypothesis that views the Earth as a coherent system of life, self-regulating, self-changing, a sort of immense organism (Lovelock, 1988). But the analogy is also drawn more specifically - "..... most currently used materials will have to be recycled, just as living organisms have recycled the water, oxygen, and carbon dioxide for the past billion years. In following suit with our man-made refuse, we will finally be returning to that steady-state environmental balance known to all other organisms on our planet" (Wagner, 1974).

Of course, it is legitimate to draw analogies, if they are useful, but there are hazards to insisting on pressing an analogy too far. If the analogy between natural and human-industrial systems is taken as a general metaphor for a preferred way of working, that is fine. But if the analogy is based on a flawed perception of natural systems, or if the aim is to emulate natural systems too literally, then the lessons learnt from natural systems may be damaging and may even lead to a breakdown of the will to achieve the initial goal. Poets have, literally, gone mad striving for the unattainable.

So it is instructive to look at the main features of materials cycling in natural and semi-natural ecosystems. What is really going on? By understanding this it might be possible to avoid a whole bevy of well-meaning and partially insightful people chasing off after a will-o-the-wisp, and it all ending in tears at bedtime. Into the bargain, we may be able to set some desirable but more realistic goals for human society.

2. NATURAL SYSTEMS: FEATURES AND FALLACIES

In natural and semi-natural ecosystems materials cycle - from the abiotic environment through organic material and, eventually, back to the abiotic environment again. This cycling would not take place without external inputs of energy, fixed and transformed by the living elements of the system as energy flows through it. A supply of energy is a prerequisite of the system and its ability to keep materials cycling within it.

It has been customary to consider the cycling of materials in natural and semi-natural systems at two levels: (1) the 'global' circulation of a single element; and (2) the cycling of a single element or group of elements in a specific natural or semi-natural ecosystem such as a temperate or tropical forest, an ocean or a stream. The 'global' circulation of a single element - often referred to as a 'biogeochemical cycle' - has come much in prominence in the last decade, with interest focusing, for obvious reasons, on the carbon and sulphur cycles.

Whilst there are common features of these cycles, such as the transfer of elements from one system reservoir to another: from the atmosphere to the soil; uptake by living organisms (or directly from the atmosphere); molecular incorporation and re-arrangement; deposition in sediments or litter; decomposition and molecular simplification and release; and renewed uptake, different biogeochemical cycles exhibit different functional characteristics. Some biogeochemical cycles are relatively well-balanced with uptake and release being generally in equilibrium - they are cyclic. The nitrogen cycle is such a cycle with a huge reservoir, only potentially available to living organisms, in the atmosphere in a gaseous form. On the other hand, another essential element, phosphorus, exhibits acyclic characteristics and severe 'shortages' build up, often severely limiting primary productivity. Whereas loss of nitrogen from the atmosphere, through the cycle, to sediments is balanced by volcanic action replacement, most phosphorus is in an insoluble form as organic or inorganic solids, widely dispersed in water and soils, or concentrated in localised deposits of organic origin such as are found in Morocco and Georgia, USA. Natural biogeochemical cycles should not always be viewed as beautifully balanced, perfectly functioning systems.

A number of important considerations follow at the specific ecosystem level from this tendency to an imbalance between supply and potential demand for minerals. A steady state system exhibits large variations between the proportion of an element in the component reservoirs, and these variations contrast from element to element. Take one illustration from a careful study of a mature evergreen temperate forest system (*Table 2* and *Table 3*).

The percentage of a major nutrient contained in the vegetation of this ecosystem varies from as little as 2 per cent (P) to nearly half (K). Soil as a reservoir represents from as much as 98 per cent (P) to less than half (K) while litter represents an insignificant elemental reservoir (P) to over 10 per cent (Ca).

Table 2. Elemental partitioning between the major components of a 36 year-old *Pseudotsuga menziesii* forest ecosystem (kg/ha). Source: Cole, Gessell and Dice (1967)

	Nitrogen (N)	Phosphorus (P)	Potassium (K)	Calcium (Ca)
VEGETATION:				
Trees	320	66	220	334
Ground vegetation	6	1	7	9
LITTER	175	25	32	137
SOIL	2809	3878	235	741
TOTAL	3310	3970	494	1221

Annual flows or transfer proportions are even more instructive. Annual uptake by vegetation from the system varies some 60-fold, from 0.2 per cent (P) to 12.5 per cent (K); the soil suffers from varying degrees of element loss, some 5 per cent of total per annum in the case of calcium, about 3 per cent of potassium and over 1 per cent of nitrogen. Losses from the system may be as high as 18 per cent of the annual uptake (Ca) or 'use'. However, expressed as a proportion of total stocks in litter and soil, annual losses are a small proportion, ranging from 0.5 per cent for calcium to 0.0005 per cent for phosphorus.

Such variations in the material flow characteristics of forest systems can be explained by known differences in ion species mobility (high in the case of potassium), which varies from form to form for an element (contrast the behaviour in ecosystems of NO_3^- and NH_4^+). This mobility is greatly affected by physico-chemical features of the system such as pH and eH (a measure of the oxidation-reduction state). In conclusion, nutrient or element cycling varies much in the detailed features of the system including element differences, system compartmentalisation, rates of transfer, use, accumulation and system leakage.

Table 3. **Element flows between components of a 36 year-old *Pseudotsuga menziesii* forest ecosystem (kg/ha/annum). Source: Cole, Gessell and Dice (1967).**

	Nitrogen	Phosphorus	Potassium	Calcium
UPTAKE BY VEGETATION	38.8	7.2	29.4	24.4
RETURN BY VEGETATION	15.3	0.6	15.0	15.7
ACCUMULATION:				
Forest	23.5	6.6	14.4	8.7
Litter	11.6	-0.4	5.3	1.1
Soil	-34.6	-6.3	-19.9	-11.5
SYSTEM LOSS	0.6	0.02	1.0	4.5
Percent soil inventory taken up by vegetation per annum	1.4	0.2	12.5	3.3

In relation to materials cycling, the next feature of natural and semi-natural ecosystems to note is that ecosystems are not static entities in the long-term. Many ecosystem types progress through pioneer, building, mature and degenerate phases. As they do so, features of materials cycling in the ecosystem become modified. Long before the advent of European man, temperate and dry deciduous forest systems in the North American continent went through a cycle, even in the climax phase, of the gradual build up of nutrients in what was often a deep and dry litter layer that accumulated over decades. This immobilised the source of nutrients in the system to dangerous proportions, restricted water penetration to the soil and mitigated against the germination and establishment of propagules and seedlings to replace gaps in the tree cover due to tree ageing and death. A rich literature has documented the part that naturally occurring forest fires play in functionally and structurally renewing ageing ecosystems, not least by releasing nutrients to the system again in more accessible forms (Goldammer,1990; Kozlowski and Ahlgren, 1974) .

Materials cycling is not just a matter of mass-balance considerations; the relative distribution of materials and the rates of transfer between system compartments are also of significance. *Figure 3* and *Table 4* illustrates the relationship of sinks or reservoirs in an ecosystem and the transfer rates of materials between them. *Table 4* which provides data for these stocks and flows in four different natural systems. Sinks and transfers in each system are normalised on the 'available nutrients' sink (X_5) in each system.

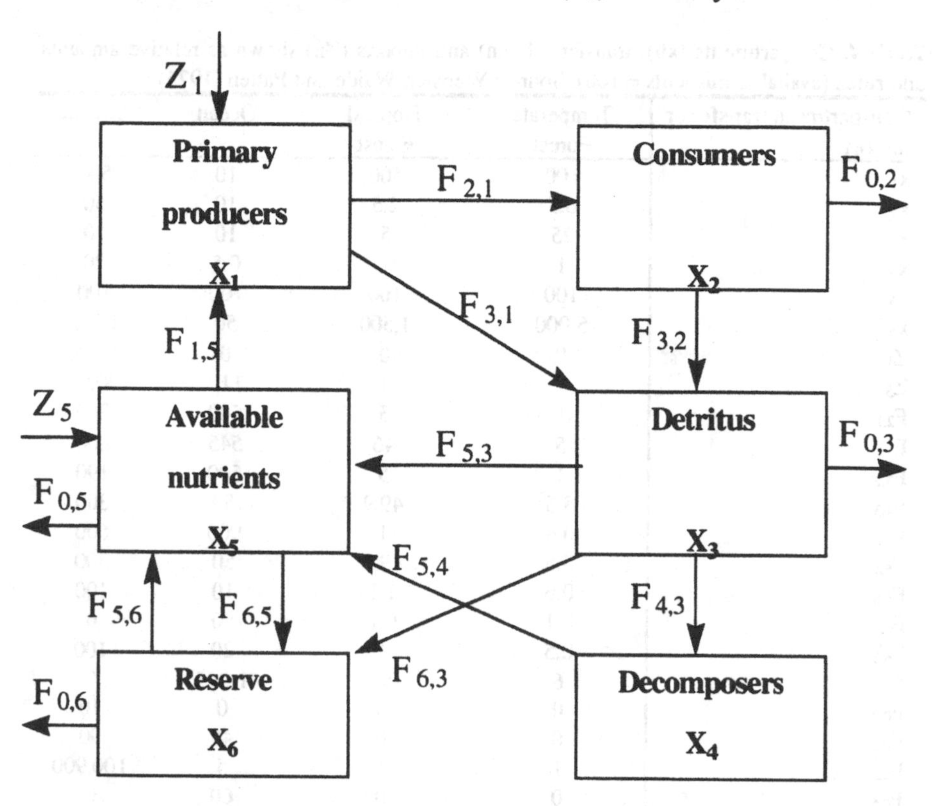

Figure 3. Compartments (Xn), transfers (Fn,m) and imports (Zn) of materials in relation to a notional ecosystem. Source: Webster, Waide and Patten (1975)

Variations such as those shown in *Table 4* have tremendous implications for the effects of disturbance of ecosystems by human activity, and their resilience under such conditions. For instance, clearance of the primary producers (X_1) of a tropical rain forest ecosystem means that the equivalent of one-third of the total nutrients in the reserve (X_6) have left the system; this compares with only 2 per cent in the temperate forest. Forest destruction, or the cropping of an area following clearance, results in a proportionate level of removal of nutrients from such different forest types. Once a major component has been removed or seriously disturbed, the time required for a build up of nutrients to levels approaching that in the original forest may be considerable. For instance, in a tropical forest, recovery of nutrient levels to those in the original forest levels may take 40 years or more (*Table 5*).

Table 4 also gives information about the differing rates of cycling of materials in natural systems. Relative to the available nutrients, the total nutrient transfers are highest in the stream system, and lowest in the temperate forest system.

Table 4. Compartments (xn), transfers (Fm,n) and imports (Zn) shown as relative amounts and rates (available nutrients = 100). Source: Webster, Waide and Patten (1975)

Compartment, transfer or import	Temperate Forest	Tropical Forest	Ocean	Stream
x_1	100	500	10	500
x_2	0.5	2.5	10	50
x_3	25	5	10	10
x_4	1	1	0.5	20
x_5	100	100	100	100
x_6	5,000	1,500	50	1,000
Z_1	0	0	0	1,000
Z_5	1	1	110	100,000
$F_{2,1}$	1	5	500	200
$F_{3,1}$	5	46	545	800
$F_{3,2}$	1	5	500	190
$F_{4,3}$	5.5	49.9	50	300
$F_{5,3}$	0.4	1	900	600
$F_{5,4}$	5.5	49.9	50	300
$F_{5,6}$	0.6	1.1	10	100
$F_{6,3}$	0.1	0.1	50	0
$F_{6,5}$	0.5	1	20	100
$F_{1,5}$	6	51	1,045	0
$F_{0,2}$	0	0	0	10
$F_{0,3}$	0	0	45	90
$F_{0,5}$	1	1	5	100,900
$F_{0,6}$	0	0	60	0

This brief discussion has illustrated the enormous variability in stocks and flows of minerals and nutrients across natural and semi-natural systems. This variability is in respect of the relative size of nutrient compartments from ecosystem to ecosystem, the rate of cycling between compartments, differences in stocks and flows of one element from another and between species of a single element, the degree of overall loss from systems (leakiness under equilibrium and disturbance conditions), and the age and developmental stage of the ecosystem.

Table 5. Change with age of major nutrients in the standing crop of vegetation of a tropical forest following clearance and cropping (kg.ha-1). Source: Nye and Greenland (1960)

AGE (yrs.)	FOREST	Nitrogen	Phosphorus	Potassium
5	Secondary	572	31	454
18	Secondary	706	107	605
40	Mature Secondary	1,614	112	673

3. HUMAN ACTIVITY: RATES AND REHABILITATION

Some studies have attempted to apply a 'natural system' estimation of the quantitative features of a cycle to human-controlled systems (see Denaeyer & Duvigneaud, 1980). It may be useful to take the lessons that can be learned from the movement of materials within ecosystems and apply them to human activities. Materials are used in the huge variety of products deemed necessary for health, happiness and well-being. But the cycling of materials in natural and semi-natural systems is complex and varied.

Ecosystems exhibit features of accumulation and of shortages in terms of the materials cycling. Not all cycles are balanced and in equilibrium. Some are relatively resilient to perturbations, including those by human activity, other cycles less so. In natural ecosystems that accumulate materials in a relatively inaccessible sink a need exists for regular, though infrequent, catastrophic events (such as forest fires) which release materials for more rapid recycling in the system once again. Within a relatively stable human society, based on durable goods and infrastructures, not given to 'replacement activity', it might also be expected that large amounts of substances would be 'locked up'. These stocks of materials could over the longer term form the basis for large-scale and routine cycling of materials (cf. Frosch, Chapter 3).

Natural and semi-natural ecosystems, even at relative equilibrium, can be 'leaky' (see $F_{0.5}$, $F_{0.6}$ in *Table 4*). Methods have evolved to 'top-up' the system requirement for materials. In human systems it is difficult, and would be too expensive, to recover materials used that become widely dispersed in the environment: lead from leaded gasoline; zinc from tyres; and nitrogen that, in the oxidised form, now 'floods' the environment (Vitousek, 1997). All ecosystems require an input of energy for the materials cycle to operate. Industrial recycling operations also require inputs of energy, for collection, for sorting, for clearing for storage and for treatment. There can be significant energy costs in recovery and recycling. In human systems, materials uses subject to dissipation could best be managed by input reduction leaving the less dissipative uses to be managed by recovery and recycling.

Few natural systems are in a continuous and unchanging equilibrium state. Age and development bring about alterations in materials cycling rates and in the structure of many ecosystems. This is also a characteristic of human systems, although there has been insufficient study of the way in which relatively stable materials use in human systems subtly changes and shifts. Stability in natural and semi-natural ecosystems, in relation to materials cycling, is largely *rate determined*: alter the rate of materials transfers between components by intervention in the system and it is difficult to re-establish a stable system. Generally 'fatigue' sets in and the system manifests a greater degree of leakiness. Many 'restored' ecosystems exhibit such features (Chadwick, 1974).

Most natural and semi-natural ecosystems are not really 'self-contained' within the boundaries normally distinguished from them. They are affected by other, seemingly, self-contained system attributes and quite severely by attributes of the system not directly linked with materials cycling.

4. CONCLUSIONS: LESSONS AND LESS

For many materials, the rate of use in society now vastly exceeds anything generally encountered in natural ecosystems. As stability is largely a rate determined characteristic, it is to be expected that many of these materials uses are unsustainable. Frequently the first indicator of unsustainable materials use are signs of 'system fatigue'. Just as natural or semi-natural ecosystems are rarely completely self-contained, nor is the use of materials in the economy isolated from other societal activities. It is not immune to other interests and priorities - from considerations of energy use, public acceptance, demand and supply for goods and services, and associated economic considerations. While the efforts to greatly reduce the throughput of materials utilised, to deliver a service for perhaps a four or ten times reduction in materials intensity are to be welcomed (Weizsäcker, Lovins and Lovins, 1997; see Chapter 4), actions need to be taken in relation to the system of materials uses as a whole, rather than through efficiency improvements linked to final demand alone. But we should never do nothing because we can only do a little.

REFERENCES

Chadwick, M.J. (1974). The cycling of minerals in disturbed environments, in Chadwick, M.J. and G.T. Goodman (Eds.) *The Ecology of Resource Degradation and Renewal*, Blackwell Scientific Publications, Oxford, 3-16.

Cole, D.W., S.P. Gessell and S.F. Dice (1967). Distribution and cycling of nitrogen, phosphorus, potassium and calcium in a second growth Douglas fir ecosystem, in Young, H.E. (Ed.) *Symposium on Primary Productivity and Nutrient Cycling in Natural Ecosystems*, University of Maine, Orono, Maine, 197-232.

Denaeyer, S. and P. Duvigneaud, P. (1980). L'Ecosystème Urbs: (comparison Bruxelles - Charleroi), in Duvigneaud, P., S. Denaeyer and Brichard, C. (Eds.) *Ecosystèmes Cycle du Carbone Cartographie*, Le Ministère de l'Education Nationale et de la Culture Française, Bruxelles.

Goldammer, J.G. (Ed.) (1990). *Fire in Tropical Biota*, Springer-Verlag, Berlin

Kozlowski, T.T. and C.E.Ahlgren (Eds.) (1974). *Fire and Ecosystems*, Academic Press, New York.

Lovelock, J. (1988). *The Ages of Gaia*, Oxford University Press, Oxford.

Nye, P.H. and D.J. Greenland (1960). *The Soil Under Shifting Cultivation*, Commonwealth Agricultural Bureaux, Farnham Royal.

Vitousek, P.M., J.D. Aber, R.W. Howarth, G.E. Likens, P.A. Matson, D.W. Schindler, W.H. Schlesinger and D.G. Tilman (1997). Human alteration of the global nitrogen cycle: sources and consequences. *Ecological Applications* 7 (3), 737-750.

Wagner, R.H. (1974). *Environment and Man* (Second Edition), Norton, New York.

Webster, J.R., J.B. Waide and B.C. Patten (1975). Nutrient recycling and stability of ecosystems in Howe, F.G., J.B. Gentry and M.H. Smith (Eds.) *Mineral Cycling in Southeastern Ecosystems*, U.S. Energy Research and Development Administration, Washington, D.C.

von Weizsäcker, E., A.B. Lovins and L.H. Lovins (1997). *Factor Four: Doubling Wealth - Halving Resource Use*, Earthscan, London.

Chapter 3

Towards the End of Waste

Reflections on a New Ecology of Industry

ROBERT A. FROSCH
Senior Research Fellow, John F. Kennedy School of Government in Harvard University, Boston.

Key words: waste, recycling

Abstract: **All wastes can be conceived as potential resources for reuse. There are many ways in which recovery could be achieved for the multitude of wastes generated in industrial economies. A framework for characterising recovery and reuse strategies which takes account of environmental and money costs, and political acceptability is proposed. This suggests that waste recovery should be considered in product design, recovery of a greater range of materials should be enhanced, that wastes which are not recovered for direct reuse should be concentrated for future use, and that stress should be placed on reducing fugitive emissions from future industrial ecosystems. A metals case study showing high materials productivity is presented.**

1. INTRODUCTION

I want to think simply and abstractly about all of industry, considered as a single block. That block takes in materials and energy, transforms them into products and wastes, and then excretes the products and wastes. At the end of their useful lives, the products may become wastes and may also be excreted, that is, "disposed of" as waste. Increasingly, we are concerned with decreasing the amount of waste to be "disposed of," and thus with changing the nature of manufacturing processes and products. In this essay I will rethink the ecology of industry as a problem in the present and will consider the future flows of materials within and among industries. Where might we try to go? How might we try to get there?

By any measure, wastes abound in modern economies.[13] In the United States, the material wastes from manufacturing, mining, oil and gas extraction, energy generation, and other industries currently are on the order of ten billion metric tonnes per year, though a large fraction of this is water. US air emissions of materials, such as the carbon in carbon dioxide, are approximately two billion metric tonnes per year. The end products of industry annually turn into about two hundred million tonnes of municipal solid waste in the United States.[14]

Traditionally, waste is whatever material is left, to be disposed of later. The emerging field of industrial ecology shifts our perspective away from the choosing of product designs and manufacturing processes independent of the problems of waste. In the newly developing view, the product and process designers try to incorporate the prevention of potential waste problems into the design process.[15] Industrial ecology notes that in natural ecological systems organisms tend to evolve so that they can use any available source of useful materials or energy, dead or alive, as their food, and thus materials and energy tend to be recycled in a natural food web.

Even with the design of products and manufacturing systems to minimise waste, some waste energy and material, whether from manufacturing or from products at the end of their useful lives, will be inevitable. The second law of thermodynamics (in the simplest of terms, the impossibility of converting all the heat from a reservoir of energy into useful work) ensures that there will at least be waste energy from a process; waste energy frequently appears as waste material, or it may be carried away as heat in a material. We do not know how to design processes that are perfectly economical with materials, and, in any case, this may be impossible to do.

Nevertheless, the idea of industrial ecology is that former waste materials, rather than being automatically sent for disposal, should be regarded as raw materials - useful sources of materials and energy for other industrial processes and products. Waste should be regarded more as a by-product than as waste. Indeed, as part of the design process for manufacturing end products, wastes might be designed to be resources for new products. The design optimisation process would include the generation of waste, the design of waste, and the cost consequences of alternatives, for example, in reuse or disposal. This practice occurs in some parts of some industries, but is not widespread.

The overall idea is to consider how the industrial system might evolve in the direction of an interconnected food web, analogous to the natural system, so that waste minimisation becomes a property of the industrial system even when it is not completely a property of an individual process, plant, or industry.

I will approach the problem by trying first to abstract the essence of industrial systems in an ecological sense, subsequently considering how we might characterise and graph future states of industrial ecosystems. Next I will contemplate the potential fates for wastes and how they might be balanced. Then I will look at some of the evolutionary trends of industrial ecosystems and the properties that might mark or lead to attractive states. Finally, I will examine how we might choose policies that are likely to lead toward attractive outcomes. My approach is not to forecast but to postulate future states and sets of states, inquiring as to how they might be reached via attractive, or at least tolerable, routes. I am thinking about coupled states of industry and the environment external to industry; my scale is industry, society, and the world. When I use "we," I mean "we, the society," in whatever way a social decision may be reached to do, or not to do, something technically attractive.

2. ABSTRACTING THE INDUSTRIAL SYSTEM

Let us consider industry, indeed, the whole of humanity and nature, as a system of temporary stocks and flows of material and energy. Materials are seen as what they fundamentally are: elements (atoms) in the sense of the periodic table, compounds (in the sense of atoms bound together into molecules by energy embodied in chemical bonds), or mixtures of elements and compounds (perhaps several of each). In this sense, everything is a dance of elements and energy. For this discussion, energy is considered at the chemical but not the nuclear level; elements the elements that they are. For example, plastics - polymers - are seen less as specific materials than as collections of carbon atoms, hydrogen atoms, and some others (for example, nitrogen, oxygen, and chlorine) bound together by the energy of chemical bonds.

The processes of industry re-sort atoms into various collections or mixtures of elements bound by energy. Every step in manufacturing - product creation, product use, and product disposal; - is a more or less transient event, a temporary (possibly long-lived, but temporary) use of some set of atoms and energy. (We hark back to the world of Democritus and Lucretius!) In this sense, products are just way stations in the flow of materials and energy - a temporary storage of elements and bond energy. Thus the whole sequence of events from mining or extraction to disposal is seen as a sequence of rearrangements of elements and energy.

From this point of view, reuse of materials produced as "wastes" in the course of production, or as "wastes" at the end of product life, is only another re-sorting of the elements and energy. The point is trivial, but the general way it puts all parts of the process into the same simple framework may suggest some useful lines of thought.

Clearly, the roles of energy and energy cost determine what may be done. Energy binds materials into compounds, but energy may also take compounds apart. Energy is required to drive the processes of negentropy (i.e., those that increase order or pattern, particularly those with value to humans) that separate mixtures back into the elements and compounds from which they were mixed or that assemble atoms that have been distributed in a diluted way into collections of compounds (molecules) or other atoms. Thus, the availability and cost of energy will fundamentally determine which of these processes is economically useful for transformations and when.

At each stage in a process that produces "product" and "waste," we have some choices that determine the material forms of each and we may link the series of choices with later sets. We postulate a universe of material/energy paths through the production, life, and dissolution of a product or set of products. We can also consider each path to be a sequence of transformations from one material/energy embodiment to another. (I generalise on a larger scale the chemical engineer's view of life.) We can view the whole of material industry as a network of such paths or transformations, connected at each end (extraction of materials and disposal of products) to the environment external to the process and product, and at places in the middle (disposal of incidental waste).

3. PLOTTING STATES OF THE INDUSTRIAL SYSTEM

Each network of paths or transformations may be considered a "state" of industry. Given some idea of impacts and costs of alternative paths, we could, in principle, choose among them. We would need somehow to rate the environmental impact of so complex a thing as the total industrial system. For purposes of this general discussion, I will assume that we can rate total environmental impact, and we will use only one measure or graphical axis to do so. I also assume that cost needs only one measure or axis. Cost implies dollars, but it should really be thought of as cost in the most general sense: total effort, including capital, work, energy, economic opportunity costs, and all other forms of effort, somehow calibrated in dollars. To help choose we might then create a two-dimensional graph of these states of industry by plotting cost versus environmental impact, placing each alternative state in its appropriate position.

With only one cost and impact for each state, I am implicitly considering only the total social economic cost and the total environmental cost without examining the question: Upon whom do the elements of cost and impact fall? Alternatively, we could struggle with a multidimensional space in which many axes represent different kinds of environmental impact, various kinds of costs attributed to different actors, and political and cultural variables. Such a multidimensional approach could, in principle, take account of regional differences, local politics, and other concerns. For the immediate purpose, however, I will continue to illustrate the ideas with the two dimensions of environmental impact and cost.

We must then ask about objectives. In particular, what principles should we use to choose preferred states, representing networks of paths through the sequence of material transformations? We can visualise the two-dimensional graph in which each state appears as a point, resulting in a cloud of points. Given the usual uncertainties in impact assessment and cost estimation, the "points" could represent statistical haloes, but at this level of generality, visualising points is fine. I presume we would like to choose among states as close to the origin as possible: least cost for least impact. Various states will in fact apportion the costs and impacts to different actors, but if we were to work outward from the origin, we would be looking for solutions that are "optimum" in some useful sense, even including the political acceptance of choosing among states near the origin on the basis of where the costs and impacts fall. However, a moment's thought about the multi-dimensional, more realistic possibilities and the uncertainty haloes of the points would reveal that detailed optimising principles are far from obvious.

For some given scenarios of industrial technology and industrial organisation, we might expect the possible states, the points on the graph, to have some systematic relationship or lie on a curve. For example, I would expect the attainment of very low levels of total industrial waste or "zero waste" to be very expensive for industry; extremes are frequently costly to attain. Symmetrically, absent internalisation of environmental impact costs, simply "throwing away" wastes might be expected to have high impact and low cost to industry, so these state points would be in that region of the graph. A standard economic interpretation might expect the curve connecting possible states of similar technology and industrial organisation to be hyperbolic in character, like a standard demand curve. In that sense, the graph of effort versus impact can be thought of as a conventional tool of the economist.

4. FATES FOR WASTES

To find future states desirably located near the origin of the graph, it seems reasonable to examine those that imply both the minimisation of waste in the production process and the use of wastes from various production processes across industry as input materials. Finally, we search for "best" states by some optimisation process that simultaneously considers total cost and impact. This leads us to a scenario or, rather, a set of scenarios.

Let us assume that manufacturers do an excellent job of minimising process waste under some set of economic and regulatory pressures. Some process waste will still remain to be dealt with, as will materials available as "waste" from products arriving at the end of their useful lives. Therefore, as a baseline, let us assume a future state in which an "optimum" overall balance has been struck between limiting the creation of waste during production and reusing "wastes" that are produced as raw materials in other productive processes or that result from products at the end of their lives. This "optimum" has been chosen to lead to a minimum amount of total waste impact on the environment at minimum total cost. It would imply the more complete recycling of valuable and more easily recoverable materials (metals for instance)in 'clean recycling' systems. For these materials the 'fund' of 'active' materials would remain reasonably stable, with the small proportion of virgin inputs to replace marginal losses to dissipative waste streams. However, increased cycling of materials will always meet technological and energetic limits, and some waste remains with which the overall system must deal.

The scenario must employ some policy for dealing with the remaining waste. The current policy, generally speaking, is to dispose of it: destroy it, if chemical (i.e., take it apart into simpler compounds or into elements through reprocessing or incineration), or "bury" it. In the general spirit of provisioning for the long run, waste for which no immediate reuse possibility exists could be stored against future need. Suppose we consider sites that are "contaminated" with hazardous materials to be "filing cabinets" for potentially useful materials. Would this approach cost less or more than the current sites or systems for "disposal"? The care of potentially valuable materials may be more economical than their "disposal." Such a shift in how we view "waste sites" would require more thoughtful characterisation and labelling of wastes and perhaps better technologies for packaging them.

Filing cabinet storage makes sense for many elements, particularly those with volatile market prices, and for compounds that involve elements (particularly more exotic minerals) that are likely to be of future use. What to do with organic materials lacking an immediate prospect of reuse is less clear. For many such materials, storage against later reuse as chemical feedstocks might be sensible. For example, pesticides are complex organic chemicals. In the spirit of petrochemical cracking and transformation of crude oil, such a mix of organic chemicals might be a good feedstock from which to extract energy and/or derive simpler compounds suitable for chemical synthesis. Excess bond energy might be available to help power the cracking process. However, the variable streams of available wastes might make it hard to maintain a stable process. The technical challenge is then two-fold: first to design chemicals which are simpler and easier to un-build; and second to design chemical processing systems taking into account significant variability in the stream of input materials.

Many materials might be more beneficial only as sources of energy - the energy embodied and stored in their chemical bonds. The extraction of this bond energy should be viewed as a matter of chemical processing, not just burning, and therefore as requiring the same level of process control as other chemical processing. Such control has not always been practised with "burning" to get the chemical energy back. Incineration without energy generation as part of the process seems silly unless it is clear that the price of the power and the avoided costs of other disposal of the material do not amortise the costs of adding power generation equipment and its operation to the incinerator. The "cogeneration" alternative clearly becomes more attractive if the price of energy rises.

5. TRENDS AND PROPERTIES

The state described above might not cost the minimum; other states nearby in the imaginary graph might achieve nearly similar impacts for a lower total economic cost to society, presumably with a different distribution of costs among the players. Even in a highly abstract and simplified picture we see complex possibilities for choice. If we go again to more dimensions, in which various kinds of environmental and other impacts and various kinds of costs appear as axes of the display, the problem is yet harder.

Before discussing the postulated state further, some other states with resemblance to it bear description. These may or may not be in the same part of the graph with regard to costs and impacts. An earlier state from our industrial/environmental history may be described as one in which direct manufacturing costs were minimised as a way of minimising product cost, but environmental impacts were generally regarded as external to the industry. We have recently been in a state in which direct manufacturing costs are minimised, but regulation and, increasingly, social obligation require industry to take responsibility for disposal of waste in a way that has a low environmental impact. We now see demand for movement towards a state in which manufacturing combines cost minimisation with low or zero production of waste.

The historical sequence implies next the state in which waste production during manufacturing is combined with the reuse of wastes within industry, as is the possibility of varying the materials in a product and the processes of production in order to change the nature of the waste materials, making them easier for someone to use as process input materials. Incomplete consumption of material in manufacturing and reuse is also implied, therefore leaving some waste to be disposed of at some cost.

The popular term "sustainability" presumably implies maintaining the global stock of available materials for as long as possible while at the same time preserving the general environment in liveable condition for as long as possible.[16] This latter condition should hold true for disposal and storage as well.

The availability of materials implies the small disposal - that is, the reuse - of elements that have limited availability. For this exercise I consider only Earth as available. If we make the rest of the solar system or the rest of the universe available, the principle remains the same, but the numbers and time scales change.

Preserving the environment for as long as possible against poisoning by the disposal of materials would require that we not disperse materials that harm the environment, even in concentrations so low as not currently to seem troublesome. Dispersal at the level of one part in a million, if continued for a thousand years, can dangerously accumulate to one part in a thousand. Although this principle omits the question of the economics, choices, and technology of future generations, it does suggest that good technologies for generating negentropy at the lowest possible direct energy cost would be welcome. Although the second law of thermodynamics tells us what the lowest energy cost must be, it does not prevent us from using technologies in which part or all of the energy cost is "free" to us, that is, where energy is used that is not otherwise directly available to us or used by us. Solar energy can be such a source. So can some or all of the energy used by organisms - for example, micro-organisms, because they may extract energy from sources we do not use - provided we can figure out how to have them be our collectors.

These considerations suggest that, in general, producing *concentrated* wastes that could be useful as someone else's raw material is likely to be more interesting in the postulated state than producing *diluted* wastes. This finding reverses the wisdom with which sanitary engineers began the twentieth century, "The solution to pollution is dilution". Technologically the problem is that many waste streams are complex mixtures and compounds, often with specific characteristics. One waste stream is different to another, even if it arises in a similar process. New analytical and reprocessing technologies will be needed to cope with the panoply of waste streams which are available from industrial and domestic sources. A shift to a strategy of concentrating wastes also suggests that a weakness of the alternative industrial ecosystems may be the small fugitive spills and emissions over long periods of time. Such a state is likely to require ever-better technology and practices for control.

Let us discuss some of the other properties of the industrial system that we have developed as our example. It will require a fairly widespread, large-scale, market-enabling information system to describe stocks and flows of materials in the industrial system, including information on those materials available for purchase or delivery and those that may be contracted for on a longer-term basis. Information on potential receivers, or buyers, of particular materials will also be needed. I am unaware of any publication, comparable to the weekly slick-paper tabloid *Chemical Marketing Weekly*, that specialises in chemical *wastes*, although that publication probably does include some materials that originate as wastes.

Beyond straightforward information leading to direct transfers of material, perhaps through brokers of materials, the brokering of more detailed adjustment possibilities would be needed. Adjustments in the products or processes of possible suppliers or users might generate "wastes" for manufacturing systems designed to use them. This process must occur for the production of particular materials now, but not much seems to happen to adjust wastes.

Specialised open-market commodity exchanges of the "pit" kind might develop in waste commodities. A computer net exchange system now operates among dealers in used auto parts, and this concept might be extended to other "waste" materials. More generally, the Internet portends cheap ways to link otherwise unconnected buyers and sellers to create markets and to search large, poorly structured databases for highly specific items.

To work for the reuse of significant quantities of materials, the system would have to be based upon realistic economics, in which the alternative processes and product materials of buyers and sellers made financial sense, including process and product costs, information and transportation costs of the various alternatives, and possible final disposal costs, whether by alteration or "landfilling". For some materials lacking realistic alternative uses, fresh incentives might make recycling economically viable. These could take forms such as taxes on their disposal in order to force reuse to be economical. The idea is that for some necessary materials, the least environmental impact might occur when a limited supply of them circulated, as opposed to flowing them through the system to "disposal".

Box ItemThe Industrial Ecology of Copper

Metals, including copper, have been important to much industrial activity and the reuse of products and recycling of materials becomes increasingly important when we consider sustainability issues. In order to investigate how the copper processed in industry actually leads to leakages into the environment, the team of researchers interviewed metal businesses located around New England, examined official reports and conducted telephone questionnaires with other companies. They thereby obtained information both at the level of the individual firm and at the level of the system.

By analysing the available data, the researchers concluded that average efficiency of a copper using firm in the study is 95.09% while that for lead is 95.93%. This high number was possible because 62% of the copper was being used by only 2 of the 64 copper using firms in the manufacture coils from copper ingots. Annually, about 1% by weight of the copper used by the 64 firms in Massachusetts is lost in emissions or landfills. This does not include the efficiency of the scrap system in recovering the copper from used products. The materials that leave the system include some of the sludge from the polishing and finishing of products, some swarf (metal powder), some metal mixed with foundry sand and some baghouse dust recovered from the air flow. The metals in these materials are frequently recovered and sold or reused.

From the analysis, it appears that there are two key driving forces: the ordinary cost-consciousness of business (the value of metal) and the evaluation of potential liability. They may conserve the copper even beyond what would be economically optimal because of the risk of future environmental liabilities, and its potential monetary costs. The existence of complex, somewhat draconian regulatory systems provide a useful strong pressure to find ways to deal with `wastes' that will protect the company from regulatory intervention. This provides a useful motivation for reuse and recycling. However, the detailed regulatory treatment of recycling, as a form of disposal, limits its efficiency. These details may, generally for reasons of scale, lead a firm to dispose of material rather than send it for recycling. Scrap dealers and secondary smelters are key actors in the circulation of metal.

Based on the Paper entitled: "The industrial ecology of metals: A reconnaissance", Robert A. Frosch, William C. Clark, Janet Crawford, T. Ted Tschang and Audrey Webber, delivered at the World Conference on Copper Recycling

In cases where "waste" material had no immediate use or was not available in sufficient quantity for an immediate buyer, public information on the contents of the filing system in landfills, with realistic systems of storage and retrieval costs, would need to be maintained. Thus, as materials came into marketable value, as happens today with copper scrap for smelter material as the market price varies, their availability would easily be known, and they could re-enter the active materials market. Both the transportation and the filing-cabinet storage systems will require considerable attention to safety, and the resulting costs would automatically be part of the trading system.

Such a system would likely require a new kind of regulation for waste and hazardous-waste materials, one that recognised movement and disposal by use in an industrial process as equivalent to what is now considered final disposal. In addition, the current tendency towards an infinite chain of liability - in which no one's liability is transportable with the material, and in which liability does not die, even when the material is transformed into something else - might need somehow to be altered. Cyclicity of materials use cannot be encouraged if liability is not transferred.

The new state of the industrial ecosystem that I have described assumes a reasonable supply of waste material even after a mature system of waste prevention is developed in the manufacturing industry. Many materials will not be internally reprocessed for reuse by the manufacturer because, for example, it may make no sense for a hard-goods manufacturer also to be a metal smelter or a maker of polymer feedstock. Waste material from sheet metal and semi-fabricated metal parts manufacturing will also remain available as material to be reprocessed. Not everything will be amenable to internal reprocessing.

6. CHOOSING POLICIES

How can we decide which policies are likely to lead in the right direction, where the end states reached would be "better" and well chosen? Nearby states are likely to resemble the state I have postulated (high cyclicity, low waste, waste concentration), differing perhaps in the proportions of waste minimisation and reuse, or in the specific configuration of the industrial networks, which nevertheless all have roughly the same amount of material reuse. Other states may differ in their network properties, or represent different kinds of environmental impact making up the same or a different, quantitative index of total impact, or represent different total societal costs or allocation of costs to different actors.

We must keep in mind that policies will not lead to a particular chosen state. The model of the future will be too crude, and systems do not respond by behaving exactly as policymakers desire. Actors in the system respond by doing what is in their interest, or what they perceive to be in their interest, so while systems move in a direction that the policies help determine, they tend toward a family of possible future states, rather than to some specific state determined by the set policy. We need some process for deciding which policies are likely to push the system in the direction of a set of states that would be desirable or contain mostly states that are desirable.

A procedure that might work in principle would be to sample the graph of states, looking for those that are considered desirable or acceptable in terms of their position in relation the cost/impact axes. Perhaps, if we are ambitious, we might look for states that are desirable or acceptable in terms of the finer structure belonging to other axes that we collapsed into our overall indices of cost and impact; these might include regional differences, assignment of cost to other actors, and institutional differences.

Given that we find a set we like, we would then consider the set of policy initiatives most likely to move industry in the direction of each state. Especially interesting are the families of policies that are common to moving in the direction of various kinds of states that are desirable or acceptable, policies that also do not appear likely to push industry in the direction of undesirable sets.

My model for policy choice among industrial ecosystems is statistical mechanics, which has developed very successfully to study systems consisting of a large number of interacting elements - particularly systems in which the large number of elements and possible interactions present an otherwise almost insuperable challenge to understanding the behaviour of the whole system.

Technical questions abound about this scheme, for example, about the relationship of families of policies to the resulting state sets to which they lead; about the degree to which simple measures of state desirability reasonably represent the complexity of real world variables; and about whether the policy sets leading to generally desirable states as described by the simple measures remain robust in the face of the underlying complexity. In looking at such policy choices and deciding whether they push the right way, discussions must also examine some of the ways in which a policy set might perversely or unexpectedly push industry towards undesirable states.

The way of arriving at end states need not be by a process of long-term governmental, or large-scale, collective, detailed planning and detailed regulation. My interest is in defining states and looking at policies that may foster a move in their direction, not an attempt to get to them by compulsion. I have emphasised provision of information, freedom to contract, and profitable commitments.

Experience suggests that systems with multiple tight connections between many elements tend to be brittle and easily become unstable if a few links are broken or a single actor is removed. Systems with loose connections are much more robust, especially if the network is of single and no more than double connections. The intention here is to suggest a direction for industrial development that would be realised by standard market mechanisms with the usual relatively loose, two-party network of connections or transactions. If this development is achievable, the issue of special brittleness of the new system need not arise to any greater degree than it customarily appears in any set of market relationships of manufacturers, suppliers, and customers. Clearly, the issue of robustness would need to be considered as a new system begins to develop.

7. CODA

I have sought to suggest a framework for thinking about materials and their flows in the context of industrial waste, about the balancing of costs and environmental impacts in possible future states of industry, and about a method of policy examination. I feel that this approach, while abstract, contains elements that make further discussion and elaboration of its possibilities worthwhile.

I believe that ten billion or so healthy people cannot prosper on Earth without a manufacturing industry and large, complex materials flows. A simple agrarian society will not be efficient or effective enough to support likely future human numbers. Yet, vast reductions in waste seem possible if we begin to re-conceive the ways we understand and operate industrial ecosystems. Jumping at solutions to particular environmental problems as they arise is easy but has not carried us nearly far enough. Little effort seems to have been given to the general, long-term, large-scale global waste problem in an analytical way. We need to stimulate more such thought.

ACKNOWLEDGEMENT

Reprinted with permission from Technological Trajectories and the Human Environment, copyright 1997 by the National Academy of Sciences; Courtesy of the National Academy Press, Washington, D.C.

---

[1] See Iddo K. Wernick, Robert Herman, Shekhar Govind, and Jesse H. Ausubel, "Materialisation and dematerialisation: Measures and Trends", *Dædalus* 125 (3) (Summer 1996).

[2] Energy wastes are also huge. Even in the most efficient economies, perhaps 10 percent of the energy extracted and generated from primary sources actually serves the end user. On the connection between material and energy wastes, see Nebojša Nakicenovic, "Freeing Energy from Carbon", *Dædalus* 125 (3) (Summer 1996).

[3] See Robert A. Frosch, " Industrial Ecology: Adapting Technology for a Sustainable World", *Environment* 37 (10) (1995): 15-24, 34 - 37; Robert A. Frosch, "industrial Ecology: Minimising the Impact of Industrial Waste", *Physics Today* 47 (11) (1994): 63-68; and Robert A. Frosch, "Industrial Ecology: A Philosophical Introduction", *Proceedings of the National Academy of Sciences USA 89* (1992): 800-803.

[4] For a general discussion of the meaning of sustainability, see Chauncey Starr, "Sustaining the Human Environment: The Next Two Hundred Years", *Dædalus* 125 (3) (Summer 1996) and Robert M. Solow, "An Almost Practical Step Toward sustainability", *Resources Policy* 19 (3) (1993): 162 - 172.

Chapter 4

Dematerialisation

Why and How?

ERNST ULRICH VON WEIZSÄCKER
President, Wuppertal Institut für Klima, Umwelt, Energie, Wuppertal.

Key words: **factor four, factor ten, ecological rucksacks, MIPS**

Abstract: **The provision of services to end consumers requires the movement of large masses of matter and energy. Each service carries an ecological rucksack. Sustainable development depends on a radical improvement in the materials and energy productivity of economic activities. Many opportunities for such improvements exist today, but the broader goal of a factor ten reduction in materials intensity will take many decades to achieve.**

1. INTRODUCTION: WHY SHOULD WE LOOK AT MEGATONNES?

We humans move more earth than do volcanoes and the weather combined (Schmidt-Bleek, 1994: 37). By moving things around we do a lot of damage. We overstress the earth's capacity to safely absorb the billions of tonnes that we return as waste or as overburden. The avalanches of matter may turn out to be the greatest threat to the global environment. `In the past, environmental protection has targetted nanograms. Now its high time for us to look at the megatonnes,' is one of the favourite sayings of Friedrich Schmidt-Bleek, a pioneer, along with Robert Ayres, in making materials - toxic or not - a major issue of environmental policy. A near exponential growth in world production of metals can be observed over the past two centuries (see *Figure 4*).

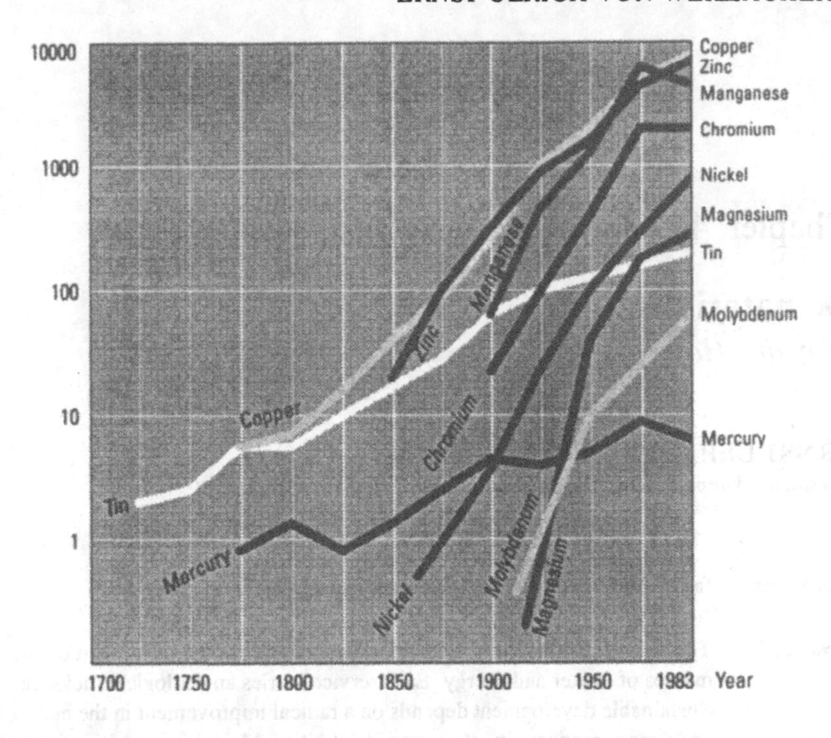

Figure 4. **Exponential growth in the use of metals by our civilisation has been the pattern for decades. Note the logarithmic scale (Ayres and Ayres 1996: 7).**

The question is whether the moving around of metals and other materials is causing any harm to the environment. Historically, the attention given to this question entered through the back door: what we noticed first in the industrialised countries was the waste problem. Landfills have been the cause of endless quarrels in and around American and European cities. For a while incinerators were welcomed as a solution; but soon they themselves became the targets of fierce environmental opposition. Recycling of waste materials was the logical next step, but again, recycling plants were unpopular in many neighbourhoods, partly because they were themselves polluting the environment. Progress remained rather slow. World-wide, the avalanches of waste kept growing.

From an economic growth point of view there are also benefits to this. Waste management has become a flourishing multi-billion dollar business creating more than a million jobs in the OECD countries. However, those who had to pay the bill for those jobs began to have a strong incentive for reducing their costs. Waste prevention also became a major focus for corporate management and for hundreds of able consultants. Recycling, re-manufacturing, dematerialisation acquired the status of a profitable activity.

On the other hand, technological progress was rapid in the fields of prospecting, mining, processing and shipping of material resources. With technological progress all these activities became more efficient and ever cheaper. Productivity improvements in resource extraction and processing have meant continued growth in the amount of virgin materials being used. Since we live in linear economies, where most material is rapidly transformed into a waste, the waste problem has grown. With the front door further open, more material has been passing through the back door.

Since 'back door' problems have their origin in the abundance and low cost of virgin resources and in the inefficiency with which we use these materials, it is reasonable to look at the ecological problems that occur at the early stages of materials cycles. So let us pay attention to the 'front door' problems. The most obvious problem is depletion of finite mineral resources, although I will focus on habitat and ecosystem destruction. Metal mines, quarries, roads leading to the mines, tailings often full of toxic metal residues - pose a substantial danger to the habitats of plants and animals. As the megatonnes shifted around are increasing and the network of roads and railways become more extensive, the dangers for biodiversity keep growing too.

Habitat destruction may be only a part of the problem. The chemistry of soil and surface waters may also undergo unfavourable changes as a result of massive physical disruptions caused by resource extraction. Indeed, as Wilhelm Ripl (1994) has shown, earth movements and human-induced changes of water flows have begun seriously to erode the cation concentrations in soils and watershed, leading to acidification. Ripl holds that the causes for soil and water acidification in South Scandinavia lie more with earth movements than with acid rain. ***Figure 5***, a qualitative drawing by Ripl indicates a dramatic loss of cations in recent decades. Nearly all the reserves built up in the ice age have been used up in just a few decades.

Earth movements have another effect on acidification. Peter Neumann-Mahlkau (1993) reports that more sulphur dioxide is created by the mobilisation of atomic sulphur through soil movements than by all industrial combustion processes taken together. Mining, processing and shipping of material resources is not only the source of local environmental degradation but also of greenhouse gas emissions (Young, 1992).

2. ECOLOGICAL RUCKSACKS

The main cause of the huge avalanches of matter is the low concentration of ores in the earth's crust. Each one of the 'raw materials' that is in use in our civilisation, has a history of mining, refining and shipping involving the transformation of many more tonnes. Metals have been extracted from ovens

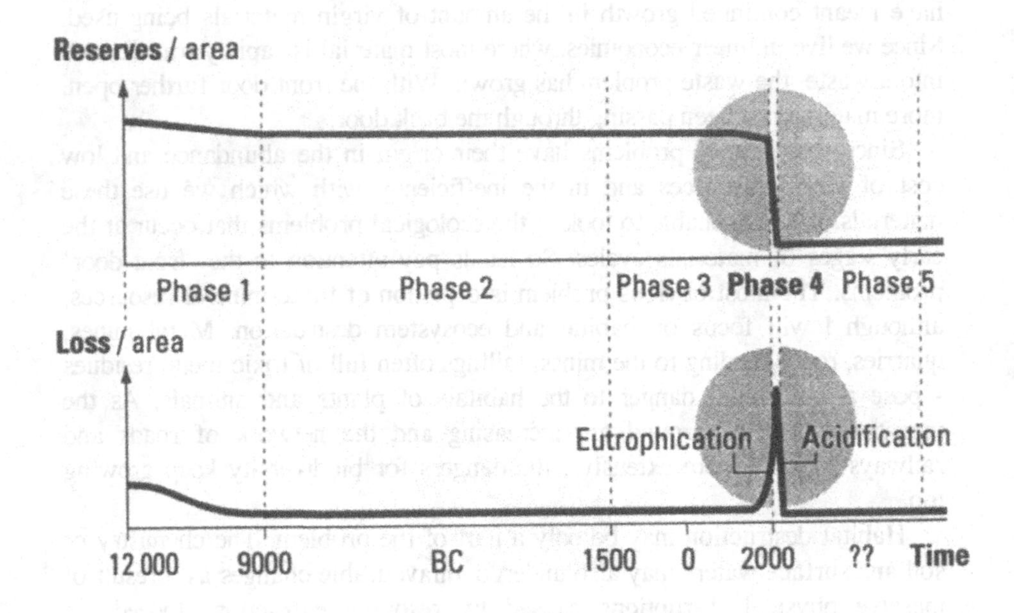

Figure 5. Loss of cations in Scandinavian soils. A large stock of cations was built up in Scandinavian soils and was maintained over the post-glacial millennia. A hundred years of earth movements and drainage have drained so much cation off the soil that lasting acidification followed with a severe danger of desertification.

containing mostly stones and dust. In the case of gold and platinum the ratio between metals and overburdens is roughly 1:300,000. Often these leftovers, overburdens or tailings from metal mining are poisonous to plants and animals. Cleaning the tailings is not a trivial matter and will hardly ever to restore already contaminated aquifers, watersheds, habitats and landscapes.

Referring to the mass movement on the mine and further down the road towards the manufacturing of commercial metals, Schmidt-Bleek (1994) speaks of the `ecological rucksack' carried by the metal. *Figure 6* shows the relations between the metal (and other mined products) obtained and their respective ecological rucksacks.

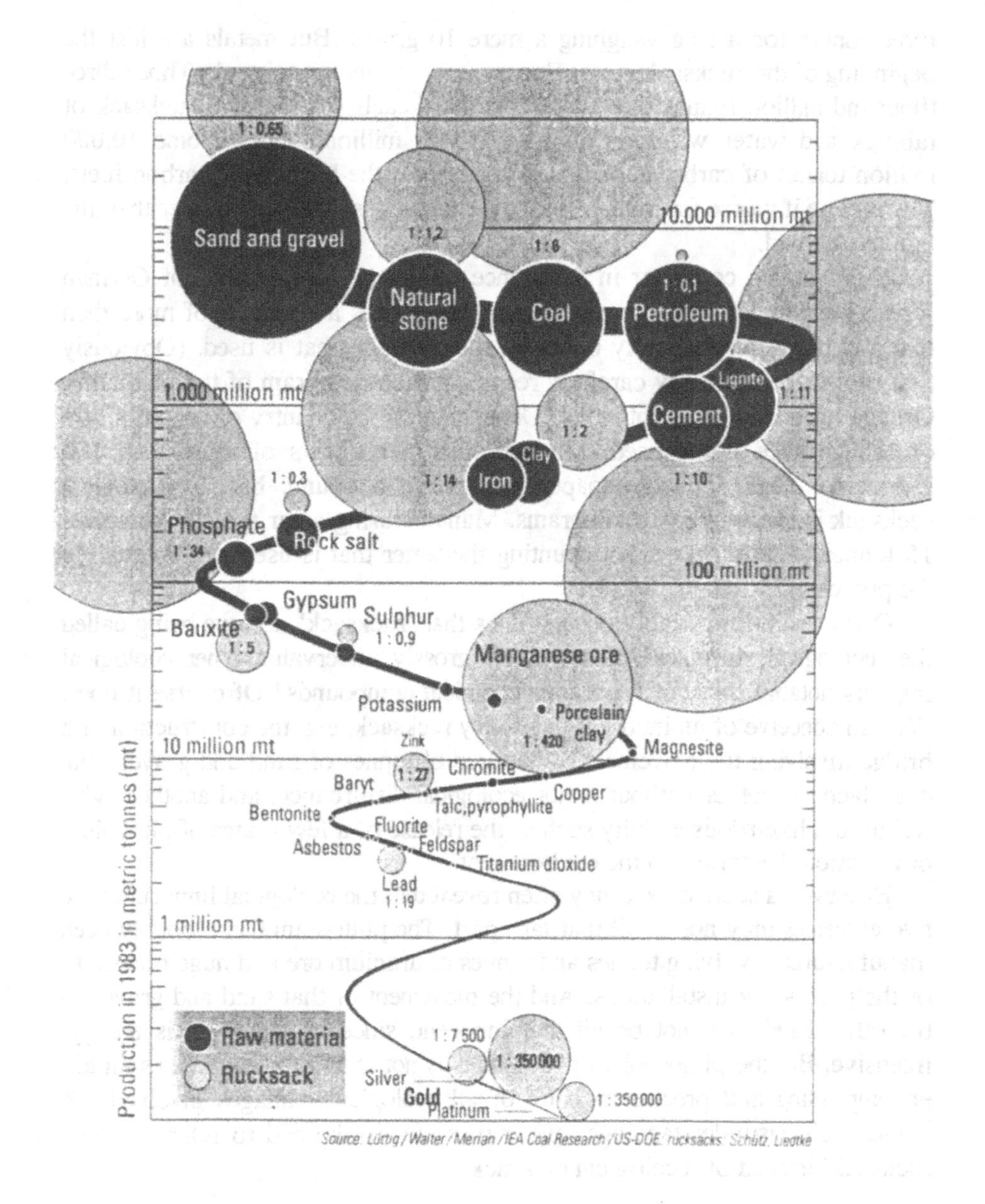

Figure 6. The ecological rucksacks of metals. A kilogram of metals may involve tonnes of materials to be moved and processed. Schmidt-Bleek (1994) calls this the ecological rucksack. Some examples are given of ecological rucksacks typically carried by different metals. Gold is among the worst.

In the case of gold and platinum the relation is 1: 350,000. Imagine the weight of the ecological rucksack of the gold ring on your finger. It would be three tonnes for a ring weighing a mere 10 grams. But metals are just the beginning of the rucksack story. Energy, too, carries a rucksack. Those three thousand million tonnes of coal that we burn each year carry a rucksack of tailings and water weighing at least 15,000 million tonnes. Some 10,000 million tonnes of carbon dioxide are released in the burning of carbon fuels. The picture if worse for lignite. Here the rucksack is 10 times heavier than the lignite obtained.

The catalytic converter in cars, once heralded as the saviour of German forests, weighs less than nine kilograms but carries a rucksack of more than two and half tonnes, chiefly owing to the platinum that is used. (Obviously that rucksack shrinks by carefully recycling each milligram of the platinum). Orange juice isn't innocent either. Depending on the country of origin, a litre of orange juice has caused soil and water movements of more than 100 kilograms. Your daily newspaper may weigh a pound, but its ecological rucksack is likely to be 10 kilograms. Manufacturing a car typically involves 15 tonnes of solid waste, not counting the water that is used and polluted in the process.

One question invariably occurs: does that `rucksack' deserve being called the `ecological rucksack'? Does it not grossly undervalue other ecological impacts notably those of hazardous chemical compounds? Of course it does. We can conceive of an innocent but heavy rucksack, e.g. the construction of a bridge involving the movement of hundred of tonnes of sand and gravel from one place to another without major ecological disturbance; and another light-weight but hazardous activity such as the release of a few grams of plutonium or infectious bacteria into the environment.

However, a second look may often reveal that the ecological impacts of the two extremes may not be all that far apart. The plutonium has probably been manufactured involving tonnes and tonnes of uranium ore and huge rucksacks of the processing installations. And the movement of that sand and gravel on the other hand may not be all that innocent, since its transport is energy-intensive. But the philosophy of rucksacks is not meant to be taken as an all-encompassing and precise measure of all ecological damages involved. To avoid such misunderstandings, one may prefer in the end to refer to `mass rucksack' instead of `ecological rucksack'.

3. MIPS: A QUICK AND DIRTY YARDSTICK

Knowing about the limitations of the rucksack concept, one will nevertheless want to have some quantitative formula allowing comparisons and assessments. For this, Schmidt-Bleek (1994) has made a bold attempt by establishing the Material Intensity Per Service (MIPS) concept. For all types of goods and services, he and his team estimate and calculate the material flows in tons on a cradle to grave basis. They do it for the ring on your finger, for the daily newspaper, for orange juice and for autos. Although goods tend to represent the availability of services, it is ultimately services that interest the user of goods. It is the mile driven in a passenger car or the convenience of sitting in a comfortable chair or the 'service' of showing marital status by a ring that would ultimately be used as the denominator in the MIPS calculation. (The definition of a service actually turns out to be the most difficult part for Schmidt-Bleek's MIPS calculation team!) Longevity of goods has a positive influence on MIPS. For many goods longer life leads to a decrease in the MIPS of the respective services.

MIPS is sometimes taken to be a definitive and generalised yardstick for the ecological impacts of services. This is overdoing the case for MIPS. Other and more refined yardsticks such as toxicity, land-use and greenhouse gas emissions, are available. However, it is extremely valuable to have something truly simple for the `quick and dirty' assessment of ecological impacts. Estimating mass inputs to a product or a service is a good, simple proxy measure.

Using - for a first approximation- the MIPS yardstick we can now proceed to look for strategies to make rucksacks lighter, i.e. to `dematerialise' our economy. As Amory Lovins and I have shown in Factor Four (Weizsäcker, Lovins and Lovins, 1997), a quadrupling of material productivity can be reached in many cases. The book features twenty examples of quadrupling energy productivity. It also provides twenty examples of reducing material needs, i.e. dematerialisation. They include efficient water use in paper manufacturing, households and agriculture; waste reduction in industry; a fourfold durability of furniture and appliances; information substituting for journeys and mail; plastic not requiring mechanical sorting for recycling; and products using materials with small `rucksacks'.

Energy productivity turns out to be an essential component in reaching high levels of cradle-to-grave material productivity. First, because energy itself has a substantial ecological rucksack, and second because energy in most cases is the driving force for material movements of all kinds. Hence, the twenty examples in Factor Four of quadrupling energy productivity help support the goal of a quadrupled resource productivity, while also helping to solve other ecological problems such as the greenhouse effect.

4. THE FACTOR TEN CLUB

A factor four in material productivity would make ecological rucksacks lighter by that factor which gives much relief from some ecological problems.

The trouble is that a factor of four may just not be enough for ecological sustainability. Schmidt-Bleek thinks a longer term aim of a factor of 10 improvement in energy and materials productivity should be adopted.

Looking at the total impact of human interference on the biosphere, Schmidt-Bleek, like others (e.g. Rees and Wackernagel, 1992; Weterings and Opschoor, 1992) comes to the conclusion that material turnovers should be reduced by some 50% at least on a world wide basis. Given the fact that per capita consumption is something like five times higher in OECD countries than in developing countries, and that further increases in world population are unavoidable, he says that sustainable levels of material flows will not be reached unless the material intensity of the OECD countries is reduced by a factor of ten. This is then the material target of sustainability.

Based on these considerations, Schmidt-Bleek has taken the initiative of founding the 'Factor Ten Club' of prominent environmentalists subscribing to that goal.[1] The principles of the Club were laid down in the Carnoules Declaration in October 1994. The Declaration calls for an efficiency revolution de-subsiding resource use and a new perception and definition of welfare. It is acknowledged that the reduction of material turnovers in many cases could lead to reduced economic turnover, i.e. reduced GDP. Hence, one would expect resistance to resource productivity from old-fashioned economists and from political camps that are exclusively concerned with economic growth and the labour market.

5. FACTOR FOUR VERSUS FACTOR TEN

People keep wondering why the author of Factor Four is also a member of the Factor Ten Club. What is the logic in it? The answers are threefold:

Factor four may be seen as a first big jump towards Factor Ten. If a factor of ten may realistically take a hundred years to reach, a factor of four may take only forty years.

Energy efficiency, so it seems, cannot realistically be increased by a factor of ten, but a factor of four may indeed be possible. Think of refrigerators, cars or light bulbs. From today's average models in use, a step-jump improvement by a factor of four in energy efficiency may be the boldest thing that is imaginable today. On the other hand, reducing material intensity by a factor of ten may not be outlandish at all when thinking of a car with a doubled life-time and a careful reduction by a factor of five of the total mass movements involved in the manufacturing process; the re-manufacturing of the car body alone would yield such a factor if compared to the production from virgin materials.

The Factor Four book was meant as an encouragement to engineers, managers, end-consumers and opinion-leaders. For this purpose it had to provide an abundance of simple examples. Simplicity has a price. For a refrigerator or a steel pylon or the freight capacity of a rail track, more than a factor of four would just not be imaginable. On the other hand, for a complex process such as the daily food diet or the cradle-to-grave optimisation of traffic is considered, a factor of ten may be possible. But such complex systems are much more difficult to transform, and frequently require institutional and behavioural changes, as well as technological changes.

6. CONCLUSION

Bold steps towards increased resource productivity are available and they are ecologically imperative. But can they become part of a realistic strategy in a world preoccupied with unemployment and growing inequality? Yes they can. Resource productivity in a time of growing demands and shrinking resources should be an economic imperative too. And it will help to de-emphasise labour productivity which should be good for saving and creating jobs.

What may be beneficial at the macro-economic level is not necessarily profitable at the level of the firm. Economic and fiscal policies are available, however, to make it profitable. Perverse subsidies have to be removed. As André de Moor and Peter Caldway (1997) have shown, world-wide some US$ 700 billions are spent annually for ecologically undesirable activities in the fields of water, energy, auto use and agriculture alone. Ecological tax reform can also help spur resource productivity and reduce the need for labour rationalisation.

If economic and fiscal policies have a new vision of technological progress and are adjusted accordingly, no contradiction exists between resource productivity, high employment rates and economic welfare. We shall have to join forces with economists, business managers, engineers and many others in society to make resource productivity and dematerialisation happen.

REFERENCES

Ayres, Robert, U. and Leslie W. Ayres (1996). Industrial Ecology: Towards Closing the Materials Cycle, Cheltenham: Elgar.

Dieren, Wouter van (ed.) (1995). Taking Nature into Account. New York: Springer.

Moor, André de and Peter Calamai (1997). Subsidising Unsustainable Development. Undermining the earth with public funds, Toronto: Earth Council (ISBN: 0-9681844-0-5).

Neumann-Mahlkau, Peter (1993). Acidification by pyrite weathering on mine waste stockpiles, Ruhr District, Engineering Geology, 34, 125-134.
Rees, William and Matthis Wackernagel (1992). Ecological footprints and Appropriated Carrying Capacity in A.M. Jannson et al. (eds.). Investing in Natural Capital, Washington: Island Press.
Ripl, Wilhelm (1994). Management of watercycle and energy flow for ecosystem control: the energy-transport-reaction (ETR) model, Amsterdam: Elsevier.
Schmidt-Bleek, Friedrich (1994). Wieviel Umwelt braucht der Mensch? Basel: Birkhäuser.
Weizsäcker, Ernst von, Amory Lovins and Hunter Lovins (1997). Factor Four. Doubling Wealth, Halving Resource Use, London: Earthscan.
Weterings, R. and J.B. Opschoor (1992). The Ecocapacity as a Challenge to Technological Development, Rijswijk: Advisory Council on Nature and Environment.
Young, John E. (1992). Mining the Earth. Worldwatch Paper No. 109. Washington: Worldwatch Institute.

---

[1] Among the Club members are Jacqueline Aloisi de Larderel, Director of UNEP's Industry and Environment Programme, Leo Jansen and Claude Fussler who have written chapters for this book, Herman Daly, formerly at the World Bank, Ashok Khosla, President of Development Alternatives, India; Jim Mac Neill, the former Executive Director of the Brundtland Commission; Hugh Faulkner, former Executive Director of the Business Council for Sustainable Development; Richard Sandbrook, Executive Director of the London based International Institute for Environment and Development; and Wouter van Dieren who has produced the new report *Taking the Earth into Account* (Van Dieren, 1995) for the Club of Rome; Robert Ayres has been a founding member of the Club; I am also proud of being a member.

Chapter 5

Analytical Tools for Chain Management

HELIAS UDO DE HAES, GJALT HUPPES & GEERT DE SNOO
Centre of Environmental Science, Leiden University.

Key words: Company chains, ecolabelling, LCA, SFA, MFC, energy analysis

Abstract: **A range of tools exists for analysing the materials and energy flows through chains of economic activities. These include life cycle analysis (LCA), materials flow analysis (MFA) and substance flow analysis (SFA), among others. These frameworks, their appropriate use, and their strengths and weaknesses are discussed. The chapter makes clear that these tools must be used within the context of participative decision- and policymaking processes.**

1. INTRODUCTION AND OVERVIEW

Chain management refers to the management of chains or networks of economic processes, in contrast to the management of single processes. Chain management is discussed here in the context of environmental management since it creates additional possibilities for influencing economic processes directly and indirectly through other processes in the chain. Thus, by purchasing environmentally friendly products, consumers will positively influence upstream processes, dealing with the production of given products. Chain management may also offer better possibilities for pollution prevention and resource conservation. For instance, the way in which products are manufactured may aim at the prevention of emissions during a later stage of the life cycle, or may facilitate recycling of the product or its materials. Finally, a more encompassing chain approach may help to avoid a possible shifting of problems, from one area to another area, the shifting of environmental burdens from one societal group to another, or the shifting of risks to the future. The aim of chain management is to reduce or avoid such unwanted consequences of environmental management.

Chain management may focus on different objects. Managing each object requires the adoption of specific policy instruments and the use of specific analytical tools (See *Table 6*).

Chain management of product stewardship can focus on product systems including all processes from cradle-to-grave related to a given product. It thus focus on the function delivered by the product. The term "product" is used in the broad economic sense: it also refers to services, like for instance waste management or a holiday trip. Chain management may start with quality aspects of the product itself, in as far as these are determined by upstream processes. An important example concerns the hormone and pesticide residues or germs in meat due to upstream processes. It may also include the burdens to the environment of the different processes involved. A typical policy instrument is ecolabelling, which aims to include the environmental performance of the product during its whole life cycle. Other instruments include green product marketing by companies, green purchasing policy of governmental agencies or firms ("green procurement") and technical standards (c.g. "no cadmium pigments"). The analytical tool for this type of chain management is Life Cycle Assessment or LCA for products in the sense of both technical goods as well as immaterial services (cf. Heijungs et al., 1992; Consoli et al., 1993; ISO/14040, 1996).

Table 6. Chain management: Objects, policy instruments and analytical tools

object of chain management	policy instruments	analytical tools
product systems	green marketing, ecolabelling, purchasing policy, admission requirements	LCA, MPC
materials/substance/energy metabolism company chains within region cradle to grave	taxes, subsidies, (emission permits (eco-audit), (certification), green marketing agreements, covenants	MFA/SFA, Energy Analysis, environmental IOA (environmental accountancy), (company ecobalance)

Chain management can also focus on the metabolism of materials and energy in the socio-economic system in a given region; and it may also include the management of the flows in the environment. The relevant policy instruments are subsidies or tax measures aiming at influencing the material use of a given region as a whole. The analytical tool which supports this type of chain management is Material Flow Accounting (MFA). It analyses flows and accumulations both in the physical economy as well as in the environment. MFA can investigate total mass via bulk materials like "plastics" or "fibres", groups of related substances like chlorinated hydrocarbons or nitrogen compounds, or single elements like cadmium or mercury. The tool which deals with the latter two types of objects (single elements and groups of related substances) is generally denoted as Substance Flow Analysis (SFA). Other tools which belong to this family are Energy Analysis, dealing with energy carriers, feedstock energy and energy dissipation in a given region, and the different types of environmental Input-Output Analysis (cf. Schrøder, 1995).

There can also be combinations of the tools mentioned above. Thus, an LCA study can be directed at focal points of a SFA; the tool for this is called "Material-Product-Chain (MPC) analysis". This tool also (cf. Kandelaars et al., 1996).

Chain management can focus on the environmental functioning of a single company. Here policy instruments include the environmental audit and, based on that, environmental certification, either in the framework of EU-EMAS or ISO-EMS. A major incentive will be the public image of a company. Analytical tools which support these are environmental accountancy of a firm, or what might be called "company ecobalance" (cf. Braunschweig and M. Müller-Wenk, 1993).

Chain management may also look at companies in a chain or network perspective in terms of "horizontal" relationships within a region and "vertical" relationships in a production-consumption chain. The first aims for instance at an optimal allocation of waste flows within a region for further processing. The second option implies that companies will include the environmental performance of their suppliers in their own environmental management decisions, again in terms of the functioning of the supplying companies as a whole. In EMAS or EMS terms this concerns the "indirect effects". Policy instruments which aim to implement these are trade agreements between companies and covenants between companies and authorities. For instance, a company may aim to chose EMAS or EMS certified companies as suppliers. It is as yet an open point whether there is one specific analytical tool which can support this type of chain management.

In the next two sections the analytical tools of the first two types of chain management will be discussed: LCA and MFA. In the following section the possibilities and limitations of the different tools are discussed, together with the possibilities for their combined use. The chapter concludes with a discussion on the relevance of procedural requirements, aiming at an interaction between the technical study and external experts and stakeholders.

2. LIFE CYCLE ASSESSMENT, A TOOL FOR THE ANALYSIS OF PRODUCT SYSTEMS

2.1 Short history

Life Cycle Assessment (LCA) originated in the early seventies and studies were performed in countries like, Sweden, the UK, Switzerland and the USA. The products which got primary attention were beverage containers (cf. Udo de Haes, 1993). During the seventies and the eighties numerous studies were performed, using different methods without a common theoretical framework. These applications led to confusion and discredit for the 'cradle-to-grave' framework because LCA was directly applied in practice by firms in order to substantiate market claims. The research results differed greatly, although the objects of the study were often the same, thus preventing LCA from becoming a more generally accepted and applied analytical tool.

Since 1990 a number of changes have occurred. Under the co-ordination of the Society of Environmental Toxicology and Chemistry (SETAC) exchange between LCA experts has increased, and efforts are being made to harmonise the methodology. The SETAC "Code of Practice" is the outcome of this process (Consoli et al., 1993). Since 1994, the International Organisation for Standardisation (ISO) also plays a role (Technical Committee 207, Subcommittee 5). Whereas SETAC has primarily a scientific task focused at methodology development, ISO has the formal task of methodology standardisation. In 1996 the general guideline had reached the level of a draft international standard (ISO/14040).

The LCA methods are becoming sophisticated, in line with the development of software and databases. Independent peer review of the results of an LCA study, which is of crucial importance for the credibility of the results. This section describes the state of the art of LCA methodology, present developments and future perspectives.

2.2 Definition and applications

In ISO/14040 LCA is defined as follows: "LCA is a technique for assessing the environmental aspects and potential impacts associated with a product by compiling an inventory of relevant inputs and outputs of a system; evaluating the potential environmental impacts associated with those inputs and outputs; and interpreting the results of the inventory and impact phases in relation to the objectives of the study". Products provide services, but in the following we will speak of a product as pars-pro-toto for all objects of LCA.

The following points are of special relevance. The reference for the study is the function which is delivered by a product. This means that ultimately all environmental impacts are related to this function, being the basis for comparisons to be made. The product, which delivers this function, is studied during its whole life cycle; all processes related to the product during its whole life cycle are together called the "product system". These processes are studied using a systems analysis approach, employing quantitative, formalised mathematical tools. A clear distinction is made between objective and normative parts thereby ensuring transparency.

LCA can be applied at different levels. LCA can be used at an operational level for product improvement, design and comparison. The latter is, for instance, important in relation to ecolabelling programmes or purchase preference schemes (cf. European Commission, 1997). LCA can also be used at a strategic level by companies or authorities. Companies may use LCA as a source of guidance for business strategies, including decisions on which types of products to develop, which types of resources to purchase and which type of investments to make for waste management. A recent example concerns the proposal of "overall business impact assessment" (OBIA), being an LCA for a company as a whole (Taylor, 1996). Authorities may use LCA for comparing and contrasting environmental policy options, for instance in the field of waste management, of energy policy or of transportation.

The use of LCA in comparative studies and, in particular, competitive ones, makes the highest demands on methodology and data. A particular example concerns the use of LCA in ecolabelling programmes. These often include studies which meet with public criticism (see chapter 22). This is much less the case with internal applications aiming at the improvement of products or new product design, or studies which aim to improve the general environmental strategy of a firm.

2.3 Technical framework

In order to make LCA a tool for comparative purposes a first step concerns standardisation of a technical framework and of terminology. The ISO framework consists of the following four phases - goal and scope definition (see 2.4), inventory analysis (see 2.5), impact assessment (see 2.6) and interpretation (see 2.6). From *Figure 7* it is apparent that LCA is not a linear process, starting with the first and ending with the last phase. Instead it follows an iterative procedure, in which the level of detail is subsequently increased. The phases are described below:

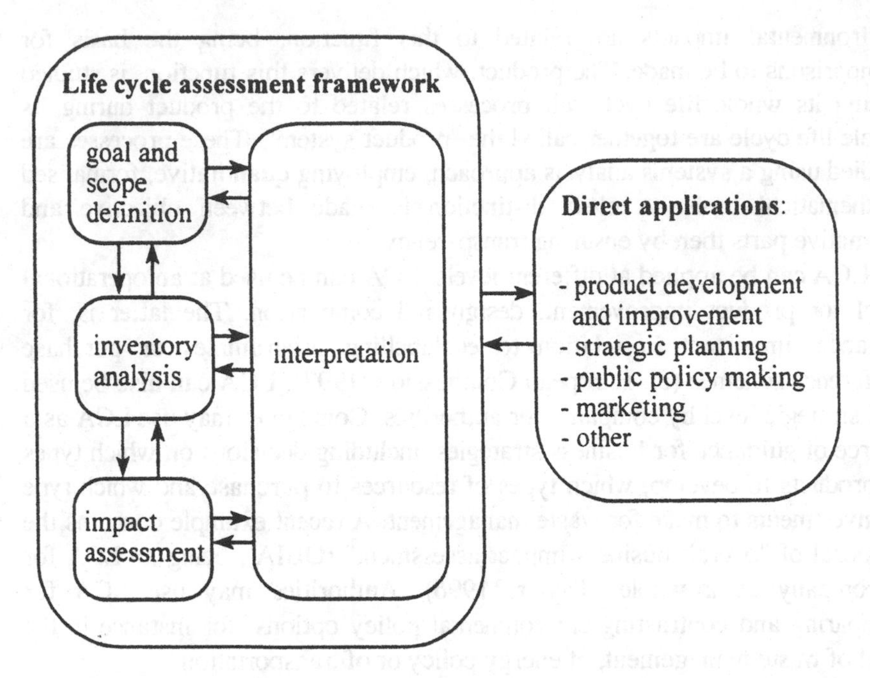

Figure 7. Technical framework for Life Cycle Assessment (source: ISO/140401996)

2.4 Goal and Scope Definition

The Goal and Scope Definition phase includes the following elements: specification of the purpose and scope of the study; definition of the functional unit; establishment of the level of detail required for the application at hand; and a procedure for ensuring the quality of the study.

Specification of the purpose and scope of the study includes the choice of the products which will and which will not be taken into account, which can be a point of serious debate. Further, the functional unit has to be defined. The functional unit is the central, quantitative measure of the function to be delivered. The definition should be undertaken with great care. It would be incorrect, for example, to compare one milk bottle with one milk carton, as the former is used many more times than the latter. "The packaging of 1000 litres of milk" may be a better reference basis. Still better may be "the packaging needed for the consumption of 1000 litres of milk", because then losses from breakages will be taken into account. In comparing different types of paint the unit of "1 litre of paint" will clearly be insufficient. "The paint needed for painting 1 m2" is already better; still more precise is "the paint needed for 1 m2 over a period of 10 years". The latter takes also the longevity of the painted surface into account. The definition of the functional unit also determines which type of alternatives can be taken into account. "1 km of transport" could possibly include all transportation modes; "1 km of car transport" clearly does not. The functional unit has no significance in absolute terms; it only serves as a reference value. In public comparative applications, for so-called "comparative assertions", stakeholders should be included in the specification of the function, the identification of the products which sufficiently fulfil this function, and the definition of the functional unit.

The reliability of a study depends not only on its level of detail but also on the procedure of quality assurance. This can be performed by an internal or an external, independent panel of experts. For comparative assertions the latter is a strict requirement.

2.5 Inventory Analysis

The Inventory Analysis is the most objective and time consuming part of the study. It includes: drawing a flow chart of the processes involved; definition of the boundaries between product system and environment; specification of processes and data gathering; allocation regarding multiple processes; and compilation of the inventory table. If a study is only performed up to the inventory analysis, it is called an LCI, i.e., a Life Cycle Inventory.

The flow chart identifies all relevant processes of the product system with their materials and energy relationships. For a flow chart the system boundaries between the product system (as part of the physical economy) and the environment have to be defined. This implies that the flows that cross the boundaries ("elementary flows" in ISO terminology) have to be defined. *Figure 8* shows the potential for confusion on this point in relation to forests and other biological production systems. Is wood a resource coming into the physical economy? Or is the forest already part of the economy and are solar energy, CO_2, water and minerals to be regarded as the elementary flows passing the boundary between environment and economy? Similarly, is a landfill to be regarded as part of the environment or still as part of the physical economy? In the first case all materials which are brought to the landfill have to be regarded as emissions into the environment; in the latter case this will only hold for the emissions from the landfill to air and groundwater. In order to make the results of different studies comparable there is a need for harmonisation here. A criterion for determining the boundaries may well be the degree to which the processes involved are steered by human activities. A forest plantation can be regarded as part of the socio-economic system. But wood extracted from a virgin forest will have to be regarded as a critical resource taken from the environment. Likewise a landfill, managed without any control measures should be regarded as part of the environment, with all discarded materials to be regarded as emissions. If the landfill is a well controlled site, separated from ground water and with cleaning of the percolation water, one may well regard this as part of the product system with only the emissions from the landfill to be considered as burdens to the environment.

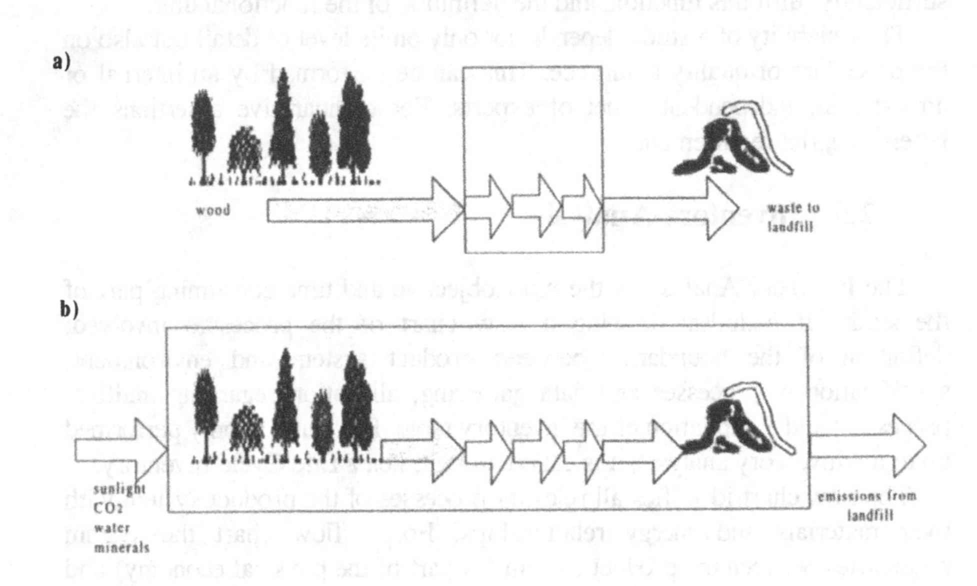

Figure 8. Two ways of defining system boundaries between physical economy and environment in LCA; a) with narrow system boundaries, b) with extended boundaries.

Given the system boundaries, the next step is the specification of the processes involved which includes a cut-off procedure determining which processes have to be analysed with respect to their impacts and which can be left out. In part it is a choice which has to be made on a case-by-case basis and will be solved in an iterative process. But there are also some general principles. The most important one relates to the impacts of basic utilities (electricity, transport, water supply, waste management). This principle states that the functioning of the capital goods concerned (the production of electricity, the driving of trucks) should be taken into account, but not the production of the capital goods (the building of the electricity plant, the manufacturing of the trucks).

The next step is data gathering about the different inputs & outputs of the process. Data are often obsolete, variable, or secret. The variability of data is in part a problem and in part an issue of interest. To start with the latter: if firms have different processes for the production of the same product, this may be the reason for awarding or not awarding it an ecolabel and thereby steer product improvement. The problem lies in the variability of the so-called "background data", i.e., data about commonly used materials and basic utilities for which averaging is necessary. A guiding principle for this averaging should be the factual size of the market: if steel has a world market, then a world average should be aimed at; if electricity has an EU market, then the average for this market should be the basis for the averaging.

The next issue on the boundaries between the product system under study and other product systems concerns simple decisions about which processes belong to which product system and which do not. The problem lies in processes which are part of more than one product system, the so-called "multiple processes" (see *Figure 9*). How should the environmental impacts of these processes be allocated to the different product systems involved? Until now the procedures used have been rather arbitrary, using rules based on mass or energy data, although they may have had great influence on the final results of the study. There are three basic types of multiple processes: co-production, combined waste processing and recycling.

Co-production means that one unit process produces more than one functional output. The question is how should the environmental burdens (the elementary flows from and to the environment) be allocated to these different functional outputs? Traditionally this is done on a mass basis. However, the example of diamond production which goes together with the production of a great bulk of stones as a by-product shows that this may not be equitable: all burdens would be allocated to the stones and not to the diamonds, although the latter are the reasons for the existence of the mine. Another principle concerns allocation on the basis of economic value, as the key steering factor for all production processes.

With combined waste treatment the problem is that emissions from an incineration plant will contain a broad spectrum of materials, which will definitely not be included in a great deal of the burned wastes. Allocating the emission of cadmium to the waste management of a PE bottle again is not equitable. The procedure should begin here with a causality principle linking as far as possible materials to different fractions of the waste. As this is not always possible, allocation on the basis of physical characteristics or economic value may be undertaken.

With recycling we can distinguish between closed loop and open loop situations. In the former there is no allocation problem, because there is only one product at stake. Generally loops will be partly or totally open: the wastes from one product system will be used as a secondary resource for another. This is a multiple process for which an allocation rule has to be defined. In present practice often a "50% rule" is used, giving an equal share to the two product systems involved, but more sophisticated logic may be applied. One may also want to allocate part of the resource needs for product system A to product system B because the latter also makes some use of the resources;

and the wastes from product system B to system A because system B also solves the waste problem for system A.

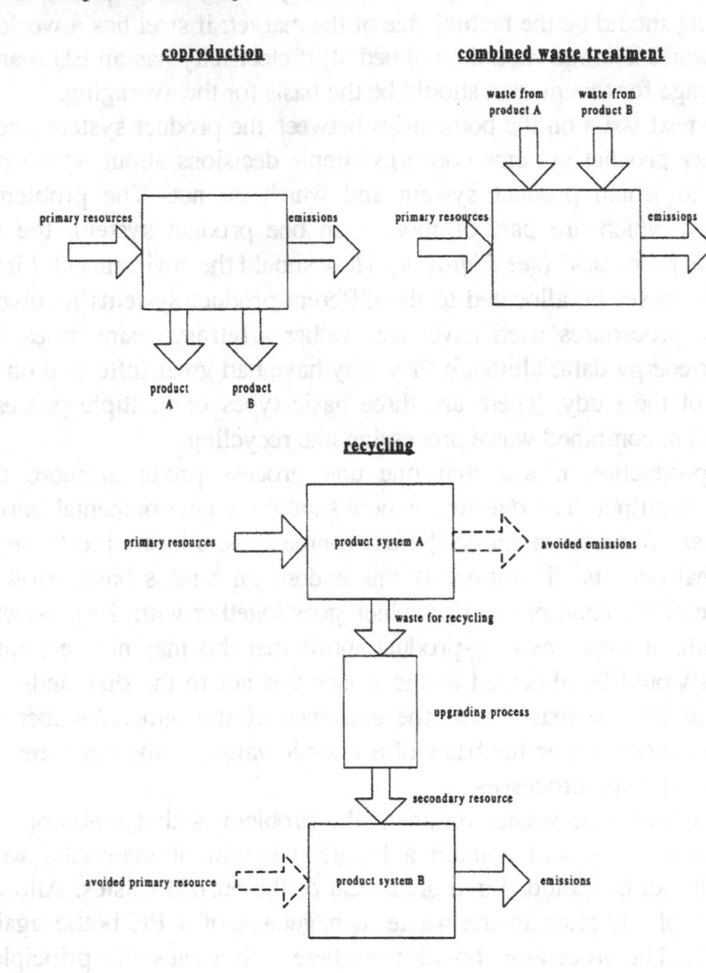

Figure 9. Allocation of environmental burdens in multiple processes; horizontal arrows indicate flows from and to the environment, vertical arrows flows from and to other product systems

One consistent allocation framework should be developed for all multiple processes. This is a key issue of the ISO standardisation of the inventory methodology (cf. ISO/CD 14041). Such a procedure does not imply a specific choice but rather a priority of different options to be checked on their applicability one after the other. This preference order consists of the following steps:

- check whether allocation can be avoided by dividing processes into sub-processes;

- check whether allocation can be avoided by expanding the boundaries of the system (the "avoided burden approach");
- apply principles of physical causality for allocation of the burdens; and
- apply other principles of causality, for instance economic value.

The inventory analysis concludes with the compilation of the inventory table, the total list of elementary flows (the extractions and emissions) connected with the product systems investigated. In the computation process care must be taken that loops of flows are taken into account properly; for instance electricity production requires steel and vice versa. Apart from the quantitative entries, the inventory table may also include issues which cannot be dealt with in a quantitative way but which have to be considered in the final appraisal of the results.

2.6 Impact Assessment

In ISO/14040 life cycle impact assessment is defined as the "phase of life cycle assessment aimed at understanding and evaluating the magnitude and significance of the potential environmental impacts of a product system". This phase interprets the burdens included in the inventory table in terms of environmental problems or hazards and it aggregates these data for practical reasons; a list of 50 or one hundred entries cannot be dealt with in decision making.

In the seventies, impact assessment was conducted in an implicit way, by defining a number of broad, inventory based parameters, which were thought to be indicative of the total spectrum of impacts. Examples of such parameters include net energy consumption, total input of resources and the total solid waste output (Hunt et al., 1974). A more recent example is the Material Input Per Service unit method (Schmidt-Bleek, 1994) in which the total material input is quantified (see Chapter 4). These approaches are time-efficient and can lead to robust results. However, all types of impacts may not be sufficiently covered by such a small number of inventory based indicators. Hence, since the mid-eighties, methods are being developed for impact assessment as an explicit LCA phase on its own.

The Swiss (BUS, 1984) and Dutch (Druiff, 1984) "critical volumes approach" directly related the outputs of the inventory analysis to some reference values (e.g. standards), to be followed by an addition of the resulting figures. This addition takes place in "volumes polluted air" and "volumes polluted water". Although practical, this procedure did not take scientific knowledge about environmental processes adequately into account, has to deal with large differences between standards and the problem that for a great number of impacts no standards are available (including resources, physical damage to ecosystems and emissions of CO_2 and CFCs). In the same period the so-called EPS-system was developed in Sweden (Steen and Ryding, 1993), based on economic valuing of the inventory output, in which impact assessment is also performed in one single step.

For transparency reasons, the ISO procedure (ISO/CD 14042) is a stepwise approach that separates the scientific and normative elements as much as possible. The successive elements are selection and definition of environmental categories, classification, characterisation and weighting

In the selection and definition of environmental categories a "problem theme approach" is generally followed for defining impact categories as proposed by CML in the Netherlands (Heijungs et al., 1992). The categories are defined here on the basis of the resemblance to environmental processes. Table 7 presents a list of impact categories as a structure for the analysis and as a checklist for the completeness of the impacts considered. This list also indicates the relevance of the different impact categories for three "general areas for protection", in fact three domains of societal values. A further analysis of the whole web of cause-effect chains may well lead to a more consistent list of impact categories.

Table 7. Impact categories for life cycle impact assessment, together with the spatial scope of the categories; Source: Udo de Haes (1996)

Impact categories	Scale
Input related	
abiotic resources (deposits, funds and flows)	global
biotic resources (funds)	global
land	local
Output related	
global warming	global
depletion of stratospheric ozone	global
human toxicological impacts	global/continental/regional/local
ecotoxicological impacts	global/continental/regional/local
photo-oxidant formation	continental/regional/local
acidification	continental/regional/local
eutrophication (include. BOD and heat)	continental/regional/local
odour	local
noise	local
radiation	regional/local
casualties	local

Pro Memoria: Flows not followed up to system boundary input related (energy, materials, plantation wood, etc.) output related (solid waste, etc).

Classification implies that the extraction and emission data (the so-called "elementary flows") are quantitatively assigned to the relevant categories. Thus, SO_2 has to be assigned to at least acidification and human toxicity.

For characterisation of the extraction and emission data different methods can be used; their aim is both quantification and aggregation of these data within the given environmental categories. These should include a measure of both the fate and effect of the substances. The fate aspect involves the distribution over and persistence within the different environmental media. Here stationary multimedia models may well form the basis. The effect aspect gives information about the sensitivity of defined indicators in the cause-effect chain. These indicators can be chosen at the level of the endpoints, such as is the case with human and ecotoxicity. But they can also be chosen earlier in the cause-effect chain, such as is the case with global warming and ozone depletion. Here the indicators are climate forcing and depletion of the stratospheric ozone, respectively.

Further refinements are possible. One refinement concerns possible information about spatial differences in sensitivity. This may be relevant for categories like acidification and eutrophication. Another refinement concerns the inclusion of anthropogenic background levels relevant for the emission of toxic substances. The starting point for risk assessment is that emissions in below-threshold situations are without effect and should therefore be discarded. However, from a prevention perspective, emissions below threshold levels occupy the environmental buffer capacity and should therefore also be taken into account.

On the basis of information thus obtained, "equivalency factors" can be developed for the different environmental categories. These enable the adding up of substances within a given impact category, e.g., the global warming potentials (GWP) and the ozone depletion potentials (ODP). The result of the characterisation element is usually called the "impact profile". However, such equivalency factors cannot be defined on scientific knowledge alone, but will to a small or large extent also be based on value judgements. In fact, this is also the case with the well accepted GWP values. The solution should be sought in authorisation by an international scientific community, together with regular updating of the defined factors.

Figure 10 shows the example of an LCA study up to the level of characterisation. It consists of a comparison of three types of gas distribution nets, composed of either nodular cast iron, PE or PVC. The proponent of the study, the Dutch Gas Company, wanted to know whether the economically preferable option of plastic tubes for new gas pipes implies an increased burden on the environment, compared to the traditional iron pipes. The scores presented in *Figure 10* are normalised in relation to the total yearly burdens on the given impact categories in the Netherlands. The major contribution concerns smog formation, mainly caused by gas leakages through pipe junctions during the use phase. In this respect there is no indication of a difference between the three types of materials. In all other respects iron contributes more than the two types of plastic; the latter two show comparable scores. The conclusion was that the use of plastic pipes instead of iron ones is also advisable from an environmental point of view. However, since the leakages far outweigh other types of impacts, the improvement of the pipes junctions was of critical importance.

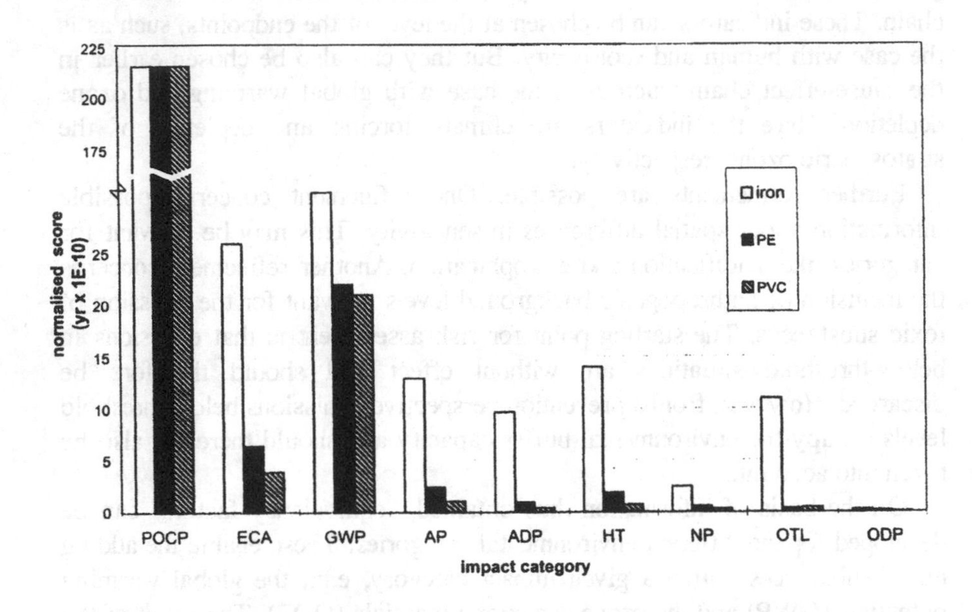

Figure 10. Normalised characterisation results of an LCA-study, comparing different types of gas distribution nets in the Netherlands (Van den Berg et al. 1996) POCP concerns the contribution to the yearly photochemical oxygen creation; ECA to aquatic eco-toxicity; GWP to global warming; AP to acidification; ADP to abiotic depletion; HT to human toxicity; NP to eutrophication; OTL to odour; ODP to stratospheric ozone depletion.

In the weighting element the environmental categories are compared with each other, with respect to their relative importance. This may include a formalised weighting procedure, resulting in one environmental index. The weighting can be done on a case by case basis, or on the basis of a generally applicable set of weighting factors. For the latter, three different lines can be distinguished, which are in part interconnected and which may to some extent be combined: a monetary approach, in which a translation into monetary values is being performed; a distance-to-target approach, in which the weighting factors are in some way related to given reference levels; and a societal approach, in which the weighting factors are set in a authoritative procedure, comparable to the setting of standards.

2.7 Interpretation

During the interpretation the results are related to the goal of the study as defined in the beginning. This includes the performance of sensitivity analyses and a general appraisal. A sensitivity analysis checks the reliability of the results of the LCA study with regard to data uncertainties and methodological choices. The further the results of the inventory table are aggregated in the following phases, the greater the need for such a sensitivity analysis. The interpretation can also start a new run of data gathering if the goal appears not to be reached satisfactorily.

3. MFA: TOOLS FOR SYSTEMS ANALYSIS OF MATERIALS AND SUBSTANCES

3.1 Short history

The first plea for the tracing of materials through the physical processes in society came from Kneese et al. (1972). Since the late seventies, a number of case studies have been performed. From the beginning of the eighties materials balances are increasingly included in environmental statistics reports. In recent years, more attention has been given to the development of modelling tools for the analysis of materials flows in society (e.g., Brunner and Baccini, 1992; Bringezu, 1993; Baccini and Bader, 1996; Van der Voet, 1996). A growing number of studies is being performed in various countries. These different MFA tools exhibit large differences in scale, ranging from companies, countries, watersheds and continents to the global level; but also in other respects such as the type of material flows under study, ranging from the analysis of individual substances to a modelling of total mass flows. So far, no standardisation of either purpose or methodology of material flow studies has been attempted. This is understandable because it has mainly been used for analytical purposes, and not for comparative assertions, as in LCA. However, as the field is becoming more structured, the need for comparing the results of different studies increases and this calls for standardisation of the methodology. This mainly implies terminology, technical framework and guidance about which type of method should be used for which type of application. It should not prescribe methods themselves. The following overview can be seen as a step in this development.

3.2 Definition and applications

MFA concerns the analysis of the flows and accumulations of materials through the physical economy and its environment. MFA encompasses a family of tools depending on the type of material analysed. Bulk-material flows are studied using Bulk-Material Flow Analysis (B-MFA; cf. Bringezu, 1993); substances or groups of related substances are studied using Substance Flow Analysis (SFA; cf. Van der Voet, 1996). MFA belongs to the still broader family of "environmental accounting", to which also Input-Output Analysis belongs in as far as this tool incorporates physical flows.

MFA can be used for different purposes:

- It can serve as a support for data acquisition and function as an error check procedure since inconsistencies in the aggregated numbers can be traced back to errors in contributing separate figures. Likewise it can also help in the identification of missing flows;
- It can be applied for the analysis of trends and their causes and can identify major problem flows to the environment and stepwise trace their causes to their origins in society. Hidden leaks from processes in society can be traced in this way; the degree to which material cycles are closed can also be assessed;
- It can be used to identify and predict the effectiveness of potential pollution abatement measures as a basis for priority setting. The modelling can elucidate the possible shifting of problems, for instance caused by a redirection of substance flows. Scenario studies are a more complex application of this type of use of the tool; and
- It can be used as a screening tool to identify issues for further investigation by other tools.

3.3 Technical framework

A well defined technical framework for MFA should consist of a number of phases which structure the analysis. Inspired by the technical framework of LCA (ISO/DIS 14040), the following phases can be distinguished: goal and system definition, inventory and modelling, and interpretation. In LCA the last of these phases is split into a technical phase (Life Cycle Impact Assessment), and an evaluative phase (Interpretation). Such a distinction does not seem necessary for MFA as yet. The following sections briefly discuss the content of these three phases.

3.4 Goal and System Definition

In this phase, the object of the study (i.e., the type of material, substance or substance group), the aim of the study (in line with the applications as described in section 3.2) and the system concerned need to be defined. The definition of the system should cover the system boundaries; possible subsystems; the inputs and outputs of system and subsystems; and the possible entities within the system and subsystems.

The system consists of all unit processes related to the metabolism of the chosen substance(s), within given boundaries. The definition of the system boundaries depends on the goal of the study and the boundaries (political boundaries in the case of administrative units, or ecological ones as is the case with the analysis of a watershed region). Other types of boundaries can also be envisaged. Thus one may perform an MFA study for an industrial sector or for one firm; or one may investigate all flows of a substance connected with the consumption in a given region. In the latter case one should also include flows outside that region into the system, related to the given consumption (the "ecological rucksacks"), thus shifting from a regional to a functional approach and taking a step in the direction of LCA.

With respect to the definition of subsystems it is usual to make a distinction between the society (or socio-economic system) and its environment. Within society the attention is focused on the physical economy, i.e., the material basis of the economy, in contrast to the financial economy. Common alternative terms for the physical economy are "antroposphere" and "technosphere". We prefer the terms "physical" and "money economy" because of their simple and clear wording; these terms refer to two ways of analysis rather than to two distinct subsystems. The (physical) economy can be further divided into economic sectors, and these in the final units of analysis, the unit processes. Alternative terms for the environment are "biosphere" and "ecosphere", which we do not prefer because they rather relate to subsystems of the environment. The environment can be further subdivided into a number of subsystems or environmental media: lithosphere, soil (pedosphere), hydrosphere, atmosphere and biosphere. The lithosphere concerns that part of the environment which has no interactions with the other components for the given substance. It is sometimes regarded as a third subsystem.

In LCA, the definition of the borderline between society and environment is critical because the core issue concerns the transgression of substances over this borderline either as the extraction of resources or as the emission of hazardous substances. In contrast, in MFA, the definition of this borderline may be somewhat less critical, because in these studies the accumulations within the economy (i.e., flows which do not pass the borderline) are a point of attention. Thus they will be considered anyhow, independent from the precise definition of the borderline between the two subsystems. However, for reasons of comparability between studies standardisation is also desirable here and one may learn from the relevant discussions within the LCA community.

For the definition of inputs and outputs of the system as a whole and of the different subsystems, one could follow the analogy principle for the two subsystems involved. Thus both subsystems will have their own inflow and outflow, exchanges with the immobile stocks in the lithosphere, and their interactions between them in the form of emissions and extractions. The last point concerns the definition of the processes within the subsystems, the flows between these processes and the accumulations which take place at the level of these processes. In the processes the flows are redistributed either into other flows or into stocks.

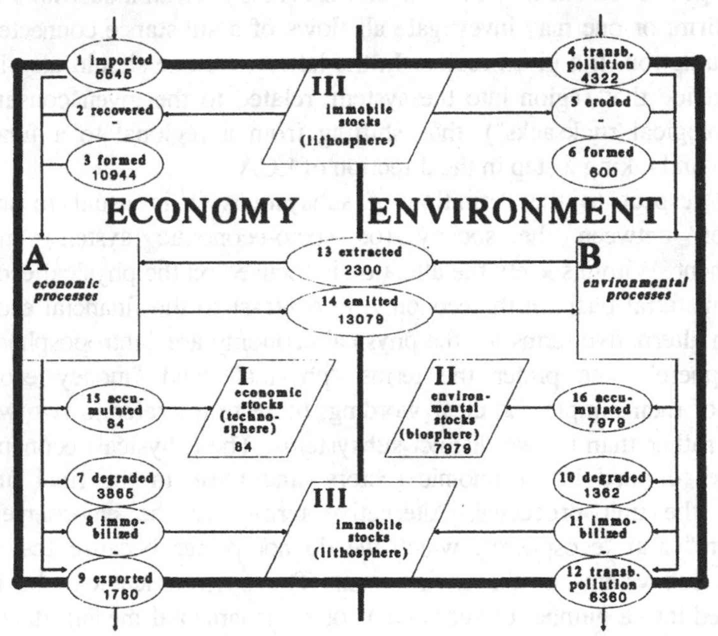

Figure 11. Substance Flow Diagram for nitrogen-compounds for the EU in 1988 (Van der Voet et al. 1996; numbers represent kilotonnes per year)

Note: Economic imports (1), recovery from immobile stocks (2), technical formation (3) and extraction from the environment (13) are inputs to the economic processes; technical degradation (7), immobilisation within the economy (8), economic export (9) and emissions to the environment (14) are outputs from the economic processes; transboundary pollution (4), erosion within the given region (5), formation within the environment (6) and emissions (14) are inputs to the environment; environmental degradation (10), immobilisation in the environment

(11), transboundary pollution (12) and extraction from the environment (13) are outputs from the environment in the given region; and, finally, accumulation within economy (15) and environment (16).

The points can be put together into a substance flow diagram (See *Figure 11*), taking the substance flow of nitrogen-compounds for the EU as an example. The two core boxes are the economic (A) and the environmental processes (B). Both have a number of input and output flows. We can already observe that the total emissions to the environment are much larger than the total extractions from the environment, that the total accumulation within the economy is much smaller than the total accumulation within the environment, that production within the system is about twice as large as the economic import, and that the environment subsystem is a net exporter of nitrogen. It should be borne in mind, however, that this is only an overview. Depending on the aim of the study the processes of course have to be analysed and presented in more detail.

3.5 Inventory and Modelling

This phase concerns the computation using different types of modelling of the flows and stocks for the given year as a (a) bookkeeping system, (b) stationary model or as a (c) dynamic model.

In a bookkeeping system, a flowchart with all stocks, flows and processes, both in society and in the environment is developed first. Then for the given period of time empirical data are collected and attributed to the flows and stocks. For society, statistical data on production, consumption, waste management and trade are linked to data regarding the content of the substance in the relevant products and materials; for the environment, monitoring data will predominantly be used. The unit processes serve just as points for redistribution of flows. In- and outflows are balanced per process, unless accumulation occurs. The overview thus obtained can be used as error check, to find missing data or to identify present or future problem flows, for example, by signalling a large accumulation in society or in a specific environmental medium. It can also be used as a monitoring instrument to register changes over time as a result of societal developments or specific policies, by drawing up this overview once in every few years (as is, for instance, done by the Dutch Central Bureau of Statistics for nutrients and metals).

In a static (stationary) model, the processes are formalised in such a way that the outputs can be formally computed from the inputs; or vice versa, the inputs can be computed so as to satisfy a given set of output values. Stationary modelling describes a stationary situation, apart from possible changes in the immobile stocks and from changes outside the given system. In principle, data from the bookkeeping overview can be used to calibrate the mathematical equations of the processes. It is preferable, however, to use data regarding the distribution characteristics of the processes themselves, if available, in order to avoid the inclusion of coincidental factors in the equations. The core point here is that one consistent mathematical structure is developed, which renders it possible to specify relations between the different flows and stocks within the system. In this way, specific problem flows can be analysed with regard to their origins. The effectiveness of certain developments or measures can also be estimated by comparative static analysis, which is not possible with the bookkeeping approach.

Figure 12 presents an origin analysis for N-compounds (presented earlier in *Figure 11*). The origins of three N-related problems are investigated: atmospheric deposition on agricultural land leading to acidification and eutrophication, groundwater pollution with nitrates and eutrophication of the North Sea. These emissions are analysed at the level of the economic sectors producing these emissions and at the level of the ultimate origins. *Figure 12* shows that for atmospheric deposition and groundwater pollution, the agricultural sector appears to be the most important; for the eutrophication of the North Sea it is the households, with agriculture and industry following with equal contributions (*Figure 12a*). At the level of the ultimate origins (the level where the substances enter the system), for all three problems the production and import of ammonia for fertiliser production appears to be the most important factor (*Figure 12b*). These two levels should be clearly separated, because the causes at the two levels are often mixed, yielding inconsistent results.

Using origin analysis, for example, on cadmium (Cd) reveals its inelastic supply since Cd is not extracted independently, but is a by-product of Zinc (Zn) (Van der Voet et al., 1994). As long as Zn-ore is extracted and purified, an amount of Cd will come on the market. As a consequence, the intake and recovery of postconsumer batteries will lead to an accumulation of Cd within the economy, probably leading to lowering of its price and to increased use. Policy measures to deal with this problem thus call for a decrease in the use of virgin Zn and storage of Cd-waste (see the ideas of Frosch in Chapter 3). The situation is similar in relation to mercury (Hg), which is now, in addition to continued extraction in countries such as Russia and Spain, also produced as a by-product of natural gas (IEM and ERM, 1996).

An MFA study can also be set up as a dynamic model, in which the process equations also include time as a variable. In this way, not only the long term equilibrium of a certain regime can be calculated, but also the road towards this equilibrium and the time it will take to reach it. For a real scenario analysis, this option is the most suitable. Time differences between production, use and waste management processes can thus be analysed. However, it also has the largest data requirement: a complete overview of stocks and flows for the initial year, and the relations between the different flows and stocks with a time specification have to be included. This may limit the accuracy of the projections. Consequently, a dynamic approach is not to be preferred automatically over the more robust static approach.

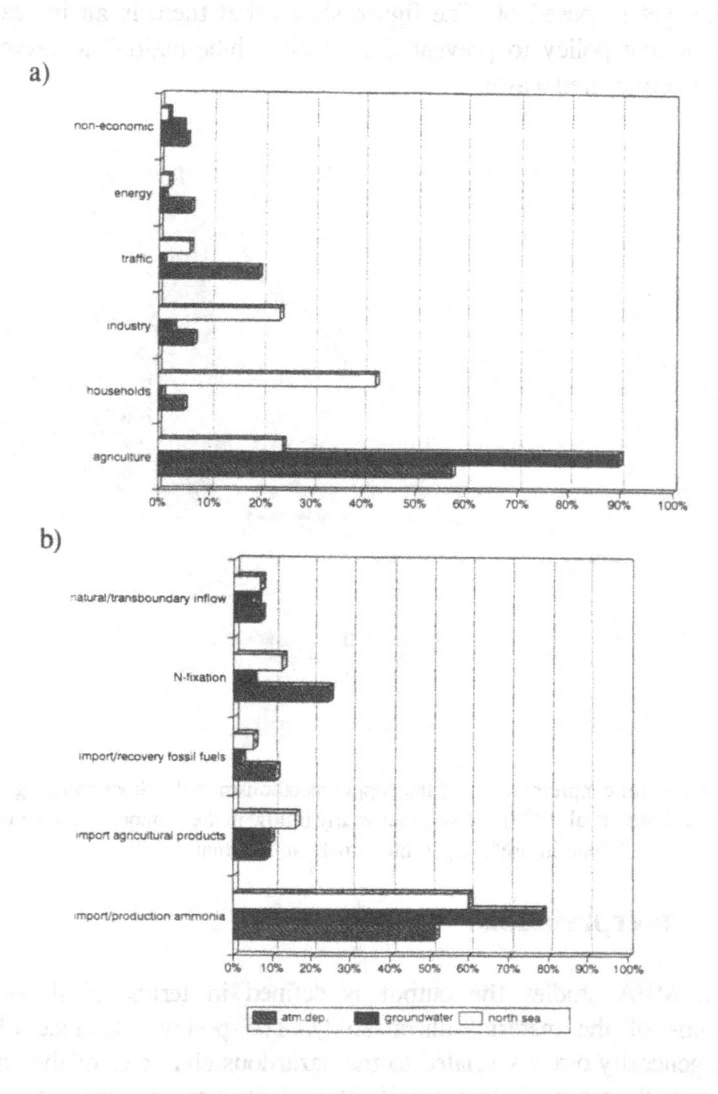

Figure 12. Origin analysis for N-compounds for atmospheric deposition on agricultural land, groundwater pollution & eutrophication of the North Sea for EU in 1990 (Van der Voet et al. 1996); analysis at the level of a) economic sectors, b) ultimate origins.

Figure 13 presents an example of such dynamic modelling (derived from Lohm et al., 1997), in relation to the Copper (Cu) metabolism of the Stockholm region for the period 1900-1995. The lowest line represents a - supposed - constant inflow of copper in this region in tons per year. The middle line represents the "stock-in-use" in the area, which is about in a steady state. The continuously increasing upper line represents the "stock-in-hibernation", which is that part of the economic stock which is not in use any more but not yet disposed of. The figure shows that there is an increasing need for an active policy to prevent the "stock-in-hibernation" to leave the economy as uncontrolled waste.

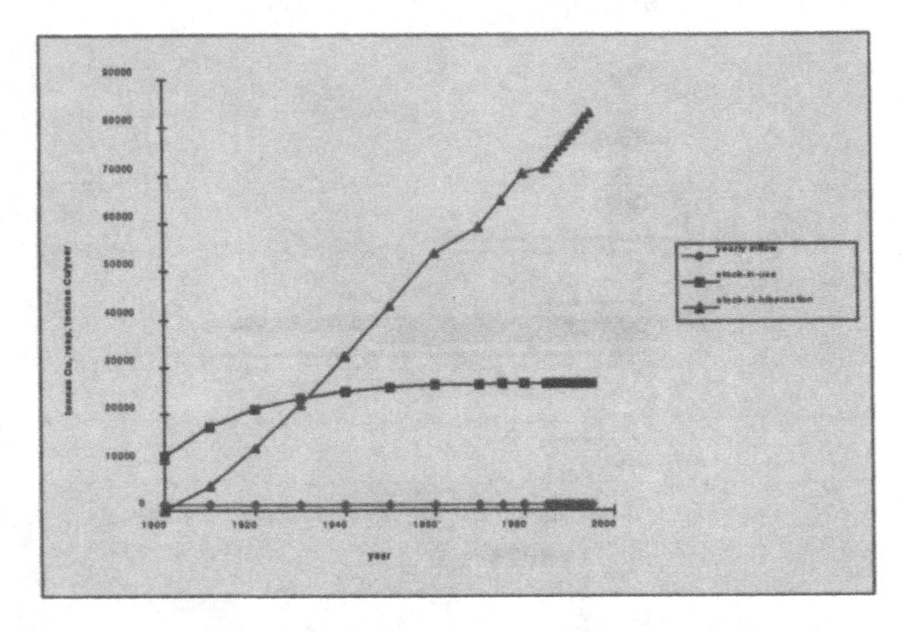

Figure 13. Dynamic representation of the copper metabolism in the Stockholm region (derived from Lohm et al. 1997). Lower line: yearly inflow in the economy; middle line: "stock-in-use"; upper line "stock- in- hibernation"

3.6 Interpretation

In most MFA studies the output is defined in terms of flows and accumulations of the material under study. The policy relevance of the analysis is generally directly related to the hazardous character of the chosen material and will need no further specification. However, in some cases, such as when a group of substances are studied, there is an explicit need for further elaboration of the MFA outcome. The data may have to be aggregated in one way or the other enabling further interpretation. A possible way to interpret the outcome of SFA studies involves the specification of the contributions of the substances to a number of environmental issues or environmental impact categories, such as global warming, ozone depletion, acidification and toxicity. This approach links up with the development of LCA impact assessment methodology (cf. Udo de Haes, 1996).

Figure 14 presents the results on an SFA study on chlorinated hydrocarbons in The Netherlands (Kleijn et al., in press) including the contributions of the different compounds or groups of compounds to ecotoxicity (based on toxicity equivalency factors as presented in Heijungs et al., 1992). A comparison between the three life cycle stages (production; application; transport and waste treatment) shows that the largest impacts are caused by the application stage. The chemical industry appears to be a relatively small polluter. Perhaps this is not surprising as the intentional use of (chlorine containing) pesticides is included here. Of course this does not preclude the responsibility of the chemical industry for the use of their products by the consumer. Another point of interest is that the results are based on about 99% of the chlorine flows. The results of this study have stimulated the proponent to focus further research on the last 1%, bearing possibly high risks for both eco- and human toxicity in particular also during the production stage.

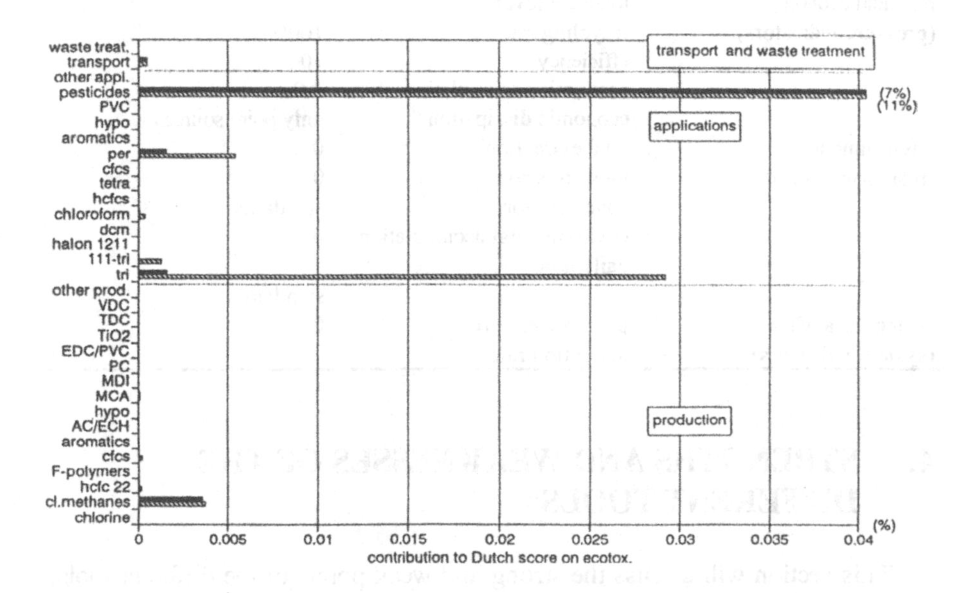

Figure 14. Contribution of emissions of chlorinated hydrocarbons in the Netherlands to ecotoxicity, normalise to the Dutch total of ecotoxic emission. Shaded bars: present situation (1990); black bars: after implementation of accepted policy measures.(Kleijn et al, in press).

If only one substance connected with only one clearly defined environmental problem is being dealt with, a further interpretation of the outcome may be needed. The total overview of flows and stocks is often too complicated to evaluate. A further interpretation can consist of a rather simple procedure, for example, by selecting specific stocks or flows as an indicator for an environmental impact category. However, it can also be more elaborate, by calculating compound indicators relative to the management of the whole chain. For example one can focus on the efficiency of processes or groups of processes in terms of the ratio between the production of the desired output related to the magnitude of the input of the process, or to the amount of the produced waste. Another example, related to resource management rather than pollution, is the fraction of secondary materials used for the manufacturing of products within the system. Table 8 gives an overview of 12 possible sustainability indicators to be derived from MFA studies, dealing with the physical economy (the "pressure indicators"), the environment (the "response indicators") and the system as a whole (the "system indicators"). The response indicators tell us whether action is required; the pressure indicators how to proceed; the system indicators about relationships with processes which are external to our defined system.

Table 8. Overview of MFA-sustainability indicators

(sub)system	criterion	reference value
physical economy	total use level	-
(pressure indicators)	recycling rate	100%
	efficiency	100%
	economic accumulation	< 0
	economic dissipation	only point sources
environment	total extraction	0
(response indicators)	total emission	0
	concentration	standards
	environmental accumulation	0
	daily intake	
		standards
system as a whole	pollution export	0
(system indicators)	disruption rate	-

4. STRENGTHS AND WEAKNESSES OF THE DIFFERENT TOOLS

This section will discuss the strong and weak points of the different tools, and how they may together offer opportunities for chain analysis as a basis for chain management.

4.1 Strengths and weaknesses of LCA

The LCA approach allows for considerable detail and specificity. The precision with which a functional unit is defined, system boundaries are established, allocation procedures specified, and equivalency factors developed for impact assessment, are aspects conducive to a sophisticated, science based analysis. However, they also introduce uncertainty in the results. The more specificity and differentiation in the method, the more relevant the results in principle, but at the same time the less certain. Take for instance the development of impact assessment as simple addition in mass terms of inputs and outputs, to the present framework of life cycle impact assessment in terms of fate and effect-based equivalency factors. Mass terms will be quite limited in their relevancy, but they are rather robust. Further specificity in fate and effect information increases the relevancy of the results, but introduces at the same time substantial uncertainty. And every further differentiation in the methodology implies additional data needs, which may not (or not yet) be available. Apart from that, methodological choices have to be made, which may also significantly influence the outcome of a study. These points presently bear the risk of a decreasing credibility of the LCA approach, in particular for so-called "comparative assertions".

There are different ways of reducing this risk. First of all there is scientific development: better databases about the environmental burdens of unit economic processes, more science based equivalency factors, etc. This is primarily the task of SETAC as scientific association. In line with this, the methodology has to be further standardised, a task for the ISO subcommittee on LCA. An example concerns the priority procedure for allocation: the existing competition between different approaches has now been replaced by a standard preference order for subsequent procedures. But the conditions under which the different procedures apply are far from fixed yet.

In addition to this, there is in our opinion a need for more simple LCA approaches: approaches which are more easy to apply and produce robust results, but which still are sufficiently specific and relevant. As to data availability, a fruitful approach may be to investigate whether data from input-output databases at the level of economic sectors can be used, instead of data at the level of separate unit processes. For allocation procedures one may try to agree on a simple default procedure for initial screening purposes, going into more elaborate procedures with further detail of study. And for impact assessment one may strive for authorised sets of equivalency factors, representing the scientific state of the art and consequently including default values for numerous parameters, which can then be refined in a continuous updating process. In this way LCA should strive to become more practical, without becoming arbitrary. This can in particular be relevant for applications of LCA on higher levels of integration, with questions regarding the material basis of our society as a whole. An example concerns LCA studies comparing different life styles.

4.2 Strengths and weaknesses of MFA

Compared to LCA, MFA is often more descriptive and less directly linked with decisions. This in particular holds true for the bulk-flow analysis procedure, which for instance provides information for a discussion about the dematerialisation of the economy. Such information is not easily connected to decisions of companies or other actors in society. For SFA, with a clearer focus on substance groups or individual substances, this may be different. But such a clearer focus is obtained at a certain cost. The focus on an individual substance implies that environmental impacts related to a substitution of that substance by another substance will be outside the scope of the study. Thus with respect to detergents, NTA may possibly cause more ecological harm than the phosphates which it replaces, but that will not show up in an SFA on phosphorus. This shortcoming can be solved by supplementing the SFA study with one or more LCA studies, aiming at a comparison of the use of the two substances for the given function. This approach has been called Material-Product- Chain (MPC) analysis (cf. Kandelaars et al., 1996).

MFA studies only deal with the physical economy and not with the monetary economy. They do not include information about prices or consumer choices. One can argue that such a link can be easily included by supplementing the MFA study with complementary monetary data, linking possible economic measures to impacts on material flows. For bulkflows of materials or substances this may well be possible (see again the MPC approach). But it is difficult or impossible for flows of substances which constitute only a small or even an undesirable part of a product, because the monetary values are attached to the products as a whole, and not to individual compounds (Guinée et al., in press). Again, different tools will have to be used in combination, but here without clear possibilities for a formalised linkage.

4.3 Strengths and weaknesses of company ecobalances

The analysis of the environmental performance of companies as a whole has the advantage that it is directly linked to the level where decisions are made: the individual company. For a company ecobalance, as mentioned in the introduction, there are no clear limitations. In fact this tool can also be described as a `gate-to-gate LCA'. Measures which aim at steering a chain of companies as a whole may well be a very promising practical approach for chain management (cf. Udo de Haes and De Snoo, 1996, 1997). But it should be acknowledged that as yet there is no clear analytical tool for the chains of companies together: a reason may well be that the companies together do not constitute a system in the strict sense. Although LCA and MFA may contribute for specific aspects of the analysis of the chains, the core of the analysis may well have to deal with the socio-economic and organisational aspects.

4.4 Towards a toolbox for chain analysis

Together the different tools provide what may be called a tool box for chain analysis. However, the suggestion should be avoided that for any purpose just a bunch of tools is ready and available. Specific tools are in development for supporting specific types of chain management. It is imperative that these types should be clearly distinguished so that indeed the right tools are used for answering the given questions.

On the other hand, the different tools are not independent from each other, as they may have common technical elements. Thus mass balances, which are the core of MFA, can play a role as a quality check in LCA. Techniques being developed in the field of risk assessment increasingly are used in life cycle impact assessment and consequently improve the quality thereof. Thus developments related to one type of tool may have implications for other tools.

Finally, separate databases are in development for the different tools, not taking into account specific requirements for other tools, consequently leading to limited options for use. Some gaps are now being filled. Thus, for non-detailed LCA studies increasing use is made of input-output databases, contributing to the extension of these databases with environmental indicators. Another example concerns the inclusion in LCA databases of data on the total magnitude of processes, and not only on their input-output characteristics, so that these data can also be used for MFA purposes.

5. CONCLUDING REMARKS: RELEVANCE FOR DECISION MAKING

A common misunderstanding is that tools like LCA and MFA will produce an outcome which mechanically can be translated into a given decision. This picture is not correct, given the many uncertainties and value choices involved. Such an incorrect image may also lead to a loss of value of the approaches, in particular with external applications involving critical comparisons. High expectations of a technical solution to essentially political choices will be left unfulfilled. Instead, these tools will function optimally in such external applications only if they are embedded in clear decision making processes. A comparison can be drawn with the application of Environmental Impact Assessment, which derives its strength from its integration in a well structured decision making procedure as is the case in many countries. With tools like LCA and MFA a social input may at least play a role at three different moments, with quite different characteristics and different parties to be involved (cf. Consoli et al., 1993).

The first moment concerns major choices related to the definition of the goal, scope, general set-up and choice of tools of a study. For broad social acceptance an input may well be essential from different stakeholders in the project, possibly including governmental authorities and NGOs. Thus in the Dutch ecolabelling procedure a stakeholder workgroup, consisting of industry, trade, consumer and environmental NGOs and authorities, advises on major choices such as the definition of product group, functional unit and scope of impacts.

Secondly, there is a need for guidance during a study. Here more technical questions have to be answered, such as the choice of allocation procedures for the inventory and equivalency factors to use for the different impact categories. Although more technical, these choices may still influence the outcome of the study considerably. Especially for strategic decisions and for public applications one may think here of an independent expert panel that also conducts a quality check or review after the study. Finally, if a study involves weighting between impact categories, this task may, at a case level, again be performed by the first stakeholder panel, but may also be based on a not case-dependent procedure. By recognising that full objectivity can never be reached, procedural requirements should deal with the uncertainties and value judgements in a balanced and transparent way.

REFERENCES

Baccini, P. and H.-P. Bader (1996). Regionaler Stoffhaushalt - Erfassung, Bewertung und Steuerung. Spektrum Akademischer Verlag, Heidelberg/Berlin/Oxford.

Berg, N.W. van den, G. Huppes and J.W.B. Wikkerink (1996). Environmental life cycle assessment of gas distribution systems, GASTEC N.V. and CML-S&P, Apeldoorn/Leiden, The Netherlands.

Braunschweig, A. and R. Müller-Wenk (1993). Ökobilanzen für Unternehmungen: eine Wegleitung für die Praxis Haupt, Bern, Stuttgard, Wien.

Bringezu, S. (1993), Towards increasing resource productivity: how to measure the total material consumption of regional or national economies? Fresenius Envir. Bull. 2, pp 437-442.

Brunner, P.H. and P. Baccini (1992). Regional materials management and environmental protection. Waste Management and Research 10, pp 203-212.

BUS (1984). Schriftenreihe Umweltschutz no. 24. Bundesamt für Umweltschutz, Bern, Switzerland.

Consoli, F. et al. (1993). Guidelines for life-cycle assessment: a 'Code of Practice', Brussels : Society of Environmental Toxicology and Chemistry (SETAC).

Cramer, J. (1991). De illusie voorbij, op weg naar een brede aanpak van milieuproblemen. Van Arkel, Utrecht, The Netherlands.

Druijff, E.A. (1984). Milieurelevante produktinformatie, CML-Mededelingen no. 15, CML, Leiden, The Netherlands.

EEC (1993). EEC Council Regulation no. 1836/93. Allowing voluntary participation by companies in the industrial sector in a Community eco-management and audit scheme (EMAS). Official Journal of the European Communities (10-7-1993, No L 168): 1-17.

European Commission (1997). Guidelines for the application of life cycle assessment in the eco-label award scheme, Luxembourg.

Fava, J., F. Consoli, R. Denison, K. Dickson, T. Mohin and B. Vigon (1993). A conceptual framework for life-cycle impact assessment: Workshop of the Society of Environmental Toxicology and Chemistry and SETAC Foundation for Environmental Education, February 1-7, 1992, Sandestin, Florida, USA. Brussels: SETAC; Pensacola: SETAC Foundation for Environmental Education Inc.

Guinée, J.B., P. Kandelaars and G. Huppes (in press), Linking economic models to physical substance-flow models: possibilities and problems. Proceedings of ConAccount workshop "From paradigm to practice of sustainability", Leiden, 21-23 january 1997.

Heijungs, R., J.G. Guinée, G. Huppes, R.M. Lankreijer, H.A. Udo de Haes, A. Wegener Sleeswijk, A.M.M. Ansems, P.G. Eggels, R. van Duin and H.P. de Goede (1992). Environmental life cycle assessment of products : guide and backgrounds, Leiden: Centre of Environmental Science, Leiden University.

Hunt, R.G., W.E. Franklin, R.O. Welch, J.A. Cross and A.E. Woodall (1974). Report EPA/530/SW-91c. Environmental Protection Agency, Washington, USA.

Huppes, G. (1993). Macro-environmental policy: principles and design. Amsterdam: Elsevier.

Instituut voor Europees Milieubeleid (IEM) and Environmental Resources Management (ERM), (1996). Mercury stock management in The Netherlands. Background document for the workshop "Kwik uitbannen of beheersen?" IEM, Brussels.

International Organization for Standardization (1995). Environmental management systems - specification with guidance for use. Draft international standard ISO/DIS 14001.

International Organization for Standardization (1996). Environmental management - Life cycle assessment - Principles and framework. ISO/14040.

International Organization for Standardization (1997). Environmental management - Life cycle assessment - Goal and scope definition and inventory analysis, ISO/DIS 14041.

International Organization for Standardization (1997). Life Cycle Assessment - Life cycle impact assessment. ISO/CD 14042.

Jänicke, M., H. Mönch, T. Ranneberg and U.E. Simonis (1989). Economic Structure and Environmental Impacts: East-West Comparisons, The Environmentalist, vol. 9, pp 171-182

Kandelaars, P.P.A.A.H., J.B. Opschoor and J.J.M. van den Bergh (1996). A dynamic simulation model for material product chains: an application to gutters. Journal of Environmental Systems, 24 (4), pp 345-371.

Kleijn, R., A. Tukker and E. van der Voet, in press, Chlorine in The Netherlands, Part I: an overview. Journal of Industrial Ecology.

Kneese, A.V., R.U. Ayres and R.C. D'Arge (1972). Economics and the environment : a materials balance approach, John Hopkins University Press, Baltimore.

Lohm, U., B. Bergbäck, J. Hedbrandt, A. Jonsson and C. Östlund (1997). Metals in Stockholm, a MacTempo case study. Annex 2 to: P. Brunner et al.: Materials accounting as a tool for decision making in environmental policy (MacTempo). Vienna.

Schmidt-Bleek, Friedrich (1994). Wieviel Umwelt braucht der Mensch? Basel: Birkhäuser.

Schröder, H. (1995). Input management of nitrogen in agriculture. Ecological Economics, 13, pp 125-140.

Steen, B. and S.-O. Ryding (1993). AFR-Report 11. Avfallsforskningsrådet AFR, Swedish Waste Research Council, Stockholm, Sweden.

Taylor, A.P. (1996). Overall business impact assessment (OBIA). Proceedings 4th Symposium for Case Studies LCA. 3 dec. 1996 Brussels, Belgium: 181-187.

Udo de Haes, H.A. (1993). Applications of life cycle assessment: expectations, drawbacks and perspectives. J. Cleaner Prod. 1 (3/4), pp 131- 137.

Udo de Haes, H.A. (Ed.) (1996). Towards a methodology for life cycle impact assessment. SETAC-Europe, Brussels.

Udo de Haes, H.A. and G.R. de Snoo (1996), Companies and products: two vehicles for a life cycle approach? Int. J. LCA 1 (3), pp 168-170.

Udo de Haes, H.A. and G.R. de Snoo (1997). The agro-production chain; environmental management in the agricultural production-consumption chain, Int. J. LCA 2 (1), pp 33-38.

Van der Voet, E. (1996). Substances from cradle to grave, development of a methodology for the analysis of substance flows through the economy and the environment of a region. Thesis, Leiden University.

Van der Voet, E., L. van Egmond, R. Kleijn and G. Huppes, (1994). Cadmium in the European Community: a policy oriented analysis, Waste Management and Research, 12, pp 507-526.

Van der Voet, E., R. Kleijn, L. van Oers, R. Heijungs, R. Huele and P. Mulder (1995). Substance flows through the economy and environment of a region; Part I: Systems Definition, Envir. Sci. & Pollut. Res. 2 (2).

Van der Voet, E., R. Kleijn and H.A. Udo de Haes, (1996). Nitrogen pollution in the European Union, origins and proposed solutions, Environmental Conservation 23 (2), pp 120-132.

Chapter 6

Material flow accounts: definitions and data

IDDO WERNICK
Researcher, Columbia Earth Institute, New York

Key words: Substance flow analysis, material flow accounts.

Abstract: **The author explains that material flow accounts represent an important tool for cataloging and calculating the apparent and hidden movements of bulk materials through the economy. He also argues that MFA information can be used to design indicators that show the relation between flows and environmental impacts. MFA's can thus indicate the opportunities for policy improvement at the meta or megascale.**

1. INTRODUCTION

To investigate the environmental effects of the physical material flows used to sustain modern economies, research institutions and governments have shown interest in producing accounts of the materials used within national and regional economies. Materials Flow Accounts (MFA's) catalogue and quantify the use of bulk materials typically measured in millions of tonnes (MMT). MFA's provide gross information that indicates the scale of bulk material flows to meet the basic infrastructure needs (e.g., energy, housing, transportation, agriculture) of modern societies. Substance Flow Accounts (SFA's) focus on more toxic materials used in amounts several orders of magnitude smaller that pose chronic or acute threats to human health and the environment.

Both MFA's and SFA's should, in principle, provide a complete materials balance equating the quantities identified as inputs with those labeled as outputs in accordance with the law of conservation of mass. Analysts must consider material flows apparent to the public, like municipal solid waste, as well as hidden ones in mines, quarries, power plants, factories, and farms. This constraint requires materials accountants to gain an understanding of the industries, technologies, products that use a given material, and search for data on ever more remote and obscure flows.

Both large and small scale materials flows impact the environment. Anthropogenic flows of hydrocarbon fuels for energy and fixated nitrogen for agricultural fertilizers affect global biogeochemical cycles that govern the climate and ecosystems. Consumption of forest and agricultural products affects land use, the potential for carbon uptake in terrestrial ecosystems, and hydrological and sedimentary flows at the regional to continental level. For SFA's, more specific information on the chemical form and geographic distribution of material flows is essential to evaluating the environmental impact. These smaller volume materials flows can enter public drinking water supplies, enter the human food chain, and can also disproportionately affect certain species in natural ecosystems upsetting their biological balance.

2. DEFINITIONS

Proper accounting requires the delineation of the system under study. Government interest and the availability of data sets promote analyses bounded by state, national, or regional borders. MFA's and SFA's conducted at this scale indicate the performance of several environmental drivers within well defined geographic boundaries. National or regional accounts invite consideration of how to capture flows of commodity materials and manufactured products across political borders (i.e., imports and exports). Economic borders can also provide a boundary, because governments and industry associations collect and organize physical data by economic sector. These MFA's provide insight on the resource efficiency and pollution loads of basic industries. Individual elements and compounds form the boundary definition for SFA's.

After defining system boundaries, the MFA must choose what mass movements to include within those boundaries. For geographic based MFA's the question arises: What constitutes the internal boundary that the flow must cross to be included in the account? Establishing an internal boundary ensures that mass movements will not be counted more than once and adds conceptual clarity to the purpose and execution of the analysis. In general terms, the internal boundary can be defined as that between "nature" and the "industrial economy." In some cases this boundary is vague. For instance, in assessing materials flows from agriculture and forestry: Should one consider the harvested material or should one consider the mineral and chemical production inputs to farms and industrial forests? From this follows the question of whether to include the flow of abundant basic natural resources, such as atmospheric oxygen and water, between natural reservoirs and forests, farms, and factories. For sector based MFA's, the question of what internal boundaries are crossed forces consideration of how far upstream to follow the chain of materials suppliers and how to evaluate the indirect and hidden material flows occurring outside the sector under study.

The criteria for what to include in MFA's range from narrowly restrictive to broadly inclusive. One available option for determining what to include would account for only those materials flows associated directly with economic transactions. Other methods include ancillary material flows such as mine tailings from metal production, overburden removed to access coal seams, and eroded soil from agricultural land. These ancillary flows represent important mass movements that result from society's need for food, materials, and energy, and impact the local environment. Some MFA's consider not only these flows, but also include the translocation of earth for activities such as road building and harbor dredging. Though the massive mobilization of sand and dirt do impact the environment in various ways, the relative environmental threat posed is simply not proportional to their tonnage. The presentation of data on these flows should not obscure environmentally important information associated with smaller volume flows.

MFA's can classify information at different points in the material flow. For example, an MFA might consider the flow of salt through the industrial economy as one among several industrial minerals. Alternatively it could consider the complex flows of chlorine and sodium chemicals that derive directly from salt. In constructing materials accounts, analysts should maintain uniformity by assessing different materials flows at similar points in the production sequence. *Table 9* shows a partially disaggregated account of materials inputs and outputs in the United States economy in 1990.

SFA's must consider commercial flows of elements and compounds and must also include them as traces in other minerals and wastes. For some substances, such as lead, SFA's should consider how much of the substance in question remains contained within the industrial/social system, how much leaves national borders, how much is disposed of, and how much is irretrievably lost (see Chapter 12). For organic pollutants, the SFA must identify the time and/or place of chemical species creation and destruction

Table 9. Materials flows: U.S. 1990

INPUTS

Material Group	Apparent Consumption (MMT)		Per Capita Per Day(kgs)
Energy	Coal	843	
	Crude Oil	667	
	Natural Gas	378	
	Petroleum Products	2	21
Construction Minerals	Crushed Stone	1092	
	Sand & Gravel	828	
	Dimension Stone	1	21
Industrial Minerals	Salt	41	
	Phosphate Rock	40	
	Clays	39	
	Industrial Sand & Gravel	25	
	Gypsum	23	
	Nitrogen Compounds	17	
	Lime	16	
	Sulfur	13	
	Cement (imported)	12	
	Other	24	4
Metals	Iron & Steel	100	
	Aluminum	5	
	Copper	2	
	Other	4	1
Forestry Products	Saw Timber	123	
	Pulpwood	73	
	Fuelwood	52	
	Other	13	3
Agriculture	Grains	220	
	Hay	133	
	Fruits & Vegetables	71	
	Milk & Milkfat	64	
	Sugar Crops	51	
	Oilseeds	45	
	Meat & Poultry	42	
	Other	5	7

OUTPUTS

Destination	Amount (MMT)	Per Capita Per Day kgs)
DOMESTIC STOCK		
Construction	1677	18
Other	203	2
Total	1880	21
ATMOSPHERIC EMISSIONS		
CO_2 (carbon fraction only)	1367	15
Hydrogen	255	3
Methane	29	0.3
CO (carbon only)	29	0.3
NO_x	19	0.2
VOC	18	0.2
SO_2 (sulfur fraction only)	10	0.1
Particulate matter	6	0.1
Total	1735	19
OTHER WASTES		
Processing waste	136	1.5
Post consumer waste	276	3
Coal ash	85	1
Yard waste	35	0.4
Food waste	13	0.1
Wastewater sludge	9	0.1
Total	555	6
DISSIPATION	145	2
RECYCLED	244	3

Source: Wernick, I.K. and Ausubel, J.H., 1995.

which may occur during intermediate stages of production or after release to the environment. Analyses at this level require detailed flowsheets for standard industrial processes as well as data from operating plants to gain accurate information on the actual flows. In cases where only one side of the material input/output ledger is supplied, employing the materials balance principle along with a process description allows analysts to construct the other side.

Though generally considered separately, material flows couple to energy flows. Different finished materials require different energy inputs for their production. Comparing the energy needs for processing an equal mass of aluminum, steel, cement, and polystyrene yields an approximate ratio of 85:10:2:1. Furthermore, materials themselves contain more or less energy content as measured in joules of heat value per kilogram which affects available disposal options. Materials used for energy (e.g., fuels, wastes) form a sizable fraction of gross materials flows around the globe. These flows include the consumption and transport of hydrocarbons and fuelwood as well as the material flows associated with their extraction and processing. MFA's show that the size and complexion of future materials flows will depend on the social choice of whether to drive energy systems by harnessing chemical energy from solid, liquid, or gaseous fuels, nuclear energy from radioactive metals, or solar energy from the sun.

To capture the heterogeneity of materials and their impacts, MFA information should support the design and use of a suite of indicators to indicate the relation between materials flows and specific environmental impacts. Indicators should provide analytic insight on the impact of the flow to the local environment as well as the effects beyond. In addition to allowing for international comparisons, indicators could consider dimensionless ratios of various materials flows to gauge the efficiency of raw material conversion to finished products in a single nation or industry. Other indicators could index kilograms of consumption to variables such as population, hectares of land, and economic activity as measured by Gross Domestic Product.These combined measures serve to better indicate the relationship between materials use and social dynamics. *Table 10* shows a sample set of metrics devised for gaining insight into the environmental significance of MFA's.

Table 10. National materials flows: sample metrics

Metric		Dimensions	Formula	Environmental significance
Total Inputs		MMT and MMT/ Capita	Aggregate total consumption of material by class on absolute & per capita basis	Benchmarking national resource use
Input Composition	Fuel mix (H/C)	Dimensionless	Consumption ratio of Btu's from Natural gas:Petroleum:Coal	CO_2 emissions
	Processed metal: ceramic:glass: polymer ratio in finished products	Dimensionless	Consumption ratio of said materials in finished products and structures	Gross shifts in materials & energy use, materials efficiency and cyclicity, mining and processing waste
Input Intensities	Intensity of use	MMT of inputs/ $\$10^6$ GDP	Material consumption quantity for selected input materials/ GDP	Relationship of resource use to economic activity
	Agricultural intensity	Dimensionless	Agricultural materials consumption/ Total crop production	Materials efficiency, eutrophication of water bodies, topsoil erosion, ecosystem disruption, energy use
	Decarbonization	MMT of Carbon inputs/$\$10^6$ GDP	Carbon inputs/GDP	Relationship of carbon emissions to economic activity
Recycling Indices	"Virginity" index	%	Quantity of all virgin materials/ Total material inputs	Materials efficiency and cyclicity, mining and processing waste, energy use
	Metals recycling rate	%	Quantity of recycled and secondary metals production/ Primary production from ores	Materials efficiency and cyclicity

Metric		Dimensions	Formula	Environmental significance
Recycling Indices	Renewable net carbon balance	%	Forest growth/ Forest products harvested	Global carbon balance of sources and sinks, land use, ecosystem disruption
Output Intensities	Green productivity	%	Quantity of solid wastes/ Quantity of total solid physical outputs	Materials efficiency and cyclicity
	Intensity of use for residues	MMT/ $\$10^6$ GDP	Generation quantity for selected materials waste streams/ GDP	Relationship of waste generation to economic activity
Leak index		%	Quantity of materials dissipated into the environment/ Total material outputs	Materials efficiency and cyclicity, media contamination
Conversion Efficiency	Industrial conversion efficiency	%	Total output for an industrial sector/ Total inputs	Materials efficiency
	Process to post-consumer wastes ratio	%	Process wastes from industry/Post consumer waste	Relating hidden to apparent consequences of industrial production
Physical Trade Index		%	Mass value of net trade in manufactured products/ Mass value of net trade in raw resources	Domestic resource consumption & environmental burden caused by exports
Mining Efficiency	Mining wastes	Dimensionless	Quantity of wastes generated/ Ton of finished product	Solid wastes, acid mine drainage
	Byproduct recovery	Dimensionless	Total byproduct recovery/ Total output	Materials efficiency, solid wastes

Source: I.K. Wernick and J.H. Ausubel, 1995.

3. DATA

Data on material flows come primarily from national governments and industry associations that conduct industry surveys and maintain data sets. Consulting firms also collect data useful for analyzing commercially valuable material flows. However, due to their commercially value, the price of these data may be beyond the reach of analysts without direct economic interests.

Data from individual firms can be essential to constructing accounts, especially in cases where relatively few companies handle a given material. These data may not be released however, as companies wish to prevent the leaking of proprietary information.

Comparing numbers from different data sources requires a consistency in the data which must be checked. Typically, the data found in international data sets represent different actual flows arising from differences in the methodology and the statistical integrity of data collection in individual nations. Data reported by government and industrial organizations ignore unreported materials flows such as those in rural subsistence economies and transactions on the black market. Political objectives may also provide incentive to over- or under-report certain flows in different countries. Some governments have been accused of underreporting the extent of their agricultural land to improve crop yield estimates and some environmental non-governmental organizations use inflated estimates of waste generation to rally public opinion.

Several factors present barriers to establishing long term series of data describing the identical material flow with published data. Official statistical definitions within the same country can change in time with the evolution of the government bureaucracy and changes in industrial practice. Industry conventions for record keeping may also change to reflect the introduction of new products or processes as well as new regulatory requirements to monitor waste streams. Natural resource agencies and companies may use different measurement units than manufacturers of consumer products downstream. To maintain consistency, MFA's may need to restrict themselves to the reported data from a single source (or small set of sources) to ensure comparability across data categories and over time.

Data sources may diverge in how they aggregate information. Data on steel may include specialty steels without distinguishing between them and standard carbon steel. Data on wood products may include different products under the category "composites" depending on the agency or industry organizations reporting the data. Even if individual flows included in a catch-all category are specified, knowing their precise fraction of the total may not be. Without supplementary information, analysts are limited by the cataloging choices of the data source. If relying on more than one source, analysts may need to resort to using the most inclusive category so that numbers compared with one another represent the same material flow.

A further source of ambiguity inherent in comparing data sets, particularly for waste data, is the degree to which the data include water. This fraction can rise to 90% for some high volume industrial waste streams. Paper mills use a 50:1 ratio of water to pulp during the manufacturing process for instance, though modern mills recycle process water up to six times. The petroleum refining/chemical industry, food processing, and primary steel industry also use large amounts of process water that is treated before release. Deducing the amount of water in a given material flow might ultimately rely on somewhat arbitrary estimates of the solid to water ratio. In fact, many of the values used in MFA's depend on multipliers used to estimate material flows. For instance, the amount of soil eroded annually in a region derives from estimates of millimeters of soil eroded per hectare. Standard production losses of materials in a given industry and the metals content of power plant emissions provide other examples. Though used for whole regions and industries, the precise value of multipliers depend on geological factors, ground-level practices adopted by firms, and requirements to comply with environmental regulations.

As a result of the different methodologies employed to generate physical data on materials flows, the accuracy of the numbers relies on survey and statistical methods, initial assumptions, multipliers, and estimates. Differences in each of the analytic stages employed to quantify a given flow can and do add cumulatively to lead to disparities in outcome. MFA's need to qualify data by explicitly mentioning the possible differences arising from the various estimation procedures used to construct the account.

4. CONCLUSIONS

Two related factors motivate the current interest in conducting material accounts. Researchers view MFA's and SFA's as valuable tools for analyzing the environmental impact associated with the functioning of modern industrial economies. In addition, researchers and others see the accounts as a compact and transparent way to convey environmental information to government officials, resource managers, and industrial operators whose decisions influence environmental quality. Materials accountants must balance between the need to make MFA's and SFA's easy to understand by non-experts while not forfeiting the integrity of the very information they wish to convey. Much of the intricate work necessary for producing a quality flow account takes place behind the scenes in the preparation phase. Some qualifying information however must be included with the account itself so as not to mislead the intended audience. In deciding how to organize and present data, producing accounts becomes an art as well as a science.

The value of MFA's and SFA's will increase from greater standardization of methodology and data verification as well as improved analytic methods. Monetary data relating to economic transactions of materials enjoy the most scrutiny as they are used to compile national economic performance statistics and corporate spreadsheets. In the OECD countries, national statistical organizations have responded to the need for better materials accounting frameworks. Government statistical offices as well as some firms work with environmental analysts to facilitate more focused data collection and organization to accommodate material flow analysis. Researchers will need to refine methods for materials flows analysis to counter limits in the ability of government and industry to monitor all environmentally important materials streams.

Data on the geographic distribution of material flows coupled to information on the chemical form of inputs and outputs and environmental transport allows for the investigation of possible human and ecosystem exposure routes. The value of the raw data also increases and the analysis deepens by mapping information on materials and overlaying it with data on social variables such as spatial demographic and epidemiological patterns using Geographic Information Systems.

MFA's and SFA's are the ideal tool for quantifying and pointing out areas with opportunity for environmental improvement at the mega- or meta-scale level. Understanding the environmental relevance of MFA and SFA information will rely on a suite of methods for analyzing the accounts once they are compiled. To multiply their value many fold, flow accounts should be coupled to data on the spatial/temporal dynamics of social and economic data. To accurately relate materials flow and environmental quality, MFA's and SFA's should ultimately couple to information on environmental pathways to humans and ecosystems and the effects of materials mobilization on planetary biogeochemical cycles.

BIBLIOGRAPHY

Allen, D.T. and Jain, R.K., eds., 1992, Special Issue on National Hazardous Waste Databases, *Hazardous Waste & Hazardous Materials* 9(1):1-111.

Ayres, R.U. and Ayres, L.W., 1996, *Industrial Ecology: Towards Closing the Materials Cycle*, Edward Elgar Publishing, Cheltenham, U.K.

Bringezu, S., Behrensmeier, R., Schütz, H., 1996, Material Flow Accounts Indicating the Environmental Pressure of the Various Sectors of the Economy, Presented at the International Symposium on Integrated Environmental and Economic Accounting in Theory and Practice, Tokyo March 1996, Wupertal Institute, Wupertal, Germany

Hüttler, W., Payer, H, and Schandl, H., 1997, National Material Flow Analysis for Austria 1992, Institut fur Interdisziplinare Forschung und Fortbildung, Vienna, Austria.

Kleijn, R., Tukker, A., and van der Voet, E., 1997, Chlorine in the Netherlands, Part I, An Overview, *Journal of Industrial Ecology*, 1(1):95-116.

Moriguchi, Y., 1997, Environmental Accounting in Physical terms in Japan - Preliminary Material Flow Accounts and Trade Related Issues, Presented at the ConAccount Workshop, Leiden Netherlands, January 1997, National Institute for Environmental Studies, Environment Agency of Japan, Tokyo, Japan.

Stigliani, W.M., and Anderberg, S., 1992, *Industrial Metabolism at the Regional Level: The Rhine Basin*, International Institute for Applied Systems Analysis, Laxenburg, Austria.

Thomas, V.M., and Spiro, T.J., 1995, An Estimation of Dioxin Emissions in the United States, *Toxicological and Environmental Chemistry*, (50):1-37.

US Geological Survey, various years, Material flow accounts and analyses in the US for arsenic, cadmium, chromium, cobalt, manganese, mercury, salt, tungsten, vanadium, and zinc are available from the Office of Minerals Information at the US Geological Survey located in Reston, Virginia, USA.

Wernick, I.K. and Ausubel, J.H. 1995, National Material Metrics for Industrial Ecology, *Resources Policy* 21(3):189-98.

Wernick, I.K., and Ausubel, J.H., 1995, National Materials Flows and the Environment, *Annual Review of Energy and Environment*, 20:462-492.

World Resources Institute, 1997, *Resource Flows: The Material Basis of Industrial Economies*, World Resources Institute, Washington, D.C.

Chapter 7

Environmental research and modelling
Disciplinary tensions

HARMEN VERBRUGGEN
Professor and Deputy Director, Institute for Environmental Studies, Vrije Universiteit, Amsterdam

Key words: Multidisciplinarity, interdisciplinarity, environment-economy models, integrated models

Abstract: Disciplinary incompatibilities stand in the way of integrated environment-economy modelling. After a discussion of biases in environment-economic modelling, ten disciplinary incompatibilities are identified and briefly dealt with. A few lessons are drawn to improve multidisciplinary cooperation in environmental research.

1. INTRODUCTION

In modelling environment-economy interactions, different lines of thought, or scientific belief systems, have to be reconciled. However, the inherent incompatibilities are not easily dealt with, sometimes even not well understood. The diverging of thoughts is most apparent between natural scientists on the one hand, and economists and other social scientists on the other hand. In picturing the differences, it is necessary to caricature the scientific approaches of either side. Inevitably, injustice is done. Moreover, it should be realised that the disciplines are considered in terms of how they participate in applied environmental research. Bearing this in mind, it can generally be found that natural scientists reason in laws of nature and in causes and effects. If an effect is conceived as an environmental problem, then working on the cause and/or the cause-effect chain is imperative. New technologies are often presented as a way out. Natural scientists are inclined to transpose these rather deterministic natural patterns on societal processes.

Economists, by contrast, take societal processes and mechanisms as a starting point of their analyses. These processes and mechanism do not obey natural laws. That does not mean that there are no regularities. Social and economic laws are, however, different from natural laws, basically because of the phenomenon of the incomprehensibility and dynamic nature of human behaviour and societal processes. Hence, economic reasoning is less deterministic and allows for degrees of freedom.

It is therefore obvious that in modelling environment-economy interactions, there always appears a trade-off between the natural science and the economic perspective. Environmental sciences is not a real interdisciplinary endeavour (yet). Before dealing with a number of the most frequently occurring trade-offs in Section 3, a brief overview is presented of different categories of (applied) environment-economy models. Section 4 draws some lessons.

2. CATEGORIES OF ENVIRONMENT-ECONOMY MODELS[1]

The environmental sciences are basically concerned with the analysis of changes in ecological systems, the interactions between ecological and socio-economic systems and the formulation of strategies to manage these changes. Hence, environment-economy modelling tries to come to grips with the causes, the complexities of the interactions and possible coping strategies, preferably in a way to help economic agents and policymakers. Various approaches to environment-economy modelling can be taken. The classification presented here is not meant to be comprehensive, it only provides a background to place the trade-offs between alternative modelling approaches in the proper context. It is also not a critical assessment, because that would require insight into the objectives of the modeller: alternative modelling approaches serve different objectives. For instance, formal models typically picture abstract system interactions to analyse the fundamentals, whereas empirical or applied models try to give a real world representation. The latter category of models may be descriptive, i.e. these models try to empirically describe processes and interactions to trace actual developments, or to predict by sketching possible future development paths by means of scenario studies. Yet another objective might be the support of decision makers. Cost-benefit and multi-criteria models that evaluate alternative projects or strategies then come to the fore.

Here, the nature and the scope of the environment-economy interactions are the prime matter of concern. To that end, the following classification seems relevant:

1. environment-cum-economy models
2. economy-cum-environment models
3. integrated environment-economy models

All three types of models will be briefly reviewed.

2.1 Environment-cum-economy models

The first category of environment-cum-economy models starts from a description in physical units of relationships in the environment. Change and dynamism in the environmental system constitute the core model. The more complete models describe the dose-response relations for a particular disruption in the environmental system, taking account of sources and sinks, stocks and flows and the system's regeneration and carrying capacity. Consistency is guaranteed in natural scientific terms with respect to, for instance, biogeochemical cycles, dose-response relations, the laws of thermodynamics, and the law of conservation of mass. It is, of course, very well possible that certain stocks and flows are placed outside the system, depending on the scope and spatial level of the model. The more partial the model, the more missing stocks and flows. These models are usually driven exogenously, i.e. from outside the environmental system, by environmental pressure from economic activities, population growth or infrastructural interventions. Likewise, technological solutions and other management strategies are also external to the system. This means that the core environmental model is extended with add-ons or satellite modules for driving economic forces and/or economic aspects, notably cost, of management strategies and technology options: hence, environment-cum-economy models.

Examples of this category of modelling include material balances and a set of derived analyses like resource or material flow accounting, substance flow modelling, life-cycle analysis and resource regeration and carrying capacity/sustainable yield models. These models vary greatly in scope, complexity, spatial dimensions and time horizon. Their descriptive nature is a common characteristic. Thereupon, simulation and scenario models can be developed.

Indicators constructed on the basis of these modelling exercises typically try to capture the interaction between stocks and flows or try to measure a particular characteristic of an environmental system. Usually, there is a reference point, for instance the sustainable use of a particular environmental good or a particular state of the environment. Such indicators always reveal a distance or a loss, and make it possible to find out whether development is going in the wrong or right direction. This is true of both pressure indicators (which try to measure the pressure from society on the environment) and state indicators (indicators for the state of the environment) (Opschoor and Reijnders, 1991).

2.2 Economy-cum-environment models

The category of economy-cum-environment models starts from the economic perspective. Here, change dynamism in the economic system are modelled in various degrees of comprehensiveness and complexity. These models are turned into an instrument of environmental policy analysis by extension with environmental modules and add-ons. The economic core models are expressed in volumes and (monetary) prices, and meet the equilibrium mechanisms of economic theories. Depending on the specific type and scope of the model, these mechanisms are either based on a micro-economic welfare objective with utility maximising economic agents who try to meet supply and demand conditions in the market place, resort to macro-economic balancing of (national) economic accounts, or trade off economic costs and benefits discounted in time. These models may be static or dynamic and may analyse a partial equilibrium for a particular market or sector, or a general equilibrium by accounting for the interaction among all relevant markets. Essentially, economic adjustment mechanisms refer to changing relative prices, substitution among scarce means and elasticities. The subject of analysis is either the efficiency of resource allocation, or the levels of, for instance, income, investment and employment over time. The sophistication of the economic model dynamism is not matched by that of the linked environmental modules. These sub-modules link economic activities to various forms of environmental pressure, such as emissions and resource use. The link is usually a rather simple linear and fixed one-way relationship without feedbacks to the environmental system or to the economic system. Some economy-cum-environment models do catch environmental feedbacks by directly treating an emission or a natural resource as an economic good, not an interaction with the environmental system.

Examples of economy-cum-environment models include input-output models. In fact, input-output models take an intermediate position between both categories of models, as input-output models picture the intersectoral deliveries of materials/products in monetary terms which consistently sum up to national accounts. This intermediate position explains the popularity of input-output models for environmental applications. The fixed technical relations are more easily linked to non-monetized environmental resource inputs and output, i.e. affluents of the economic system. Moreover, due account is given to intersectoral linkages and indirect environmental effects. Elements of input-output analyses are therefore often used for material and substance flow analyses. Applied general equilibrium models, by contrast, better meet the interests of economists, but are more difficult to combine with environmental variables, because relative prices change and various substitutions take place. Macro-economic models are particularly relevant for policymakers, but the environment extensions often do not go beyond the inclusion of the cost to meet certain environmental objectives. Cost-benefit and cost-effectiveness models applied to environment-economy trade-offs also try to moneterize environmental cost and benefits. Where this presents difficulties, multi-criteria models provide an alternative by using non-monetary dimensions.

This category of models gives rise to the search for indicators expressed in monetary values, preferably in an aggregated form to reflect (changes in) overall welfare. Aggregate measures implicitly facilitate substitution and are a reflection of revealed preferences. National product or income are well-known measures of welfare, although incomplete and partially misleading. To arrive at better measures of welfare, attempts are being made to correct traditional national income measures for environmental losses, i.e. depletion and degradation of the environment. The 'true' green national accounts are still under profound discussion (Mäler, 1991; van Dieren, 1995). Suggested alternatives for a greener national product are adjusted savings measures or the concept of genuine savings (Pearce, *et al.*, 1998). These measures for net savings account for the depletion of natural resources and the accumulation of pollutants in monetary value terms.

2.3 Integrated models

The ultimate objective of environment-economy modelling is to arrive at genuinely integrated models, which duly reflect the relevant mechanisms operating in both the environmental and the economic system, inclusive of their mutual interrelationships and feedbacks. Integrated models are really a multidisciplinary endeavour. Given the high degree of complexity, these models usually come at two levels: either at a low spatial level or at the highly aggregated, abstract level of the so-called integrated assessment models. The former type of models start from a well demarcated area, ecosystem or specific environmental problem, and try to deal with environmental and economic interactions in great detail. Developments in time or different scenarios are evaluated by means of a pre-selected set of performance indicators. In contrast to these desaggregated models, integrated assessment models usually involve a high level of aggregation. An integrated assessment model needs to provide relevant information for policy and decision making and therefore needs to combine insights from various disciplinary fields (Tol and Vellinga, 1998). Integrated assessment models link disciplinary submodels to different degrees of sophistication and complexity. To keep it manageable and transparent, simpler versions of extensive disciplinary submodels are linked in a meta-model. Although intended to be truly integrated, hardly any integrated model can be found without either a natural science or economic bias.

3. DISCIPLINARY INCOMPATIBILITIES AND TRADE-OFFS

It will have become clear that disciplinary trade-offs, which have their origins in disciplinary incompatibilities, are more persistent than many environmental scientists wish. *Table 11* lists ten of the most frequently occurring disciplinary incompatibilities.

Table 11. Incompatibilities in Applied Environment-Economy Modelling

	Perspectives	
Subject	**Natural sciences**	**Economics**
1. Sustainable development	Sustainable use	Substitution among capital stocks
2. Time horizon	Short to very long run	Short to medium-long run
3. Spatial level	Environmental scale	Economic scale
4. Units of measurements	Physical units	Monetary values
5. Categories of analysis	Materials, substances, technologies	Sectors, products, demand and supply, prices
6. Object of analysis	Cause, effect, response	Welfare, cost-benefit, efficiency
7. Consistency	Natural laws	Economic theories and balances
8. Actors	-	Optimising economic agents
9. Dynamism	Assimilation, transformation, regeneration	Market forces
10. Attribution problem	-	-

3.1 Sustainable development

The interpretation of the concept of sustainable development ensures the divergence of disciplinary opinions. Natural scientists usually interpret sustainable development as the sustainable use of a particular environmental good. Resources, eco-systems and fishing grounds may be used up to the assimilative capacity or up to sustainable extraction or yield levels without jeopardising future use. Implicitly, each and every environmental good has to remain intact, or be compensated for technically or spatially. This constant natural capital concept underlies the sustainable use interpretation of sustainable development, and is often referred to as strong or ecological sustainability (Serageldin and Steer, 1994; Opschoor, 1992). Economists do not feel at ease with this rigid interpretation. Disregarding the nuances, economists generally define sustainable development as non-declining per capita human welfare over time (Solow, 1986; Mäler, 1991; Pearce, *et al.*, 1989; Pezzey, 1992). Welfare is broadly conceived and compromises income as well as poorly measurable components such as environmental services. However elusive this notion might seem for natural scientists, it implies a rather stringent condition, namely that the total stock of capital goods with which human welfare is produced must, at the very least, be maintained. And if the increasing needs of a growing world population are taken into account, then there has to be an enlargement in this total stock, or, at any rate, in the total welfare that the capital stock is able to produce. This is only feasible through technological development.

The divergence of opinions between natural scientists and economists is made more clear when the different conditions to maintain the capital stock are specified. The total capital stock that sustains present and future welfare comprises environmental or natural capital, man-made capital, human capital and what is called social capital. The latter form of capital, though a less tangible concept, is nevertheless of great importance to the attainment of sustainable development and stands for the social cohesion of a society. According to the strong sustainability concept, the sustainable use of (the components of) environmental capital has to be guaranteed. Thus, environmental capital has to be conserved, and, of course, preferably also the overall stock of capital. Economists, by contrast, only require a non-decline in the overall stock of capital, thus allowing for substitutability among the four different forms of capital. Although there are interpretational differences among economists, this weak sustainability concept constitutes their frame of mind. Their interpretations are differentiated to the extent that uncertainties and irreversibities with respect to environmental capital components are valued and future technological possibilities are assessed. These aspects might limit the substitutability of environmental capital for other forms of capital. But generally, sustainable development is for economists a search process, directed by societal risk analyses and preferences. It represents a series of value judgements and normative choices (Den Butter and Verbruggen, 1994). In contrast to this subjectivism, natural scientists feel safer with the objectivity of the sustainable use.

3.2 Time horizon

The analysis and models of natural scientists cover a wide range of time horizons, from very short up to (almost) infinity. Climatologists and earth scientists only think in time intervals of hundreds or even millions of years. Chemists and biologists can also handle very short time periods. For economists, however, the medium-long run is the longest period that they can deal with in their applied modelling in a well-founded manner. As for the long run, they simply shrug: the uncertainty is too great. Not without reason, a well-known economist once stated that in the long run we are all dead. It is only in their theoretical modelling that infinite time horizons are envisaged. These models, however, do not reflect a real time dimension, rather a succession of model periods converging to a steady state or indicating long-term growth paths. The varying time perspectives cause disciplinary tensions, for instance with respect to substance flow modelling. Economists wish to stop after several years, whereas natural scientists prefer to continue till the stocks are either depleted, saturated or certain thresholds are passed. In this connection, ecologists who are exploring long-term equilibrium situations, take an intermediate position.

3.3 Spatial Level

Demarcation of the appropriate spatial level of analysis is important for natural science and economic research alike. Economic research may have no specific spatial dimension, e.g. a sectoral analysis or a trade model for a particular product, or it may be confined to the boundaries of national states and regions. Although country and regional models can be linked, national boundaries remain decisive, simply because economic systems are nationally organised. The environmental system, on the other hand, disregards national boundaries. The transport model of emissions or the extent of an environmental good typically determines the relevant spatial level of analysis. An integrated environment-economy model requires a spatial level that covers as much of the relevant environmental and economic system as possible. Inevitably, certain stocks and flows or economic mechanisms are placed outside the system, unless, the spatial level is global.

3.4 Units of Measurement

Natural scientists analyse in physical units. These units may differ per substance and in dimension, and range from extremely small to huge units beyond one's comprehension. Sometimes a common denominator can be constructed, like, for example, an acid or carbon equivalent, but aggregation of measures in physical terms is in general not possible. This seriously complicates mutual comparisons, indicator development and trade-offs, to the frustration of policymakers. Economists are very fortunate in this respect. They pretend that everything can be expressed in monetary values, either directly with market prices or indirectly through imputed prices. Admittedly, market prices do not always reflect true scarcities, and imputed prices are often imperfect and incomplete. But, nonetheless, economists have one yardstick. This diverse nature of measurement units might also explain the different ways of looking at empirical data and figures. For natural scientists data are measured and fixed, albeit with a margin of uncertainty, whereas for economists prices and quantities are in a permanent state of change. A figure for the former is a fact, for the latter it is a mere indication. It will be clear that the linking of these different units of measurement in environment-economy models. i.e. firm and soft data, brings along specific frictions. These are further aggravated by the mismatch between categories of analysis and the so-called attribution problem.

3.5 Categories of analysis

Basically, natural scientists engaged in environmental research study the fate of substance flows, or material flows as bundles of substances, in the environment in the broadest sense, i.e. water, soil, air and all the life it contains, humans included. Apart from regulations, technological innovations constitute obvious coping strategies. Hence, the most frequently used categories of analysis in environmental sciences are substances, materials and technologies. The linking of these categories with economic categories and classifications of products, sectors and trade flows in prices and quantities is a cumbersome procedure. Economists themselves already find it difficult to link up their statistical categories, let alone with substances, materials and technologies. From a natural science point of view, economic categories should be as homogenous as possible, which would lead to an endless desaggregation. This does not correspond to reality, and is hard to digest for economists.

3.6 Object of analysis

Causes and effects, governed by (objective) laws of nature are at the heart of the natural scientists' thinking. They are eager to acquire knowledge about the working of physical, chemical and biological reality. Economists are also in search of knowledge, but in a more instrumental way. Their overall objective is to understand how scarce means can be put to efficient use so as to maximise human welfare.

Environmental scientists as applied scientists have a supplementary social objective; they pursue a minimal environmental impact. Quite often, a contrast is created between minimising and optimising. An economist can lightly discuss an optimal level of pollution, whereas for a natural scientist this optimum can only approach zero.

A similar tension arises with respect to the interpretation of the notion of efficiency. That energy efficiency expressed as exergy or the efficiency of food production in terms of proteins can be increased is not of importance as such from an economic perspective. Physical efficiency is not the same as economic efficiency. For an economist, only the relative scarcity of means counts. To be economically efficient, both the cost and benefits of a physical efficiency improvement and the relative scarcity of the good whose efficiency is improved, has to be weighed against alternative uses.

3.7 Consistency

Physical laws, such as the laws of thermodynamics, provide for consistency in natural science research. It is assumed that the economic system, as it is embedded in a larger bio-geo-chemical system of material and energy flows, is constrained by these physical laws. In that sense, economics must take a subordinate role to natural science (Ayres, 1998). Even assuming economists comprehend the consequences of these physical laws, they do not know what to do with this knowledge. Since the 18th century Physiocrats, economics has been dematerialising. Economists use their own consistency checks, such as supply equals demand, production equals income and balance-of-payments deficit equals savings surplus. Both systems, although internally consistent, do not match.

3.8 Actors

Typical for economists, and for social scientists in general, is their orientation towards human behaviour and societal processes. Economic models are governed by behavioural relationships, ideally expressing optimising economic agents. Economists rely on the supposed rationality of these actors, who are after all gifted with a free will. This rationality relates to both the present and the future. Other social scientists take a broader view on human behaviour than focusing on economic rationality alone. Natural scientists have to do without such actors, who indeed can act erratically, but can also be steered. That is perhaps why natural scientists often have no other policy options but regulations and technical solutions.

3.9 Dynamism

Economists envisage a great number of, sometimes even unexpected, adjustment mechanisms that generally do not come to the minds of natural scientists. Even more disturbing for natural scientists is that although economists do not speak with one voice, they seem to understand each other, but non-economists' outsiders are confronted with diverging views. This elusiveness of economists has everything to do with the subject of their analyses: the functioning of markets. Everywhere along the chain from cause to effect or from the extraction of raw materials to the treatment of waste, economists place markets. On the market place, supply and demand of factors of production (capital, labour, land, technology, resources), intermediates and final products (goods and services) meet, and there is always the tendency to equalise supply and demand in one way or another. This equalising tendency that drives market forces expresses itself in changing prices, changing quantities in demand and supply, technological changes and substitutions among factors of production and products. What's more, the market place is not a single, surveyable market, but consists of numerous separate markets for each discernible factor and product, which, however, are all to a greater or lesser extent interrelated. Hence, notwithstanding the fact that these markets are governed by economic laws, it is not always easy to indicate what will happen. The determinism of the natural scientists and the loose observations of economists are often confusing.

3.10 Attribution problem

One of the most troublesome problems in environment-economy modelling is the so-called attribution problem. This does not merely boil down to the linking of natural scientists' units of measurement and categories of analysis with non-corresponding economic units and categories, but goes an analytical step further (Heijungs, 1997). At the one side of the chain, there is the question of what or who drives certain economic activities. This is, for instance, relevant in the case of exports and imports. Can foreign consumers be held responsible for the environmental effects in exporting countries or should these effects be attributed to exporting industries? At the other side of the chain, how can a particular environmental impact, which is usually due to a combined transformation process of various emissions and interventions, be attributed to economic activities. From a scientific point of view, this attribution problem is intriguing. For policy makers who are looking for good cases for policy interventions, this problem is really frustrating. With respect to the attribution problem, it is not a matter of disciplinary incompatibilities in the first place. There is still a lot of work to be done for all disciplines.

4. LESSONS TO BE LEARNED

As already indicated in the introduction, this discussion of disciplinary tensions shows a black and white picture for the sake of clarity. On the multidisciplinary shop floor, important accomplishments have indeed been reached. The argument is that disciplinary incompatibilities stand in the way of genuine integrated environment-economy modelling, and that these prove to be rather persistent. In spite of this, the ideal case would still be a truly interdisciplinary approach to the problem of environmental degradation, or the pursuit of sustainable development. To come closer to the ideal case, the following lessons can be drawn.

First of all, to achieve multidisciplinary cooperation around a specific environmental problem as a first step, a multidisciplinary attitude is desirable. This means that the different disciplines have to make their scientific conceptions explicit, and try to indicate what their respective contributions can be in understanding and managing the problem. With an open mind and understanding of the disciplinary possibilities and limitations, an adequate division of labour can be brought about. The weight of the different disciplines depends, of course, on the problem definition and the research objectives.

Second, and a step further, a disciplinary sharing could take place. It is very well possible that a concept or technique from one discipline can fruitfully be applied in another discipline. Natural scientists, for instance, appear drawn to economic input-output models. In turn, the ecologists' concepts of sustainable yield and the view of the environment as scarce natural capital find their way in economics. This sharing of idea eases multidisciplinary cooperation.

Thirdly, and the best that can be achieved as it presently stands, is multidisciplinary cooperation that is structured along a commonly shared interdisciplinary framework, concept or technique. Examples that have recently been developed include the material balances approach, substance flow analysis, life-cycle analysis and integrated chain management. Still less elaborated but promising interdisciplinary concepts and approaches are the concept of sustainability, the notion of spatial scales, industrial ecology, industrial transformation and industrial metabolism, natural capital or environmental utilisation space and, of course, the pursuit of integrated assessment. These interdisciplinary frameworks and concepts, although partial, go beyond disciplinary boundaries, provide a common framework of analysis and direct efforts in the same direction. Also, at this higher level of disciplinary integration, the disciplinary trade-offs remain valid, but are less serious obstacles.

ACKNOWLEDGEMENTS
Comments from Richard S.J. Tol and Reyer Gerlagh are appreciated.

REFERENCES

Ayres, R.U. (1998). Industrial Metabolism: Work in Progress, in: J.C.J.M. van den Bergh and M.W. Hofkes (eds), *Theory and Implementation of Economic Models for Sustainable Development*, Kluwer Academic Publishers, Dordrecht.

Bergh, J.C.J.M. van den (1993). A Framework for Modelling Economy-Environment-Development Relationships based on Dynamic Carrying Capacity and Sustainable Development Feedback, *Environmental and Resource Economics*, 3,4, pp. 395-412.

Bergh, J.C.J.M. van den, and M.W. Hofkes (1998). A Survey of Economic Modelling of Sustainable Development, in: J.C.J.M. van den Bergh and M.W. Hofkes (eds), *Theory and Implementation of Economic Models for Sustainable Development*, Kluwer Academic Publishers, Dordrecht.

Butter, F.A.G. den, and H. Verbruggen (1994). Measuring the Trade-Off Between Economic Growth and a Clean Environment, *Environmental and Resource Economics*, 4,2,pp.187-208.

Chopra, K. and G.K. Kadekodi, in collaboration with J.B. Opschoor, H. Verbruggen, A.M. Groot and A. Gilbert (1998). *Operationalising Sustainable Development: an Ecology-Economy Interactive Approach for Palamau District in India*, Sage Publications, New Delhi.

Dieren, W. van (ed.) (1995). *Taking Nature in Account - A report to the Club of Rome*, Springer-Verlag, New York.

Heijungs, R. (1997). *Economic Drama and the Environmental Stage*, PhD. thesis, Leyden University, Leyden.

Kuik, O. and H. Verbruggen (eds) (1991). *In Search of Indicators of Sustainable Development*, Kluwer International Publishers, Dordrecht.

Mäler, K.G. (1991). National Accounts and Environmental Resources, *Environmental and Resource Economics*, 1,1, pp. 1-15.

Opschoor, H. and L. Reijnders (1991). Towards sustainable Development Indicators, in: O.Kuik and H. Verbruggen (eds), *In Search of Indicators of Sustainable Development*, Kluwer Academic Publishers, Dordrecht.

Opschoor, J.B. (1992). Sustainable Development, The Economic Process and Economic Analysis, in: J.B. Opschoor (ed.), *Environment, Economy and Sustainable Development*, Wolters-Noordhoff Publishers, Groningen

Pearce, D.G. Atkinson, and K. Hamilton (1998). The Measurement of Sustainable Development, in: J.C.J.M. van den Bergh and M.W. Hofkes (eds), Theory and Implementation of Economic Models for Sustainable Development, Kluwer Academic Publishers, Dordrecht.

Pearce, D.W., A. Markandya and E. Barbier (1989). *Blueprint for a Green Economy*, Earthscan, London.

Perrings, Ch. (1987). *Economy and Environment*, Cambridge University Press, Cambridge/New York.

Pezzey, J. (1992). *Sustainable Development Concept-An Economic Analysis*, World Bank Environment Paper No. 2, World Bank, Washington, D.C.

Seragelding, I. and A. Steer (eds) (1994). *Making Development Sustainable: From Concepts to Action,* Environmentally Sustainable Development Occasional Paper Series No.2, World Bank, Washington, D.C.

Solow, R.M. (1986). On the Intergenerational Allocation of Exhaustible Resources, *Scandinavian Journal of Economics, 88, pp. 141-149.*

Tol, R.S.J. and P. Vellinga (1998). The European Forum on Integrated Environmental Assessment, *Environmental Modelling and Assessment (forthcoming).*

---

[1] This section is partly based on Chopra *et al.*, 1998, Chapter 1. See also Perrings, 1987; Van den Bergh and Hofkes, 1998; and Van den Bergh, 1993.

Chapter 8

Software for material flow analysis

JOS BOELENS AND XANDER OLSTHOORN

Researchers, Institute for Environmental Studies, Vrije Universiteit, Amsterdam

Key words: modelling, LCA, MFA, SFA

Abstract: This chapter gives an impression of the state of the art of computer programmes dedicated to support Life Cycle Analysis or Material Flow Analysis. Several aspects of the programs, such as the data structure, graphical presentation, dealing with uncertainty and modelling capacities are described. Dynamic modelling is often missing in these programmes and, hence, this chapter also describes the concepts of Petri Nets and Expert Systems that can be useful for dynamic modelling.

1. INTRODUCTION

Computer-based tools play a significant role in the analysis of materials and energy flows through economic activities. A range of software tools have been developed that specifically serve Materials Flow Analysis (MFA) - also referred to as Substance Flow Analysis (SFA) - or Life Cycle Assessment (LCA). This chapter gives an assessment of these tools, beginning with an inventory of MFA and LCA software, examines the types of systems, features and options that the software market currently offers. The main focus of this chapter is on MFA software, although aspects of LCA software are also covered. A key question relates to the benefits to be gained from using dedicated software to support MFA and LCA, rather than generic software for database management and spreadsheet calculations.

In MFA and SFA, two major tasks are performed: (a) preparing the material flows accounts and (b) interpretation, evaluation and modelling of the account. Preparing material flows accounts involves finding and organising the data on stocks and flows of material. Given the often complex nature of the accounting framework (i.e. number of entries (nodes), system boundaries, temporal resolution) and the lack of appropriate statistical data, the preparation of an account is frequently costly in time and effort. The software supports these activities by structured data storage within databases, guarding consistency and providing tools for dealing with uncertainty.

The second area where computerised tools support MFA/SFA is interpretation, evaluation and modelling. The software may calculate performance indicators and provide a range of report functions to support these activities. The software allows the construction of complex models, e.g. models that apportion environmental impacts to a specific product of service in LCA, or models that describe the behaviour of systems that are made up of mutually related substance flows.

This chapter provides first an overview of available software (section 2) and discusses how software may support the preparation of the materials account (section 3). Section 4 discusses how software is used for interpretation and modelling. Section 5 briefly describes some new developments in using computer techniques that may improve the currently available software. Section 6 concludes.

2. OVERVIEW OF AVAILABLE SOFTWARE

There is an abundance of dedicated software tools available in the market. Recently, about 30 computer programmes were presented at two international workshops[1]. Almost every programme runs on an ordinary personal computer (PC). Intended users of the programmes are environmental scientists at universities and research institutes, and researchers in industry. The development of these programmes began in the early 1990s. It will probably take many years before the programmes are mature in the sense that they are generally applicable and support all relevant aspects of all types of material studies.

An impression of the state of the art of the currently available European software is presented in *Table 12* (see for details the world wide web (WWW) where developers present their programs)[2]. The list is far from complete, even for the European market. However, this is today the most mature market for this software with a number of major suppliers, all of whom are represented below. The first column contains the name of the software and the institute and country (indicated by its WWW acronym) where it is developed. The second column specifies the hardware platform, it shows that all programs, except Cara, run on a PC, mostly under Microsoft (MS)-Windows. The other columns are explained in the rest of this chapter.

Table 12. Overview of software (web sites and Rice et al., 1997)

Programme name, institute, country	platform	purpose	data-base	data quality	stocks	hier-archy	mode-ling
Aspect, TU Vienna, at	msdos	MFA	no				yes
Audit, Audit GmbH, de	windows	LCA					
Boustead, UK	msdos	LCA	no		no		
Cara, Wuppertal Institute, de	mac	MFA	yes		no		
Danedi, dk	windows	MFA	no	no	no	no	linear
Euklid, Fraunhofer Institute, de	windows	LCA	yes	yes		yes	
Flux, IVM, Vrije Universiteit, nl	windows	MFA	yes	yes	yes	yes	linear
Gabi, Product Engineering GmbH, de	dos/wind.	LCA	yes				
Kcl_Eco, Finnish Pulp&Paper Inst., fi	windows	LCA				no	linear
Kosimeus, French-German Inst. Of Env. Research	windows	MFA	yes	yes		yes	iterative
LMS/U1, LMS Umweltsysteme GmbH, de	windows	LCA	yes	yes	no	yes	linear
Netsim, Universitat Graz, at	msdos	wood pulp				yes	stat.state
Pems 3.0, Pira Intnl, UK	windows	LCA			no		
Sankey, TU Graz, at	windows	MFA/ LCA	no		no	no	no
Sfinx, CML Leiden, nl	msdos	MFA	no		no	no	linear
Simapro, Pre consultants, nl	msdos	LCA	yes		no		
Simtool, Ing.untern. fur Umweltanalyse, de	msdos	cement ind.	no			no	dynamic
Team, Ecobilan group, fr	windows	LCA	yes		no		
Umberto, IFEU Heidelberg, de	windows	MFA	yes			no	non-linear
Visioman, TU Vienna, at	wind./mac	MFA	no		no		

The price of the software is not included in the table; it varies from a few hundred German Marks to more than ten thousand German marks. Examples of the expensive category are Boustead, Euklid, Gabi, Pems and Team. In general these programmes are more sophisticated, and are supported by larger, more specific databases.

3. MFA VERSUS LCA

The third column in *Table 12* indicates whether the programme is primarily meant for LCA or MFA. From a computer science point of view tools for MFA and LCA are quite similar, both account material flows and represent production processes. The system boundaries form the main difference between both methodologies (Voet and Heijungs, 1994; Kandelaars et al., 1996). MFA focuses on substances, while LCA is directed at products, i.e., product functions and services in general. In contrast to LCA, MFA views flows within a geographic region (and eventually economic sectors or companies) and reflects a certain time period, usually a year. This implies that MFA should also take into account of stocks, otherwise mass balances are distorted by accumulation or depletion (see section below on mass balancing).

4. PREPARING A MATERIAL FLOWS ACCOUNT

MFA is a general concept having a different meaning depending on the context. Within the context of a firm's environmental management, MFA refers to gathering, organising and summarising all information with respect to the use of resources and its processing into products and emissions of a firm (or set of firms). MFAs may be part of a firm's environmental audit and as such may give an insight into a firm's environmental performance. The boundaries of the accounting framework are defined by the firm (e.g. its production facilities and logistics) itself. For environmental policy making by a government, the scope of the MFA is confined only to the area where government exerts authority. Usually, these MFAs are used to develop environmental policies that address a single pollutant, whereas MFAs of firms typically address all substances. The preparation of a materials flow account involves several steps:

1. selecting and/or identifying the (composite) materials or substances;
2. identifying/establishing the time period;
3. selecting and defining nodes (or entries) of the materials flow accounts;
4. searching and processing the relevant data and information; and
5. preparing the account that is consistent with respect to the law of conservation of mass.

The first three steps relate to defining the scope of the account. In book-keeping terms, this step involves definition of the entries of the account, while in terms of physical modelling this step identifies the nodes of the materials flows network. What nodes are distinguished depend on the purpose of the analysis and on the statistical and other information that is available for the analysis. MFA software can support the preparation of the material flow account by providing a well defined data structure and a user-friendly interface. The software can also deal explicitly with the quality of data and the consistency.

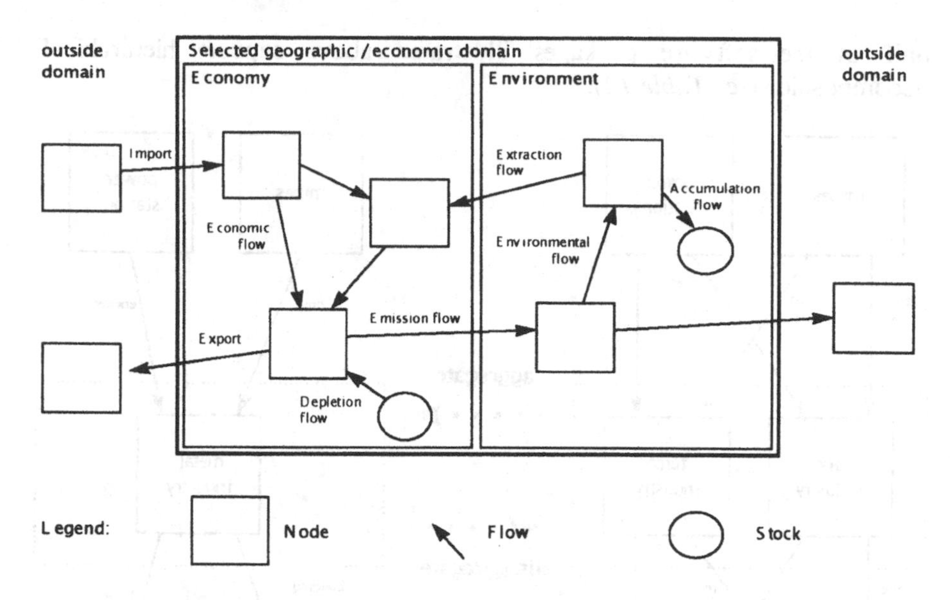

Figure 15. Overview of elements of a materials flow network

4.1 Data structure

The software represents the material flows within a network of nodes (or processes or transitions) and flows (or arcs). *Figure 15* shows the kinds of elements that can be included in materials flow network. A node may represent an economic sector (e.g., metal industry or consumers) or, on a more detailed level, technical processes. In some programmes (e.g., Flux) a node may also represent an environmental compartment (e.g., a soil). The node has inputs (e.g., raw materials) which are transformed into outputs (e.g., final products and waste). A flow transfers an amount of products or materials from one node to the other. The number of nodes within the economic subsystem can vary from a few up to more than hundred. Most of the LCA programmes supply processes with its inputs and outputs for a specific industrial sector, like Kcl-Eco does for paper production. The data can be used, in this example, by the paper industry, although the process characteristics will differ among companies and countries (see Rice, 1997).

In order to characterise a large network of flows it is necessary to be able to view the data on several levels of aggregation. For that, the data structure should incorporate hierarchic decomposition such that subsystems can be fused together into a single node and vice versa *Figure 16* demonstrates the concept of hierarchic decomposition. The very basic example shows that different branches of the metal industry are aggregated into a single node (or vice versa), flows are fused together and intermediate flows disappear. About half of the software packages illustrated above support hierarchical decomposition (see *Table 12*).

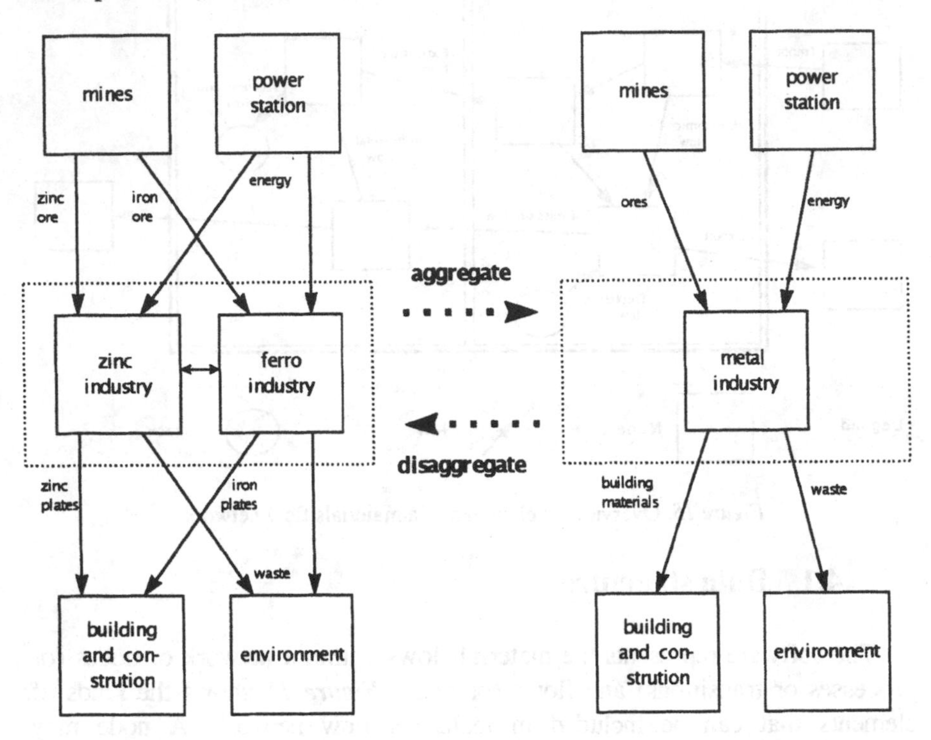

Figure 16. Application of hierarchic decomposition on a simple network

The subjects of flows are products, materials or substances. About half of the programmes (e.g. Gabi, KCL-Eco) enable definition of composition of products and materials. This facility provides an insight into the linkage between flows of several substances. For instance, zinc ore contains zinc, but also some cadmium. Since changes to the import of zinc ore affect the material flow accounts of both zinc and cadmium, these accounts should not be manipulated or modelled independently of one other.

LCA programmes and most of the MFA programmes lack a Stock entity. Within LCA stocks are of minor importance, and within MFA only the change of the stock within a time period is important. The stock itself (e.g. number of batteries within households) provides useful information, but becomes particularly interesting only when several time periods are viewed.

The more complex the data structure is, the greater is the need for a Database Management System (DBMS)[3]. DBMS is a very commonly used tool for information systems. DBase and Access are popular packages on PC's. The column "database" of *Table 12* states whether the programme uses a DBMS or not. Programmes without a database often focus on presentation of graphs (e.g. Sankey) or on calculation (e.g. Sfinx).

4.2 Human computer interface

A programme with excellent features may still fail as a tool for LCA or MFA if the human computer interface is too complex or impractical. Well designed windows, intuitive operations and unambiguous terminology are important for that reason. However, simplicity may induce functional limitations. In general a good interface is to be preferred over advanced but incomprehensible software.

A feature that supports clear representation of the structure and coherence of the data is a graphical presentation of nodes and flows. Practically all programmes include this feature. In cases where the system is very large, with for instance more than 100 flows, a graph is overwhelming and hierarchical decomposition of the graph is needed. *Figure 17* is an example of an aggregated view of a complex (about 100 nodes) material flow account of zinc in the Dutch economy. The arrows represent flows between the aggregated nodes. The blocks in the middle of the figure represent the nodes reduced by aggregation to four main types: 'Extraction' comprises all nodes that represent mining (no mining in the Netherlands); 'Production' draws all nodes together that represent production of goods that contain zinc; 'Usage refers to all nodes where (capital) goods that contain zinc are in use; Waste processing relates to nodes that process waste streams that contain zinc (e.g., waste water and municipal waste. Foreign countries and the environment represent nodes outside the system of flows and stocks in the Dutch economy. Within the main blocks of the figure additional information is inserted. The 'efficiency' (eff) of the aggregated node gives an indication of the flow to the environment in relation to total throughput. Accumulation refers to the rate of accumulation of the stock. The latter was not available in the material flow account.

Beyond graphical presentation, the programmes usually provide also a range of graphs and reports. Frequently links to spreadsheets and word processors are included. These links provide flexibility for the user in processing the output of the software.

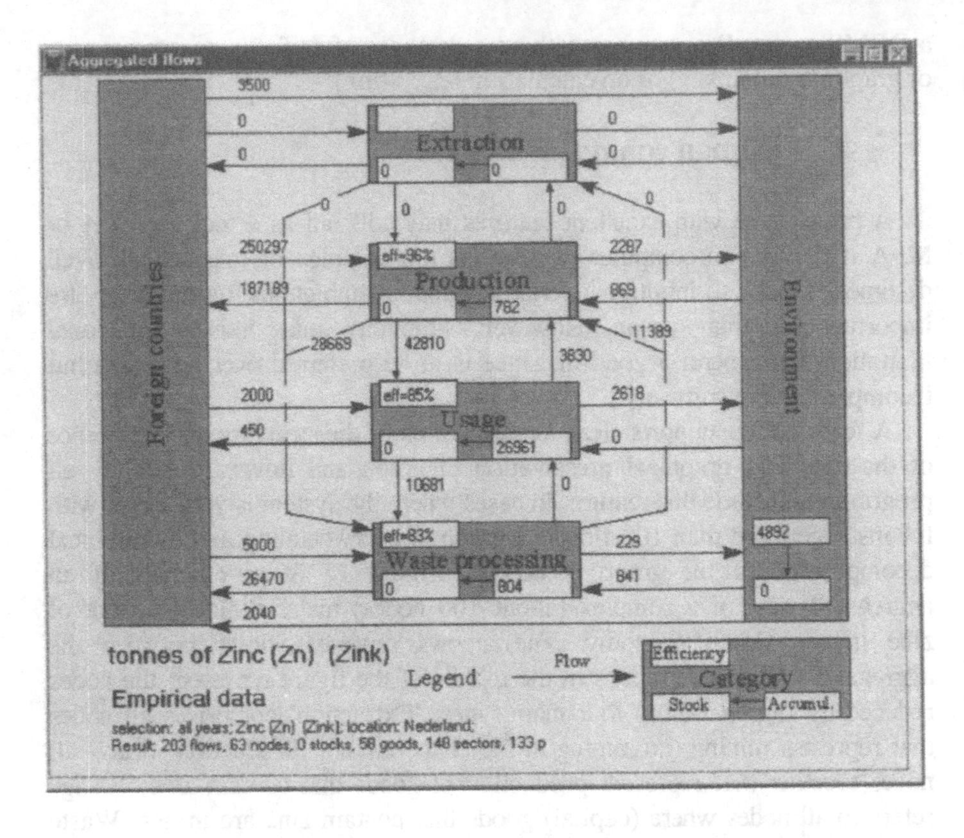

Figure 17. An example of graphical representation of (in this case aggregated) nodes and flows from the program Flux

4.3 Data quality

The lack of comprehensive, high-quality data is the major obstacle for MFA and LCA. Therefore, it is important for the software to deal with the paucity of data. Uncertainty margins may denote that the actual value is within certain stated margins with, say, a 95% probability. In the available computer programs, uncertainty information is often not available. Only Euklid, Flux, Kosimeus and LMS/U1 claim to deal with uncertainty information.

A possible reason for the neglect of uncertainty information is that they are unknowns. It is frequently hard enough to collect estimates of stocks and flows. A possible solution for the data quality problem is to use heuristic values, instead of exact uncertainty margins. For instance, in the programme Flux five classes of quality can be distinguished, varying from "very low quality" to "very high quality". The user defines the actual margins that are used for these five classes. The user can also fill in exact margins when these are known. The advantage of an heuristic approach is that it is easier to choose from a few classes of quality. Moreover, often information about quality of data is lacking completely, since perhaps only one 'measurement' is available. The heuristic approach is an easy way to deal with the reality of making rough estimates.

4.4 Mass balancing

Without the concept of mass conservation environmental scientists would have a far more complex task collecting a realistic set of data on material flows. For every node in the network, and for the system as a whole, the mass input should equal mass output (taking account of stocks), at least for the substances under study. Due to this concept it is sufficient to determine only a subset of the data; the rest of the data can be derived from that subset.

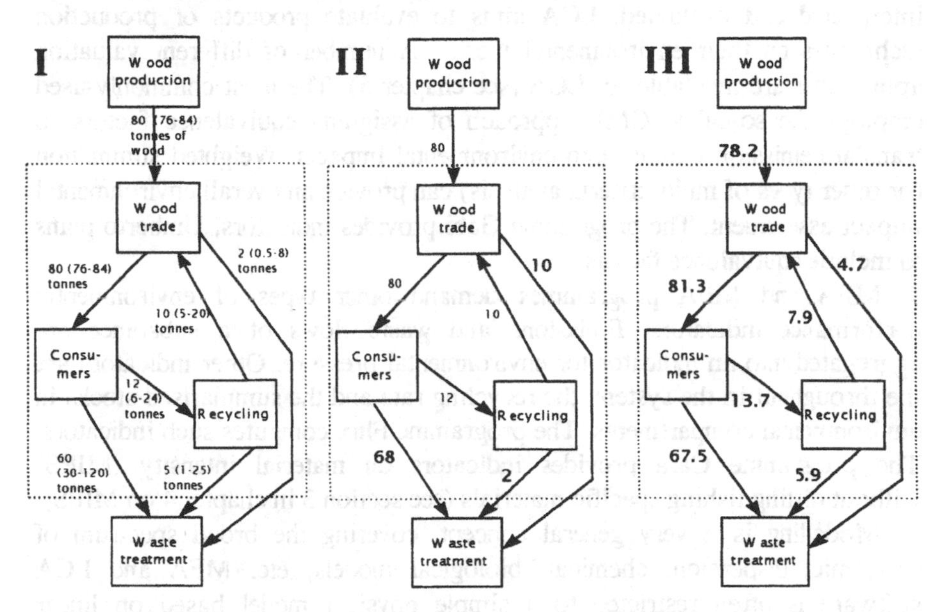

Figure 18. Example of how an inconsistent account (I) is adapted by balancing flows (II, bold digits are adapted) and by a numeric optimisation algorithm (III)

Note that changes to the stocks (i.e. accumulation or depletion) should be included as well within the mass balance. Take for instance a node that represents 'consumers'. For a relatively new consumer product with a long lifetime (more than 1 year), the input (purchase of the product), will exceed the output (waste flows) for a period of more the one year. The difference is caused by the accumulation of the stock of consumer goods within the node. Therefore accumulation should be treated as an input flow and depletion as an output flow of the node.

The concept of mass conservation also offers methods to correct a rough set of data resulting in a consistent network. One method is to appoint for every node a single input or output that is adapted such that mass-balance is preserved. The programme Sfinx applies this method. An alternative is to use a numeric algorithm that adapts all flows as little as possible, taking into account the uncertainty of the flows. The more accurate a flow is, the less it is adapted and uncertainty margins are never exceeded. The programme Flux applies this method. *Figure 18* shows an example where both methods are applied to make an account consistent. Note that the outcome of the two methods are quite different.

4.5 Interpretation, evaluation and modelling

After preparation of the materials flows account the data needs to be interpreted and evaluated. LCA aims to evaluate products or production techniques on their environmental merits. A number of different valuation approaches are available for LCA (see chapter 5). The most commonly-used employs the so-called CML approach of assigning equivalence factors to translate emission outputs into environmental impacts. Weighted summation (or other types of multi-criteria analysis) can provide an overall environmental impact assessment. The programme Gabi provides indicators, Umberto plans to include equivalence factors.

MFA and MFA programmes demand other types of environmental performance indicators. Emissions and waste flows of a substance are aggregated into an indicator for environmental pressure. Other indicators are the throughput in the system, the recycling rate and the summarised stocks in environmental compartments. The programme Flux computes such indicators. The programme Cara provides indicators on material intensity (MIPS) without distinguishing specific materials (see section 3 in chapter 4 on MIPS).

Modelling is a very general concept, covering the broad spectrum of economic, dispersion, chemical, biological models, etc. MFA and LCA software is often restricted to a simple physical model based on linear equations. LCA programmes represent a process by means of coefficients, where the production of 1 kg product involves specific amounts of raw materials, energy and waste. Several (related) products are compared on their environmental impacts. The impacts may differ due to differences within the production chain and the materials used. Modelling may comprise of experimenting with the process coefficients and the selection of production processes.

MFA models also use process coefficients, resulting in a kind of input-output analysis. Changing the size of a single flow has an effect on the rest of the network. Most programmes incorporate the relations between the input and output of a node, often restricted to linear relations. The result is a model. The user of the programme can experiment, make changes, and observe, for instance, the results in terms of environmental impacts. Several scenarios can be represented and compared with each other. Some of the programmes support scenario analysis (e.g. LMS/U1). None of the programmes currently model stocks and flows over time, incorporating scenario variables and environmental measures. There is a need for such programs, since planned developments can lead to future environmental problems that are preferably avoided.

5. CONCEPTS FOR FUTURE LCA/MFA SOFTWARE

In this section we move beyond the state of the art of LCA/MFA software, and discuss the application of new concepts. The field of mathematics and computer science may offer less familiar concepts that could support LCA and MFA. Two such concepts, Petri Nets and Expert System, are described here.

5.1 Petri nets

Currently available MFA/LCA software have certain modelling capacities but apart from Simtool, the software is restricted to supporting static modelling. Dynamic modelling is interesting since it may provide a tool for making scenarios for the (near) future. Dynamic modelling has not been developed in the context of this chapter, partly because of the complexity of the calculations, data collection and management and the human computer interface. A pragmatic approach would be to link the dynamic modelling language Stella, used by Meadows (1991) and Gilbert and Feenstra (1992), to the software. However, due to technical restrictions this is currently not possible. Another approach is to use the concept of Petri Nets.

A petri net is a mathematical concept that can be applied for (dynamic) modelling of material flows. Petri Nets are networks with nodes and arcs, just like the networks used by the LCA/MFA software. However, a difference is that two types of nodes - places and transitions - are distinguished. A place can represent a stock and a transition is a node or process that transforms input flows into output. Connections between two nodes of the same type are not allowed (see *Figure 19*). Furthermore, places may contain zero or more so called tokens, often drawn as black dots. For material flows, a token represents a unit of a product, material or substance. During modelling the tokens are shifted from place to place by means of the defined transitions. Transitions can use both linear and non-linear functions to take tokens from one or more (input) places and to put these onto one or more other (output) places.

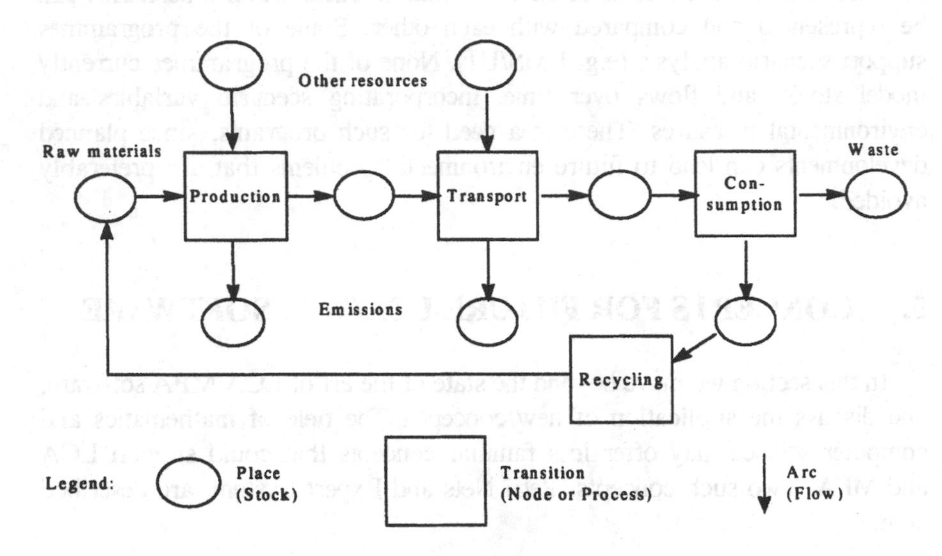

Figure 19. A Petri Net representing a production and consumption chain.

Petri Nets have existed for over 30 years and in the last decade have been put into practice extensively. Many tools and applications have been developed, as Petri Nets have become a mature tool for modelling and analysing various types of systems (van der Aalst, 1994; see also Reisig 1985 for further introduction to Petri Nets; and Schmidt and Schorb 1995 for Petri Nets in relation to MFA). One of the application domains is the field of logistics, which has a lot in common with the field of MFA/LCA.

When it comes to developing a dynamic model, the Petri Net is a promising concept to use. Petri Nets can model material flows over time in a natural way. They have a strong mathematical foundation that can deal with the complexity of dynamic modelling, for instance by formalising mass conservation and hierarchical decomposition. A second advantage is the availability of Petri Net tools which can, for instance, analyse the network structure and simulation of stocks and flows over time. Altogether, it seems worthwhile to further investigate the application of Petri Nets. Some programmes already use Petri Nets - Umberto uses Petri Nets for the graphic design, Netsim provides a link to Petri Net modelling.

5.2 Expert systems

Even though petri nets can help to develop a satisfying dynamic model, the question remains: how should values be assigned to the parameters of the model. For instance, changes in consumption patterns may induce changes to complete production chains and have side effects on other production chains. The challenge is to build scenarios and translate them into a sound and consistent set of parameters for the model. The prediction of effects of socio-economic developments and environmental policies on physical stocks and flows of materials demands expert knowledge. An expert system can be designed to operationalise such expert knowledge.

Some work has been done on developing an expert system for MFA. An expert system prototype has been developed for modelling socio-economic developments and measures for emission registration (see Langen et al., 1995 and also Brumsen et al., 1991). Although this system does not include intermediate supply of products and materials, many aspects of this system can be applied to the expert system for MFA.

Expert systems refer to a computer system or programme which incorporates one or more techniques of artificial intelligence to perform a family of activities that traditionally would have to be performed by a skilled or knowledgeable human. An expert system does not need to replace a human expert, it may play the role of an assistant (Tanimoto, 1987). The application of modelling socio-economic developments and measures is an ideal candidate for expert systems since it is a complex problem domain, high level reasoning is demanded, uncertainty is involved, the situation is dynamic and consistent solutions are required.

Figure 20 shows what the structure of such an expert system may look like. The left part shows a dynamic model for physical material flows; the parameters that control the dynamic model are put together in a separate module. The right part is the expert system; it include representations of socio-economic developments (scenario) and requirements. Requirements include consistency with respect to mass balances and stocks being non negative, but it may also include environmental targets or economic preconditions (costs).

How does the expert system work? First, the user specifies a scenario; then a set of parameters is derived from that scenario. On the basis of these parameters the account for a future period is calculated. This future account is compared with the requirements. When requirements are not met, the system adapts the scenario in co-operation with the user. These steps are repeated until the results are satisfying.

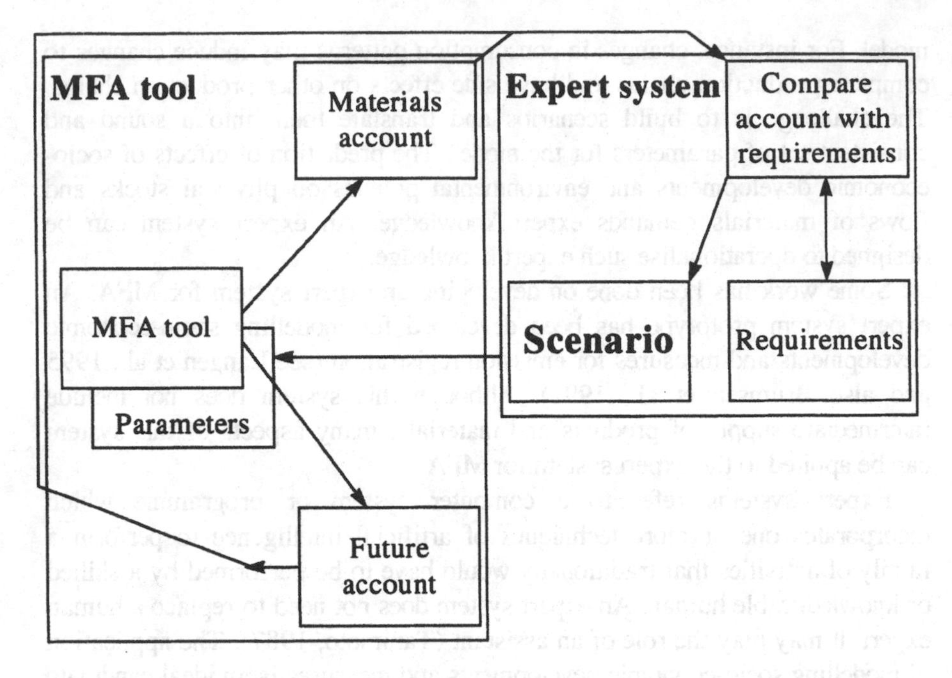

Figure 20. Possible programme structure of an expert system for performing scenario analysis

Such an expert system seems to demand a higher level of knowledge than is currently available. In spite of this, it may still be interesting to develop a prototype expert system, since the development of an expert system may induce the extension of the expert knowledge.

6. CONCLUSIONS

Currently software supporting the analysis of material flow studies is available. Such programmes support activities such as materials flow accounting, and modelling networks of flows and stocks. This chapter gives an impression of the features of the computer programmes that simplify the work of the researcher. Without these programmes a lot can be done with spreadsheets and mathematical programs, but that will often require more effort, that can probably better be devoted in making reusable programs. Why not recycle knowledge? To evaluate whether a software tool can be useful or not, and which of the tools is appropriate to support a specific piece of research, further reading and experimenting with demonstration programmes is recommended.

Many of the tools are in a prototype stage and are being developed further. A question is whether standards for LCA and MFA will arise, or whether there will be a wide diversity of programmes. Currently some efforts at standardisation of LCA data formats are being undertaken[4]. Standardisation of formats is preferable for it would imply that the exchange of LCA data was no longer complicated by the barrier of different data-formats. More intensive exchange of date may induce stronger progress in the field of LCA studies. Meanwhile, one may expect that many new programmes will be developed, some 'simple' tools which focus on a single aspect; others with extensive functionality that are complex and expensive. There is a market for these expensive programmes among firms interested in supporting eco-audits, and in embedding life cycle analysis in product design.

Currently, the ability of dedicated computer tools to translate notions of industrial transformation to predictions of physical material flows seem low. The concepts Petri Nets and Expert systems can be used in dynamic modelling. Petri Nets are interesting for their strong mathematical foundation and the natural representation of material flows. An expert system can be helpful in registering and increasing the available knowledge on the translation of industrial transformation into physical material flows.

REFERENCES

Aalst, W.M.P.v. (1994). Putting high-level Petri nets to work in industry. *Computers in industry*, Vol. 25, No. 1, p. 45-54.

Ayres, R.U., U.E. Simonis, (1994). *Industrial Metabolism, restructuring for sustainable development*, United Nations University Press, Tokyo

Brazier, F.M.T., P.H.G.v. Langen, J. Treur, N.J.E. Wijngaards, M. Willems, (1996). Modelling an elevator design task in Desire: the VT example. *International Journal of Human-Computer Studies*, Vol. 44, Special Issue on Sisyohus-VT (A.Th Screiber & W.P. Birmingham Eds.), pp. 469-520

Bruin, A.J.d., A. Reinders, (1991). *The use of knowledge based systems within the RIVM (In Dutch)*, 421504002, RIVM, Bilthoven.

Brumsen, H.A., A.A.M. Kusse, J. Pannekeet, J. Treur, A.W.S. Wenthold, (1991). *Modelling design processes for environmental decision support*, SKBS/A3/91-13, Vrije Universiteit, Amsterdam.

Gilbert, A., J.F. Feenstra, (1992). *Cadmium in the Netherlands (in Dutch)*, Vrije Universiteit Amsterdam.

TU Wien, (1995). State of the art computer programmes for materials balancing/accounting, Samba workshop 8-9 May 1995, Vienna.

Graedel, T.E., B.R. Allenby, (1997). *Industrial Ecology*, Prentice Hall, New Jersey, USA.

Hemming, C. (1995). *Directory of life cycle analysis*, SPOLD secretariat, Brussels,

Kandelaars, P.P.A.A.H., M.H. Jansen, F.J.D. Lambert, (1996). *Survey of physical flow models focusing on the environment*, TI 96-124/5, Tinbergen Institute, Amsterdam,

Langen, P.H.G.v., F.M.T. Brazier, H.B. Diepenmaat, M.P.J. Pulles, (1995). *Invenarisatie van de milieubelasting van industriele processen vanuit kennistechnologisch perspectief*, R 95/139, TNO, Delft, 54 p.

McFadden, F.R., J.A. Hoffer, (1994). *Modern database management,* 4th ed., Benjamin/Cummings Publishing Company Inc, Redwood City

Olsthoorn, X., J. Boelens, (1995). Towards a calculation and information system for substance flows (In Dutch). *Milieu, Tijdschrift voor Milieukunde.* Vol. 10, no.3, p 141-146

Preece, J., Y. Rogers, H. Sharp, D. Benyon, S. Holland and T. Carey, (1994). *Human-computer interaction,* Addison Wesley Longman, London

Reisig, W. (1985). *Petri Nets, an introduction,* Springer Verlag, New York Heidelberg Berlin Tokyo.

Rice, G., R. Clift, R. Burns, (1997). LCA Software review, comparison of currently available European LCA Software. *The International journal of Life Cycle Assessment,* Vol. 2, No. 1, p. 53-59.

Schmidt, M., A. Schorb, (1995). *Stoffstrom-analysen in oekobilanzen und oeko-Audits,* Springer Verlag, Berlin Heidelberg.

Schroder, H. (1996). Merging economic and environmental input-output analysis. *Ecological economics,* Vol. 13, No. 2, p 125-140

Singhofen, A., C.R. Hemming, B.P. Weidema, L. Grisel, R. Bretz, B. d. Smet, D. Russel, (1996). Life cycle inventory data: development of a common format. *International journal for Life Cycle Analysis,* Ecomed. Vol. 1, No. 3, p. 171-178. Landsberg, Germany,

Tanimoto, S.L. (1987). *The elements of artificial intelligence, an introduction using Lisp,* Computer Science Press, Rockville MD

Voet, E.v.d., R. Heijungs, (1994). MFA vs LCA, *SETAC Newsletter,* Vol. 4, No. 6, p. 3-4.

Voet, E.v. (1996). *Substances from cradle to grave, development of a methodology for the analysis of substance flows through economy and the environment of a region.* PhD. Thesis

---

[1] The first SAMBA workshop was at the Technical University of Vienna (TU Wien, 1996), the second at the Wuppertal Institute (WI, 1996).

[2] A number of programs are presented at http://awsunix.tuwien.ac.at/samba/samba.htm. Software Developers or sellers are invited to add (links to) information about their software on this site. A number of LCA programs are presented at http://www.ecosite.co.uk/software/ecosoft.htm. The reader is also referred to the SAMBA-L discussion list on software for MFA, which has started in 1995. You can join the discussions on the SAMBA-L list by sending an E-mail to listserv@nic.surfnet.nl. The E-mail should include the following text in the message body: subscribe samba-l <first name> <second name>.

[3] The disadvantages of not using a DBMS are uncontrolled redundancy, inconsistent data, inflexibility, limited data sharing, poor enforcement of standards, and excessive program maintenance (Mc Fadden and Hoffer, 1994)

[4] The Society for Promotion Of Life-cycle environmental management Development (SPOLD) in Brussels is working on a common format for the exchange of environmental inventory data for life-cycle (cradle to grave) assessments of products.

Chapter 9

Integrating Life Cycle Assessment and Economic Evaluation

JANE POWELL, AMELIA CRAIGHILL AND DAVID PEARCE
Centre for Social and Economical Research on the Global Environment, Norwich and London.

Key words: Economic valuation, landfills, life cycle assessment, recycling, waste management

Abstract: **Making choices on the basis of environmental assessment techniques such as life cycle analysis (LCA) requires a valuation of the impacts. A mature approach for doing this is economic valuation in which money values are ascribed to environmental and social costs associated with alternative products, processes or policies. In this chapter economic valuation is applied to a case study of waste management options including landfilling, incineration and recycling. The results show that the environmental costs of waste management options are specific to location, technology and material. Recycling is shown to be clearly beneficial in most cases.**

1. INTRODUCTION

The physical accounting of resource and energy flows in the economy is enjoying a revival in the 1990s, having already gone through an early phase of popularity in the 1960s and 1970s. Physical accounting is familiar to environmental economists since what was in many ways the seminal paper that initiated so much environmental economics, Kenneth Boulding's 'spaceship earth' essay (Boulding, 1966), firmly established that the conditions for the existence of what economists call an 'externality' or 'external effect' are potentially pervasive in any economy, regardless of its ideological basis in capitalism or socialism. Economists had long known that external effects, uncompensated third party losses in well being due to economic activity, were sufficient to distort the actual workings of the economy away from an optimum. But it seems fair to say that the empirical cognition of these externalities implied that they were fairly minor deviations. Boulding showed that, at the very least, the First Law of Thermodynamics dictated a rather different result, namely that since extracted and harvested resources had always to reappear in equal physical quantities as waste flows, the *potential* for externalities to show up almost everywhere was extremely high. Nonetheless, physical accounting can never provide a complete picture of resource use. This is because it is devoid of any value content: it does not carry with it any particular ethical implication because there is no objective function against which those resource flows are to be judged. This does not make resource accounting unimportant, it simply makes it incomplete for purposes of analysing public policy choices.

Obviously, establishing an objective function is a fairly arbitrary matter unless there is some 'meta-ethical' objective that dictates the choice of function. The way in which we use resources, for example, is likely to be very different if we think that human survival, regardless of quality of life, is some kind of moral imperative, compared to, say, maximising human well being over some defined time horizon. The consequence is that disagreements about how resources should be used are not just likely, they are inevitable. Since there is no universally agreed moral imperative, nor is there likely to be, there will be very differing views about what is a 'good' allocation of resources.

Life cycle assessment (LCA) has emerged as one procedure for engaging in physical accounting. In light of the above argument, it too must be incomplete as a technique for policy analysis. The issue of the objective function can be restated as one of selecting the weights that need to be applied to the many and heterogenous resource flows in the economy for there to be any quantifiable use for LCA. We develop this point below. What we show in this paper is that LCA can be combined with economic valuation techniques to produced a synthetic technique which we call 'LCA-EVA' for life cycle assessment combined with economic valuation analysis. We illustrate various applications of this approach. Of course, both components of this analysis will be disputed: LCA and EVA. We do not dwell on the likely sources of disputes since that would over-extend the paper. We merely demonstrate that the two techniques can be profitably combined.

2. THE PRINCIPLES OF LCA

The total environmental and social impacts of waste management systems can be determined by using the technique of lifecycle assessment combined with some evaluation methodology. Lifecycle assessment (LCA) is used to measure the environmental inputs and outputs associated with a single product or service from the mining of raw materials, through production, distribution, use and re-use or recycling, to final disposal. This is carried out in terms of raw materials, energy use, emissions to air and water and solid waste. Given some evaluation method, the total environmental costs and benefits of a product or service can be determined. This can be extended, where data exist, to include social costs and benefits as well.

There are four stages in a life cycle assessment - goal and scope definition, inventory analysis, impact assessment and interpretation of the results (Finnveden, 1996; see chapter 6 for details). The first stage defines the purpose, scope and boundaries of the study, the functional unit, key assumptions and likely limitations of the work (SETAC, 1991). In the inventory analysis the environmental impacts throughout a product's lifetime are quantified. The result is a list of pollutants and resources, and their quantities, that may have an impact on the environment. The impact assessment stage is aimed at understanding and evaluating the magnitude and significance of the potential environmental impacts of a product system (ISO, 1996). In the Interpretation stage the findings from the inventory analysis and the impact assessment are combined together. The findings of this interpretation may take the form of conclusions and recommendations to decision-makers (ISO, 1996).

As is well known, one of the main problem areas in LCA is the aggregation of the resulting environmental impacts which are usually in non-comparable units. This aggregation is carried out at the impact assessment stage. New frameworks are under constant consideration. LCANET (1996) suggests five elements of the assessment stage: definition, classification, characterisation, analysis of significance and valuation. In the definition element the impact categories are chosen. This is followed by classification where the environmental interventions from the inventory are attributed to the selected impact categories. Interventions contributing to more than one impact category are listed more than once (Udo de Haes, 1996). The aim of the analysis of significance is to gain perspective on the results for the different impact categories. This can include normalisation, in which the results of the characterisation are related to the actual or predicted magnitude of the given impact category; sensitivity analysis, related to the data; and uncertainty analysis related to the type of methods, models and assumptions (Finnveden, 1996).

The valuation part of the impact assessment requires the assignment of relative values or weights to the various impacts. Valuation methodologies in the LCA literature tend to be based on social opinion, political decisions, sustainability indicators, expert rankings, or some sort of economic valuation (Wit *et al.*, 1993). They can be related to goals and costs, or can be estimated with the guidance of either a panel of experts or a cross section of interested parties. A set of valuation factors that is widely acceptable has not been established as there is a great deal of controversy about the various weighting techniques and valuation methodologies (Powell *et al.* 1997). Moreover, as our introductory remarks suggest, any consensus that might emerge will be artificial: there is no 'true' set of weights because there is no single objective function to which all will agree.

Numerous valuation methodologies have been developed. These can be classified as: distance-to-goal techniques, scoring techniques, environmental control costs and economic damage approaches. In the distance-to-goal approach, weights are derived from the extent to which actual environmental performance deviates from some goal or legislative standard. The method ranks impacts as being more important the further away society is from achieving the desired standard for that pollutant. Examples include the Swiss critical volume approach, Swiss ecopoints (Braunschweig *et al.* 1994), and Dutch Environmental Performance Indicators (Guinée, 1994). A disadvantage of these distance-to-goal approaches is that they imply that the standards set by the political process are, in some sense, optimal. Put another way, standards must reflect *society's* evaluation of what is desirable. But political processes are themselves subject to many influences and do not themselves maximise social well being. For example, single interest groups may dictate actual standards. As such, the 'goal' component of the distance-to-goal approach has little normative significance. We argue shortly that economic approaches do carry normative significance.

The environmental control cost approach derives its weights from the expenditure necessary to control environmental damage, usually to the level of some environmental standard. There is therefore a generic link with the distance-to-goal weighting procedure. Again, the underlying rationale is that society has expressed its 'willingness to pay' for achieving the standard by implicitly voting for the expenditure required. An example of this approach is given by Tellus Institute (1992). There are circumstances in which control costs could be the relevant weights for an LCA. These circumstances would arise in contexts where there is an over-riding social compulsion to achieve a given standard. It might then be argued that the cost of achieving that standard defines the relevant cost to society. Even here, however, problems remain since it is difficult to define contexts where such overriding compulsion exists. The relevant weight is surely an indicator of the *damage* done to society, and control costs are not the same as damage costs. Indeed, if they were, the ratio of the benefits (damage avoided) to costs (of control) would always be unity, or above unity if control costs are thought to be minimum estimates of damage. But this is not credible since it would then imply there can never be a 'mistake' in setting policy targets. The existence of policy analysis techniques, such as cost benefit, programme budgeting and cost effectiveness, suggests that policy cannot always be efficiently chosen or implemented. Like the distance-to-goal procedures, then, the control cost approach suffers from having no rational grounding in the theory or practice of social choice.

Scoring techniques are an alternative approach to valuation. Weights are provided by a group of experts, or if this is controversial, by a cross section of interested parties, possibly with differing viewpoints. These can include environmental, consumer and business groups, who reflect the relevant scientific and social opinions. The opinion of the general public can also be sought. In all cases numerical weights are applied to each impact and the weighted impacts are then added. Two forms of scoring are required: first to rank the extent to which more or less of a pollutant is important, (i.e. linear relationship or thresholds), and second to rank emissions relative to other pollutants and impacts, e.g. NO_x would be weighted relative to CO_2 or particulate matter. Certainly, scoring techniques such as the Delphi technique can help make the weights more transparent, but there are serious problems in their use. First, unless expert panels come up with the same weights each time, it is necessary to convene such a panel every time a weighting decision is required. This could be impractical. Second, the relevant weights must relate to social damage, since it is in society's interests that the analysis is being done. But finding representative panels to reflect social interests, rather than the possibly narrower interests of an expert group, is complex. Nonetheless, scoring techniques constitute an advance on distance-to-goal approaches if the scoring panel is socially representative.

The weights in the economic damage cost approach are derived from explicit or implicit measures of willingness to pay to avoid the impacts identified in the LCA. The methodology is concerned with estimating the value that individuals place on market and non-market goods and services, as revealed by a consumer's behaviour (revealed preferences) or derived from their stated 'willingness to pay' or 'willingness to accept' compensation (stated preferences) (Department of the Environment, 1991). Economic values are available for a number of criteria including gaseous emissions, road congestion, and casualties from road traffic accidents, thus facilitating the inclusion of both social and environmental impacts. However some impacts do not yet have credible damage cost estimates, impacts on biodiversity being perhaps the most notable. An example of this 'LCA-EVA' approach is the UK study on landfill versus incineration of waste by CSERGE, Warren Spring Laboratory and EFTEC (1993). This paper describes some other case studies using this approach. The next section outlines the principles and offers a justification for the use of EVA.

3. THE PRINCIPLES OF EV

Economic valuation is based on the measurement of individuals' preferences. In this sense it is inherently 'democratic'. But inequality of individual votes enters in through the mechanism of recording the preferences, since what is recorded is the *willingness to pay* (WTP) for an improvement in something or, sometimes, the *willingness to accept* (WTA) compensation for the loss of something. WTP is familiar in the conventional market place: each decision to purchase a commodity records a 'vote' for that commodity, and, indeed, the economist's construction of a 'demand curve' is no more than a willingness to pay curve showing that more people will be willing pay a given price for something the lower that price is. But willingness to pay is obviously influenced by ability to pay, i.e. by incomes and wealth. Hence each market-recorded preference is weighted by the income of the individual expressing the preference. It is in this sense that market votes are not equal. Space forbids a full treatment of this issue but it is possible to further weight preferences so that this income bias is corrected. Whether such 'equity weights' *should* be used is a moot issue. It is not clear, for example, why they are rejected in the context of the workings of the market place, but required when markets are absent.

The issue for environmental impacts, however, is that many preferences occur in the context of 'missing markets', i.e. there is no market place, or at least no obvious market place. Thus peace and quiet is not traded directly in the market place, neither is air pollution. Hence the challenge is to find WTP 'as if' there was a market. Two broad approaches are used. The first is *revealed preference* whereby preferences for an environmental asset are revealed in the demand for some other asset that is traded. Examples of revealed preference techniques include:

- housing markets where the demand for housing is influenced by environmental attributes such as peace and quiet, clean air, crime-free neighbourhoods etc. It can be shown that house prices generally contain an element of price that reflects these attributes and that house prices therefore embody the WTP for these attributes. This is the *hedonic property price* approach;

- labour markets where the supply of labour is influenced by risk in the workplace. In general, higher risks are compensated by higher wages and the risk element in the wage is a measure of the WTA of the risk;

- markets for commodities which embody safety or environmental improvement. Thus markets for double glazing of windows and smoke alarms reflect the WTP for safety and environmental improvement.

The second approach relies on *constructed markets* or *stated preferences* whereby the researcher actually creates the market, usually via some form of questionnaire which either directly asks the WTP of the respondent for the environmental commodity (*contingent valuation*) or the respondent's ranking of attributes which are then linked to some other price (*contingent ranking, conjoint analysis*).

These two techniques are supplemented by the use of conventional market data in the *production function approach* whereby environmental impacts are valued at ruling market prices, e.g. crop losses from air pollution. This requires that the *dose response function* linking air pollution and crop loss is known. The resulting change in crops is valued at market (or adjusted market) prices. The production function approach has extensive application for impacts such as soil erosion, crop and forest loss from pollution, building damage and health damage from pollution.

What is the basic justification for the use of economic values? First, economic values are based on individuals' preferences. As such they purport to measure what society, as the aggregate of individuals, prefers. The use of such measures is therefore responsive to social concerns rather than to often self-selecting expert groups. Second, there is a need to integrate LCA with other equally valid social concerns, the main one being cost. Cost represents not just 'money' but the resources that are being used in a given action, e.g. a given waste disposal policy. If the resources are not used for this purpose they could be used for something else of social benefit. Hence cost is a surrogate for the sacrifice that society makes when making policy choices. A 'metric' that compares costs and benefits in the same units is therefore to be preferred to one where costs are measured in money but benefits in, say, reductions in environmental impacts scored by some expert process. The latter does not permit us to say whether the cost sacrifice is worth making or not. The economic approach, on the other hand, quite explicitly puts the impact measures into the same units as the cost measures. It is probably because most LCAs ignore cost altogether that this deficiency is not appreciated. Third, the economic approach forces a focus on damages rather than, say, emissions, since it is damage that relates to the loss of human well-being, not emissions. Again, this fundamental issue is ignored in many LCAs. Finally, the economic approach permits the integration of 'psychic' impacts, such as disamenity into the LCA, something that, in principle, LCA can do but which it rarely does. Examples of economic values for the marginal damage associated with GHGs, SO_2 and NOx are given in *Table 13* and *Table 14* below.

Table 13. Marginal damage values for greenhouse gases

	1991-2000	2001-2010	2011-2020	2021-2030
CO_2 as C	25.0	28.0	31.1	34.2
	(7.6-55.6)	(9.1-65.1)	(10.2-718)	(11.3-79.0)
CH_4	132.8	158.7	187.0	216.5
	(59.0-252.1)	(71.3-306.3)	(84.9-360.4)	(97.2-420.7)
N_2O	3561	4156	4798	5521
	(990-8921	(1172-10284)	(1380-11908)	(1571-13679)

(£1997/tonne, with ranges from 5th to 95th percentile in brackets; adapted from Fankhauser, 1995)

Table 14. Examples of unit damage costs estimates for direct effect of SO_2 and NO_x emissions

Study	receptor	SO_2 (£/tonne)	NOx (£/tonne)
Holland & Krewitt (1997)	crops	4-5	na
Calthrop & Pearce (1997)	buildings and materials	33 - 520	na
AEA technology (1997)	ecosystems	390	na
Holland & Krewitt (1997)	fertiliser effect	na	(-28)[a]
Total		427 - 915	(-28)

[a] the estimate for NOx represents a negative cost, i.e. a benefit

4. INTEGRATING LCA AND EV

As noted above, listing and measuring LCA impacts may not be very useful for making *choices* in the products, process and policy context if the impacts do not exhibit 'vector inequality', i.e. if each and every impact under option A is not superior (inferior) to each and every corresponding impact in the alternative case. This is why 'weights' are needed. Under the economic valuation approach the weights are given by shadow prices and the shadow prices reflect (marginal) willingness to pay (WTP) for avoiding (obtaining) the impact in question. Other weighting schemes are possible and need not involve economic valuation at all. However, deciding which weighting scheme to use depends very much on what the question is for:

- *What are the environmental impacts of a given product?* This question is relevant to a company and/or to a regulator. The LCA would identify and measure the set of impacts. Weighting would be irrelevant *unless* the purpose is to compare two products for some purpose: e.g. the company may wish to substitute one for the other or the regulatory may wish to proscribe one in favour of the other. Ignoring the comparability issue for the moment, the set of impacts could then be inspected to see if any of them are thought to be 'unacceptable'. What is and is not acceptable may be an internal company decision, or there may be external forces that cause one or more impacts to be highlighted. Action might then be taken to

reduce those impacts. As long as comparison is not required, weighting is not strictly required and there is no particular reason therefore to invoke arguments for or against economic valuation. Once comparability (product A versus product B) is required, then weighting is almost certainly required unless there is vector inequality of impacts. But which weights should be chosen ? It is perfectly possible to argue the case for choosing any weighting system so long as (a) it has some internal rationale, and (b) it is applied to both products in the same way. However, and this is the critical point, there will be *costs* attached to a switch of products, so that, logically, we require some 'calculus' that permits the company or regulator to compare these costs with the weighted gain in reduced impacts. If the weights are *not* prices, what this amounts to is cost-effectiveness: it costs £Z to achieve a weighted reduction in impacts of K% relative to the existing product. This is then a matter of judgement since there is no pre-ordained rule for making this choice. If the weights *are* prices, then, of course, the calculation becomes a cost benefit one - the £Z of costs can be compared directly to the economic value of the benefits of the reduced environmental impacts, say £G. If Z>G there is a presumption that the switch in product is not worthwhile. And vice versa. This link between costs and impacts is of fundamental importance. It explains why LCA has limited value as a policy-making calculus, despite claims to the contrary. The sequence is:
(i) unweighted LCA can be compared to the costs of the product change in only a very ad hoc fashion: a whole set of impacts, in different units, has to be compared with a homogeneous cost number. Hence, unweighted LCA is not very useful as soon as the comparability issue arises.
(ii) non-price weighted LCA can be compared to costs provided the cost-effectiveness context is accepted. This means that some weighted differential impacts $(\Sigma w_i.- \Delta I_i)$ is compared to a cost, say £Z. (w is the weight, I is the impact, and - indicates the difference between the two products). There can be no rule that says whether the ratio $(\Sigma w_i.- \Delta I_i)/Z$ is acceptable or not. It is a matter of judgement.
(iii) price weighted LCA, however, permits costs and impacts to be measured in the same units so that instead of relying on a judgement about cost-effectiveness, we have the rule $[\Sigma p_i.- \Delta I_i - Z] > 0$ which implies that the product change should be undertaken.

- *What are the environmental impacts of a given process?* The same observations as above apply for this question, but here comparability is likely to be a more prominent requirement. An existing technology could, of course, simply be improved. Most LCAs however are likely to focus on substitutable technologies. Choice between them will often depend on a coherent valuation of the different impacts.

- *What are the environmental impacts of a given policy?* The same approach applies except that the comparability issue becomes even more pronounced. It is unlikely that one would evaluate a policy without comparing it to another policy. In the policy context, however, governments and regulators may wish to adopt a different stance to that of the company - impacts will be weighted by what society wants rather than what the company wants. The EVA approach permits this by changing the price weights to social prices - prices reflecting the costs and benefits to society as a whole. This may well mean re-costing the company's costs.

LCA without weighting is useful in contexts where no real issue of comparability arises. The most likely case is where a company simply wishes to find out what the impacts of its product are in order to focus on reducing some or all of those impacts. Once comparability between products, processes or policies is required - and comparability will get more important as one moves from product to process to policy - then weighting is required. Once this is accepted, the case for economic valuation is very strong since comparability implies choice and choice implies cost. Hence there is a need to compare the environmental gains with the costs of action. While economic valuation is not essential for this process, it is advantageous since it avoids the need for cost effectiveness judgements.

The important point is that LCA without the link to costs is not very helpful and could easily lead to major distortions. It amounts to saying, for example, that product (process, policy) A is 'better' than product (process, policy) B regardless of the cost of making the switch. Once the link to costs is accepted, economic valuation provides a way to answer questions that other weighting systems cannot answer.

5. LCA AND WASTE MANAGEMENT

Historically LCA has been used as a tool for examining the environmental impacts of specific products, but it can be applied to services such as waste management. Relatively little LCA research has been undertaken in the waste management field, although SETAC (1991) provides some theoretical guidance on dealing with recycling within a lifecycle assessment. Several authors (Johnson, 1993, White *et al.*,1995, and Powell *et al.*, 1996) have attempted to compare the relative environmental impacts of alternative waste management options. However most have been confined to the inventory stage (Virtanen and Nilsson, 1993) and few have included a valuation of the results. As such, these studies are incomplete in the sense defined earlier. Thus, Mølgaard and Atling (1995) and Johnson (1993) have classified the inventory data into environmental impacts such as global warming and acidification, but no weighting of these impacts was attempted. A weighting system using a panel of experts was applied by Powell (1996), and Wilson and Jones (1994) employed a similar technique known as the Delphi method. One of the few studies to apply a valuation methodology to waste management techniques was that by Powell *et al.* (1996) which applies economic valuation to environmental externalities arising from recycling schemes.

6. LCA-EVA CASE STUDIES

Several LCA studies have been carried out on waste management problems including the environmental impacts of packaging, but the results are often unclear due to the problems of aggregating the inventory results, as discussed earlier. The following case studies overcome this problem by using a combination of lifecycle inventory analysis and economic valuation.

6.1 Comparative Packaging Options

This case study compares alternative packaging options for a product (CSERGE *et al.* 1995). It determines the environmental externalities associated with the production (upstream) and disposal (downstream) stages of HDPE milk bottles and gable top cartons. Upstream refers to the environmental impacts at both production and distribution stages, i.e. prior to the product and package being sold; and downstream refers to lifecycle impacts resulting from collection and recycling or disposal of the container, i.e., subsequent to the product and package being sold. For the downstream externalities five different waste management scenarios have been used:

Scenario A - 100% incineration: waste from production and consumption of both cartons and bottles are fully incinerated.

Scenario B - 100% landfill: waste from production and consumption of both cartons and bottles are fully landfilled.

Scenario C - 100% recycling: wastes from production and consumption of both cartons and bottles are fully recycled

Scenario D - "current ratios": this scenario is based on a rough estimate of the current configuration of waste treatment methods employed in the UK for cartons and bottles: landfill (90%), incineration (8%) and recycling (2%).

Scenario E - "future ratios": this scenario illustrates the changes in external costs resulting from a change in the waste management configuration to landfill 55%, incineration 25% and recycling 20%.

The lifecycle inventory analysis was undertaken using the PEMS software developed by Pira, at Leatherhead, Surrey, UK. The functional unit has been defined as 1000 two pint containers. The purpose of the study was to test the methodology rather than to undertake an in-depth comparison of the containers. Although the data set was the best available at the time it is hoped to undertake a more thorough analysis in future research.

It is assumed in this study that in scenario A (100% incineration) energy is recovered as electricity which will displace power generated elsewhere in the electricity grid. Energy savings are also made when the recycled secondary materials are used to replace primary materials. In earlier work (CSERGE *et al.*, 1993), it was assumed that the marginal source of energy displaced was old coal fired power stations. However in LCA circles, there has since been much debate about this issue and whether the displaced energy should be attributed to the 'average' source or to the 'marginal' source, and indeed, what the marginal source would be. If we are talking about a product or process generally the average source is relevant. If we are talking about discrete changes in products/processes (e.g. one extra tonne of recycling) then the right basis is the marginal source. In this analysis UK average data for displaced energy emissions have been used.

The Life Cycle Inventory (LCI) shows that when the containers are produced emissions occur in three countries: the UK, Germany and Sweden. For the economic evaluation of the LCIs, two approaches are applied. The first is the country specific, or GNP adjusted, evaluation. This implies that pollution emitted in a country during the lifecycle of cartons and bottles is evaluated at the values attributed to pollution damage in that country. The second approach uses the UK prices or unit cost damage values for evaluating the LCI results, even when they occur in Germany or Sweden. Clearly, the former approach is technically more correct. The results of both approaches are presented in *Table 15*.

Table 15. Environmental costs of milk packaging calculated by the LCI-EVA model (£/functional unit: one functional unit = 1000 2 pint containers and 2 pints = 1.14 litres)

Scenario	Country specific values		UK values	
	carton	bottle	carton	bottle
A:100% incineration	5.47	7.21	4.93	6.77
B:100% landfill	7.05	9.96	6.52	9.52
C:100% recycling	7.14	8.77	6.89	9.01
D: current ratios	6.93	9.72	6.40	9.29
E: future ratios	6.71	9.04	6.23	8.74

NB The results are given as negative externalities, i.e. as costs. Thus, a higher number means greater damage to the environment.

Table 15 shows that 'UK damages' approach produces estimates of damage that are less than the GNP-adjusted damages. This is due to the higher per capita incomes of Sweden and Germany, which mean that pollution in these countries is assumed to be valued higher than in the UK (willingness to pay is a function of income levels).

When comparing the two types of containers, the results indicate that the environmental costs associated with the production and disposal of 1000 two pint gable top cartons are less than those associated with the production and disposal of the same quantity of two pint HDPE bottles, no matter which scenario we choose to look at. The results also give some indication of which disposal method carries the highest environmental burdens. For each container type the upstream effects will remain the same no matter whether the container, after use, will be recycled, incinerated or landfilled. The variation in the results is therefore entirely a reflection of differences in the environmental costs associated with the waste treatment methods.

For both cartons and HDPE bottles, the estimates indicate that the environmental burden from incineration with energy recovery is less than those from both landfill and recycling. In fact the downstream effects, i.e. the environmental costs associated with incineration, are negative - there are environmental *benefits* in the order of just under £1.50/FU for incineration of cartons and around £2.75/FU for the incineration of HDPE bottles. This benefit can be attributed to the displaced energy effect which outweighs the pollution effect from incineration. The higher environmental benefit associated with the incineration of HDPE bottles can be explained by the higher calorific content of HDPE relative to paperboard.

For recycling and landfill, the picture is less clear cut. For cartons, the estimates imply that the environmental costs are higher for recycling than for landfill. In addition, both the 'current' waste management configuration and the 'future' scenario carry lower environmental costs than both landfill and incineration (because the inclusion of incineration reduces the environmental costs). For the HDPE bottle, however, the picture is somewhat different. Here, landfill comes out as the worst option, but recycling is the second best option with the two mixed waste management configurations lying in between. The reason why recycling fares so well for HDPE bottles lies in the fact that the materials credit for HDPE bottles is considerably higher than for cartons, and it outweighs the pollution resulting from the recycling process, thus resulting in an environmental benefit from recycling of HDPE.

With these results in mind, a comparison between the 'current' and `future' scenarios would therefore suggest that for cartons an increase in incineration would ease the environmental burden while increased recycling would increase it. For HDPE bottles, a shift away from landfill to both recycling and incineration is estimated to decrease environmental costs.

Two conclusions can be drawn from the results. Firstly, cartons appear to cause less environmental damage than the HDPE bottles. This is the case across the board for 100% incineration, 100% landfill, 100% recycling, and thus also any combinations of landfill, recycling and incineration. 'Conventional wisdom' as espoused by a number of bodies and national governments (e.g. US Environment Protection Agency, UK Department of the Environment, European Commission) declares that recycling is a better option than incineration, which in turn should be preferred to landfill. However, the ranking of environmental impacts indicates that in the case of gable top cartons and HDPE bottles, incineration should be chosen in preference over both recycling and landfill, and in the case of cartons, landfill should be preferred over recycling. It should be kept in mind that these conclusions relate only to cartons and HDPE bottles. No conclusions can be drawn about how different waste management options should be ranked generally, or about the general validity of the 'waste hierarchy'.

6.2 Lifecycle Assessment of Recycling

LCA-EVA has been applied to two recycling case studies: firstly a comparison between two different types of recycling collection schemes and secondly a comparison between the disposal of household waste to landfill and recycling. The second comparison focuses on the kerbside collection scheme alone and compares the recovery of recyclable materials and their subsequent use in new materials, with the disposal of the same waste components to landfill and the consequent use of primary materials.

6.2.1 Comparison of two recycling schemes

In the first recycling case study a comparison is undertaken between two household waste recycling collection schemes, a kerbside scheme which operates in Milton Keynes, and a 'bring' scheme in South Norfolk. The kerbside scheme collects source separated recyclable materials which householders have placed in reusable containers outside their homes. The scheme operates in addition to the normal refuse collection round. The second recycling scheme, operated by South Norfolk District Council (SNDC), comprises eleven bring recycling sites, in predominantly rural areas. In order to determine the environmental and social impacts of bring recycling a social survey was undertaken in South Norfolk to find out what distance members of the public travelled to the recycling banks, their mode of transport and whether it was a special journey. The type and quantity of recyclable materials brought was also recorded. This enabled a distance per tonne of recyclables to be calculated. Full details of the study are available in Craighill and Powell (1996).

In the kerbside scheme environmental and social impacts arise during the separate collection of materials and extensive sorting at the Materials Recycling Facility (MRF); whereas the impacts from the bring scheme occur in the delivery of the materials by householders to their local recycling centre, and the collection and delivery of the materials to the storage depot. For both schemes the transport distances, modes of transport, and loads carried have been used to determine environmental emissions, the risk of casualties from road traffic accidents and the cost of road congestion. The impacts from the transport of the recovered materials to reprocessing sites, impacts arising from re-processing and the savings within the manufacturing industry are excluded in order to carry out a location independent comparison of the two methods of collection.

The main difference between the two schemes is the distances the recyclable materials are transported. The distances were calculated from the vehicle returns and from the social survey results. In the kerbside collection scheme the collection distance is an average 13.9 km per tonne. For the bring scheme, transport is undertaken by recyclers taking materials to the recycling centres and by South Norfolk District Council collecting the materials. When only special journeys by recyclers are taken into consideration the total distance the collected materials are transported is an average 161 km per tonne.

In order to compare the two schemes an economic valuation of the environmental and social impacts was undertaken. An economic valuation of the kerbside scheme gives average values of £0.88/t for emissions, £0.71/t for casualties and £3.40/t for congestion, a total external cost of £4.99 per tonne of mixed recyclable materials. The 'bring' scheme was valued at £5.62/t for emissions, £10.93/t for casualties and £6.40/t for congestion, giving a total of £22.95/t for mixed recyclable household wastes.

In order to obtain a full economic picture of recycling the external costs can be viewed in addition to the private financial costs. The financial costs of collecting and sorting recyclable materials were provided by the relevant local authorities. The private and social costs together are estimated to be £60.00/t for the kerbside scheme and £114.95/t for the bring scheme. As a comparison, alternative financial costs of operating recycling schemes, estimated by Atkinson, Barton and New (1993), have been used; £36 to £130 per tonne for collect schemes and £18 to £28 per tonne for bring schemes. When these costs are substituted in the analysis the total economic cost is £40.99 to £134.99 for the kerbside scheme and £40.95 to £50.95 for the bring scheme.

One of the key parameters in this study is transport, which gives rise to gaseous emissions, road traffic accident casualties and congestion. To find out how sensitive the results are to the distances travelled, a sensitivity analysis was undertaken. Firstly, a 25% reduction in the average distance that householders travel to the collection banks decreases the external cost of the bring scheme by just under 5%. Secondly, a reduction in the distance to the storage depot of 25% reduced the external cost by 20%. Even when these two variables are combined they do not significantly improve the bring scheme in relation to the kerbside collection scheme. Therefore, the external cost of bring recycling is more sensitive to changes in the distance to the depot than to changes in householder distances, but the kerbside scheme still has lower external costs.

The comparative analysis of the two recycling schemes demonstrates that the method of collection has a significant effect on the overall economic impacts of recycling. The bring scheme has a higher external cost than the kerbside collection scheme. This may be explained by the rural nature of the former compared with the suburban character of the latter, resulting in greater distances being travelled. This major difference in the transport distance reflects not only the rural nature of the bring scheme, but also the inability of the collection vehicle to carry more than one material at a time. In addition, some members of the public undertake special car journeys to the recycling centres to deliver relatively small quantities of recyclables. Hopefully, this would not occur in a system of high density banks in an urban area, in which case the bring scheme could have lower environmental and social costs than the kerbside scheme.

The high cost of operating the South Norfolk scheme is surprising, as it is generally considered that 'bring' schemes are less expensive to operate than kerbside schemes. However this may be accounted for by the low rate of throughput of the South Norfolk scheme in comparison with Milton Keynes, which collects over seven times the weight of materials. Thus it is likely that considerable economies of scale are in operation, which would be reflected in both the private and external costs. To some extent the two recycling schemes represent almost opposite ends of the spectrum. One is a bring scheme operating in a rural area, still increasing in the size of its operations, with relatively little capital equipment in the way of collection vehicles and sorting machinery. The other is a well-established and highly organised kerbside collection scheme in an urban area, which has received substantial investment in equipment. The environmental costs and benefits of recycling can vary considerably with size and type of the scheme and geographical location. It is of interest to note that although the two local authorities involved in these studies co-operated fully, data such as distances travelled and energy used in the sorting operations were not collected as a matter of course. Therefore the environmental costs of the scheme are unknown to those who operate them.

6.2.2 LCA comparison of recycling and landfill

The second case study compares the kerbside recycling system in Milton Keynes with the available alternative, waste disposal to landfill. The recycling system includes kerbside collection, sorting, transport and the use of secondary materials in manufacturing in place of primary materials. The waste disposal system includes the collection of waste and its final disposal to landfill, plus the consequent use of primary raw materials in manufacture. The analysis also includes the impacts arising from materials extraction, manufacturing and reprocessing.

The environmental impacts from landfill include landfill gas generation, leachate and disamenity. Where there is energy recovery from landfill gas, credit has been given for the emissions associated with the generation of energy from fossil fuels that would be displaced by the energy recovered from the landfill gas. As some waste components are inert, it is necessary to calculate the disposal savings for each waste component individually. Credit is also given for the energy saved by the use of secondary materials instead of primary materials in the manufacturing process.

The results of the environmental impacts of the lifecycle inventory can be found in Craighill and Powell (1996). An economic valuation of the LCA of the Milton Keynes kerbside recycling scheme, including the transport to the reprocessor, is given in *Table 16*. To compare the two waste management systems, the costs and benefits from landfilling and the use of primary materials is compared with recycling and the reprocessing of secondary materials in *Table 17*. It can be seen that for most waste materials recycling is a better option. Only plastics recycling results in an overall net cost to society.

Table 16. Economic Valuation of Environmental and Social Impacts Associated with the Kerbside Recycling Scheme (£/tonne)

Material	Emissions (£/t)	Casualties (£/t)	Congestion (£/t)	Total (£/t)
Paper	1.66	1.15	3.61	6.42
Aluminium	2.42	1.88	4.23	8.53
Steel	1.87	1.31	3.76	6.94
Glass	1.49	1.02	3.64	6.15
HDPE	0.93	0.54	3.62	5.09
PVC	1.63	1.30	3.78	6.71
PET	3.09	2.37	3.85	9.31

Note: the figures are for both collection and transport. Source: Powell *et al.* (1996)

Table 17. Economic Valuation and Comparison of Environmental and Social Impacts Associated with the Disposal of Waste to Landfill and the Use of Primary Materials, with Recycling and the Use of Secondary Materials (£/tonne)

Material	Landfill and Primary Materials (£/t)	Recycling and Secondary Materials (£/t)	Net Benefit from Recycling (£/t)
Aluminium	1880	111	1769
Paper	300	74	226
Steel	269	32	238
Glass	255	67	188
HDPE	9	12	- 3
PVC	7	12	- 4
PET	14	21	- 7

Source: Powell *et al* (1996). Numbers are rounded to nearest whole unit.

The results suggest that the type of collection and sorting scheme has little effect on the relative environmental impact of recycling versus landfill disposal. To test the robustness of this conclusion a sensitivity analysis has been carried out on the main study parameters.

For the recycling scheme, a 25% reduction in the distribution distances reduces the external costs by 0.8% for aluminium, 0% for paper, 0% for steel 0.6% for glass, 2.1% for HDPE, 4.9% for PVC and 5.4% for PET, with correspondingly small increases in the net benefit from recycling. Alternatively, if the emissions from using secondary materials in place of primary materials within the manufacturing process were reduced by 25%, this results in a decrease in the external cost of recycling of between 15 to 25%. This gives an increase in the net benefit from recycling of 1.6% for aluminium, 7.5% for paper, 2.7% for steel and 8.6% for glass. In addition there is a significant effect on the net benefit of recycling plastics, improving that of HDPE by 107% (and turning it from a cost to a benefit), and reducing the net cost of recycling PVC and PET by 43% and 56% respectively.

It can be seen that the results are significantly more sensitive to changes in the data relevant to the manufacturing system, than to data concerning the collection stage.

For most materials, recycling appears to be more environmentally desirable than landfilling and often produces net external benefits. However, at present there are limited reprocessing outlets for most recovered materials. In some parts of the UK this results in recovered materials being transported long distances at significant financial, environmental and social cost, thus significantly reducing the benefits of recycling. As levels of materials recovery and demand for secondary materials increase, this situation will hopefully improve as more re-processors come on line.

The results of this study would appear to support the adoption of the waste management hierarchy. However, the public relations value of recycling may have introduced an element of 'recycling at any cost' mentality in the introduction and operation of many recycling schemes. Although this approach is justified by the environmental savings realised for the schemes in this paper, this may not always be the case. The results are sensitive to the level of transport and the manufacturing process displaced by recycling secondary materials. A thorough consideration of the environmental costs and benefits of the type of waste management scheme adopted is required for each location.

7. OVERALL CONCLUSIONS

Through the use of case studies we have shown that LCA can be successfully integrated with economic valuation techniques. It is arguable that economic valuation has a more consistent basis than other weighting techniques: it 'mimics' the market and is responsive to public preferences. Its obvious defects are that willingness to pay data are still scarce in many areas where impacts could be important. This deficiency can only be corrected by extensive further study on economic valuation and the generation of a database of such values.

REFERENCES

AEA technology (1997). *Cost benefit analysis of proposals under the UNECE multi-pollutant, multi-effect protocol*, Department of the Environment, Transport and the Regions, London

Atkinson, W., J.Barton, and R. New. (1993). *Cost Assessment of Source Separation Schemes Applied to Household Waste in the UK*, Report No. LR 945, Warren Spring Laboratory, Stevenage.

Boulding, K. (1966). The economics of the coming spaceship earth, in H.Jarrett (ed), *Environmental Quality in a Growing Economy*, Johns Hopkins University Press for Resources for the Future, Baltimore, pp.3-14.

Braunschweig, A., R.Forster, P. Hofstetter and R Muller-Wenk. (1994). *Evaluation und Weiterentwicklung von Bewertungs-metoden fur ˆkobilanzen -Erste Ergebnisse*, IWO-Diskussionsbeitrag, No 19, St Gallen, Switzerland.

Calthrop, E. & Pearce, D. (1997). Methodologies for calculating the damage from air pollution to buildings and materials: an overview. In V. Kucera, D. Pearce, & Y-W Brodin (eds) *Economic evaluation of air pollution damage to materials*, Swedish Environmental Protection Agency, pp.148-161.

Craighill, A.L. and J.C. Powell (1996). Lifecycle assessment and economic valuation of recycling: a case study. *Resources, Conservation and Recycling* vol. 17(2), pp75-96.

Centre for Social and Economic Research on the Global Environment (CSERGE), Warren Spring Laboratory (WSL) and Economics for the Environment Consultancy (EFTEC)

(1993). *Externalities from Landfill and Incineration*, Her Majesty's Stationery Office (HMSO), London

Centre for Social and Economic Research on the Global Environment (CSERGE), Economics for the Environment Consultancy (EFTEC) and Pira (1995). *Integrated Life Cycle and Economic Valuation of Beverage Container Waste*, CSERGE, Norwich and London (Report for private client).

Department of the Environment (1991). *Economic Appraisal in Central Government, A Technical Guide for Government Departments*, HMSO London.

Finnveden, G. (1996). Life-Cycle Impact Assessment and Interpretation, Theme Report within the LCANET Project, draft version 1.0. (in preparation).

Fankhauser, S. (1995) *Valuing Climate Change: the Economics of the Greenhouse*, Earthscan, London.

Guinée, J.B. (1994). *Review of Classification and Characterisation Methodologies*, Paper presented to the 4th SETAC Conference *Towards Sustainable Environmental Management*, Brussels, 11-14 April, 1994.

Holland, M. & Krewitt, W. (1997). *Benefits of an acidification strategy for the European Union*, European Commission DGXII Joule programme.

International Organization for Standardization (ISO) (1996). Draft International Standard ISO/DIS 14949; *Environmental Management - Life Cycle Assessment - Principles and Framework*, International Organization for Standardization, Geneva.

LCANET (1996). *Definition Document*, CML, Leiden University, Leiden, The Netherlands.

Johnson, C.J. (1993). *A Life Cycle Assessment of Incinerating or Recycling Waste Paper*, M.Sc Thesis, Imperial College Centre for Environmental Technology (ICCET), Imperial College London.

Mølgaard, C. and L. Atling (1995). *Environmental impacts by disposal processes*. Paper presented at the EMPA conference 'Recovery, Recycling, Re-integration', Geneva, February 1-3, 1995.

Powell, J.C. (1996) The evaluation of waste management options, *Waste Management and Research*, 14, pp 515-526.

Powell, J.C., A. Craighill, J. Parfitt and R.K. Turner (1996). A lifecycle assessment and economic valuation of recycling, *Journal of Environmental Planning and Management* 39(1), pp 97-112.

Powell, J.C., D.W. Pearce and A. Craighill. (1997). Approaches to valuation in LCA Impact Assessment, *International Journal of Life Cycle Assessment* 2(1), pp. 11-15.

SETAC (1991). *A Technical Framework for Life-cycle Assessments*. Report from the Vermont workshop. SETAC, Pensacola, USA.

Tellus Institute (1992) *CSG/Tellus Packaging Study: Assessing the Impacts of Production and Disposal of Packaging and Public Policy Measures to Alter Its Mix*, Tellus Institute, Boston, Mass, Vols. 1 and 2.

Udo de Haes, H.A. (Ed) (1996). *Towards a Methodology for Life-cycle Impact Assessment*. SETAC-Europe. In preparation.

Virtanen, Y. and S. Nilsson (1993). *Environmental Impacts of Waste Paper Recycling*, Earthscan, London.

Wilson, B. and B. Jones (1994). *The Phosphate Report*. Landbank Environmental Research and Consulting, London.

White, P.R., M. Franke and P. Hindle (1995). *Integrated Solid Waste Management: A Lifecycle Inventory* (Blackie Academic and Professional) Glasgow.

Wit, R., H.Taselaar, R. Heijungs and G. Huppes (1993) *REIM: LCA-based Ranking of Environmental Investments Model*, CML 103, Leiden, The Netherlands.

Chapter 10

Dematerialisation and rematerialisation
Two sides of the same coin

SANDER DE BRUYN
Researcher, Faculty of Economics, Vrije Universiteit, Amsterdam.

Key words: **materials consumption, energy use, economic development, evolutionary economics**

Abstract: **The thesis that a 'de-linking' occurs between materials use and economic growth during economic development (the so-called 'dematerialisation hypothesis') is discussed. This chapter argues that dematerialisation is not a persistent trend in industrialised economies, but occurs during periods of rapid structural and technological change. Evidence suggests that periods of 'rematerialisation', when materials use is re-linked with economic growth, follow periods of dematerialisation. A theoretical explanation based on the idea of evolutionary patterns in materials use is proposed.**

1. INTRODUCTION

The use of materials and energy undoubtedly has economic origins and environmental consequences. The consumption of materials and energy is therefore an important interface between the economy and the environment and analysis of the patterns, causes and effects of materials and energy consumption have gained considerable interest in environmental economics. Such analysis can be conducted on the level of individual products, firms or nations. The latter orientation has resulted in empirical work investigating the 'stylised facts' of the consumption of materials and energy as the economy of a country develops. One of the positions that has been put forward is that in the process of on-going economic growth the economy is 'de-linked' from its resource base, so that rising per capita incomes would be associated with a declining consumption of resources and associated pollution. It is nowadays common to depict the relationship between resource use and income as an inverted-U curve; this implies rising levels of resource use in early stages of economic development, but declining resource use in subsequent stages. With growing environmental awareness, this process of dematerialisation (and 'depollution') has received considerable scientific and political interest. After all, the implication of an inverted-U curve between resource use and economic development would imply that economic growth might be compatible with improvements in environmental quality.

This chapter discusses the stylised facts about resource consumption and its relationship to economic development and economic growth. Sections 2 and 3 provide a historical overview of the various contributions in the literature since the beginning of the 1960s on the patterns of resource consumption in combination with their environmental consequences. They show that some partial evidence that the inverted-U curve may not represent the actual development of aggregated throughput exists. The evidence shows that an N-shaped curve is more likely (similar to the inverted-U curve but with a subsequent increase in resource use for developed economies). Section 4 assesses the factors underlying the changes in resource use and Section 5 discusses how these factors determine the throughput-income relationship over time. It will be argued that recent advances in economic theory do suggest that N-shaped patterns occur. The policy as well as the scientific implications of these findings will be discussed in the concluding Section 6.

2. FROM LIMITS TO GROWTH TO THE INTENSITY OF USE HYPOTHESIS

Until the late 1960s, the consumption of materials, energy and natural resources was believed to grow at the rate of economic growth. This gave rise to growing concerns about the earths' natural resource availability, which was most firmly put forward by the Club of Rome's "Limits to Growth" study (Meadows et al., 1972). This report predicted a linear and rather deterministic relationship between economic output and material input. Because of worldwide economic growth, mankind is likely to face widespread resource exhaustion, which in turn would negatively affect economic and population growth, human health and welfare.

The arguments put forward by the Club of Rome can be seen as a restatement of the views of the nineteenth century philosophers Malthus and Ricardo. They predicted that scarcity of natural resources (including land) would eventually result in diminishing social returns to economic efforts which effectively puts a limit on economic growth. The result would be a steady state, with a constant population, bounded by the carrying capacity of the earth.

The position that economic growth in the long run would be limited by resource scarcity was examined and tackled most forcefully by Barnett and Morse (1973:11). They state: "Advances in fundamental science have made it possible to take advantage of the uniformity of energy/matter, a uniformity that makes it feasible, without preassignable limit, to escape the quantitative constraints imposed by the character of the earth's crust. A limit may exist, but it can be neither defined nor specified in economic terms. Nature imposes particular scarcities, not an inescapable general scarcity". What they hint at is obvious: progress in human knowledge opens up new substitution possibilities and advances the technology of extraction, use and recycling, which ensures that resource scarcity does not become a permanent constraint to economic activities.[1] Simon (1981) has in this respect referred to human knowledge as 'the ultimate resource'.

The 'limits to growth' have not only been disputed in economic theory. Substantial empirical work following the Report to the Club of Rome has found increasing evidence of a 'slackening' of world material demand since the 1970s (Tilton, 1986, 1990). *Table 18* underlines this development. Between 1951 and 1969 the consumption of most refined metals increased exponentially: annual growth rates were often higher than 5%. A doubling of metals consumption took place every 15 years. Predictions of future demand for materials by Meadows et al. (1972) and Malenbaum (1978) depicted lower but still rather high growth rates for the next decades. However, if these predictions are compared with the actual developments of the world materials demand we see what statisticians would call 'a break in series'. World growth rates of metals between 1973 and 1988 have approximated a modest 1% per annum. A doubling of consumption would then occur only every 70 years.

Table 18. Annual world growth rates in the consumption of refined metals

	Actual[iii] 1951-69	Meadows[ii] 1971-	Malenbaum[i] 1975-85	Actual[iv] 1973-88
Iron ore	6.2	1.8	3.0	0.8
Copper	4.7	4.6	2.9	1.2
Aluminium	9.2	6.4	4.2	1.7
Zinc	4.9	2.9	3.3	0.7
Tin	1.7	1.1	2.1	-0.5
Nickel	5.0	3.4	3.1	1.7
GDP	4.8	NA	3.5	3.0

Sources:[i] Malenbaum (1978); [ii] From Meadows (1972), mean estimations from the US Bureau of Mines; [iii] From Tilton (1990); [iv] GDP: Estimation based on UN, Statistical Yearbook, metals World Resources Institute (1990) 'World Resources 1990-1991'.

Explanations for the slackening of world materials demand were first put forward by Malenbaum (1978) in a theoretical sketch which has later become known as the 'intensity of use hypothesis'. According to Malenbaum, the demand for materials is derived from the demand for final goods; consumer durables such as automobiles and disposables such as beer cans. Since material costs form only a small proportion of the total costs of these products, the prices of materials have an insignificant influence on demand. Instead, income is the explanatory factor in materials consumption. Malenbaum predicted non-uniform income elasticities over time and across countries because of the different characteristics of the composition of final demand associated with different stages of economic development. Developing countries with an economic structure relying on subsistence farming typically have a low level of materials and energy consumption. However, as industrialisation takes off, countries specialise first on heavy industries to satisfy consumer demand for consumer durables, such as houses, and infrastructure, and therefore materials consumption increases at a faster rate than income. The subsequent induced shift towards service sectors will result in an associated decline in the demand for materials. Hence Malenbaum depicts the relationship between materials demand and income as an inverted-U-shaped curve (the line IUS in *Figure 21* with the turning point at a).[2] Technological change has the effect of shifting the relationship between materials demand and income downwards since technological improvements in materials processing, product design and product development implies that the same economic value can be generated with less material input (the line IUS' in *Figure 21*). Late developing countries therefore follow a less materials-intensive development trajectory.

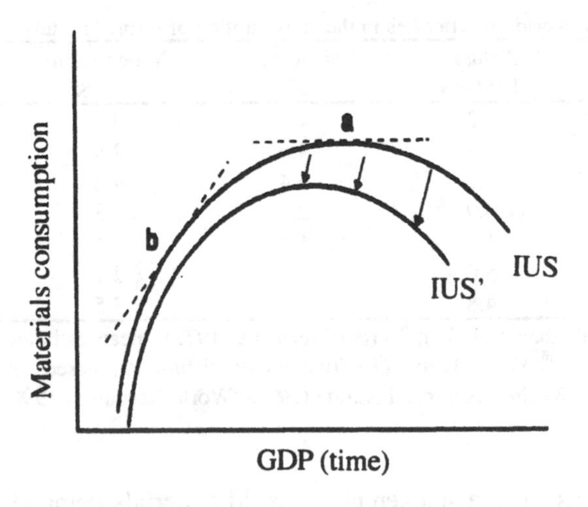

Figure 21. The 'intensity of use' hypothesis and the influence of technological change

3. CONTINUING DEMATERIALISATION?

The 'intensity of use' hypothesis has been found to hold in a number of case-studies for specific materials and energy (cf. Bossanyi, 1979; Chesshire, 1986; Williams et al., 1986; Tilton, 1990; Valdes 1990; Goldemberg, 1992; Nilsson, 1993). These show that dematerialisation, defined here as a reduction in energy and materials consumption, has occurred in a wide range of developed countries. But Labys and Wadell (1988) have suggested that such conclusions can be misleading. They argue that dematerialisation may more adequately be described as 'transmaterialisation'. The demand for materials typically follows a Schumpeterian life-cycle from introduction, via growth and maturity, to saturation and decline. Whereas the intensities of copper and iron ore in the US economy peaked during the 1940s, new peaks are currently recorded for polyethylene, platinum and ceramics. Because the collection of statistics for the consumption of new materials lags behind in the introduction and growth-stages, studies using statistical data often observe the saturation and decline stage of materials demand which may not reflect overall dematerialisation but rather materials substitution, or transmaterialisation.

Whereas transmaterialisation may be a purely descriptive phenomenon in resource economics, the environmental implications are not neutral. Resource consumption has consequences for the environment by virtue of the mass balance principle (Ayres and Kneese, 1969). There would be no reason to assume that environmental pressure decreases due to dematerialisation if only the composition of the materials and energy consumed changes but not the absolute level. Moreover, due to transmaterialisation new substances may enter the environment with serious negative impacts. For example, the impacts of DDT, CFCs and PCBs on human health and the environment were understood long after their market introduction.

Therefore, environmental economists have been investigating ways to construct indicators that represent a better overall picture of the pressure materials and energy consumption exert on the environment. Such indicators may be indicative of the "throughput" of the economy, defined by Daly (1991: 36) as the (entropic) physical flow of matter and energy from nature's sources, through the human economy and back to nature's sinks. A crucial issue in the construction of a throughput-indicator is how to add the various types of materials and energy into a single and uniform indicator. Several methods have been proposed, all of which may be critical to the results of empirical applications. For example, Moll (1993) investigates dematerialisation developments for several materials together in the US economy and finds that when aggregated over mass, the US economy dematerialises after 1970. But when aggregated over volume (in m^3) no dematerialisation trend can be found. Moll defends the use of aggregation over volume with the notion that mass in itself does not represent a function to consumers. Other aggregation schemes that have been proposed are: 'net energy' and entropy (Ayres and Schmidt-Bleek, 1993).

A more simplified method of aggregation has been employed by Jänicke et al. (1989) who investigated the developments of throughput in 31 OECD and communist economies between 1970 and 1985, where throughput has been defined as the equally weighted level of energy consumption, steel consumption, cement production and weight of freight transport on rail and road (as a general measure of the volume aspect of an economy).[3] These proxies of throughput may capture to a large extent the environmentally relevant physical realities of the economies under investigation. The results of this analysis are given in *Figure 22* where the arrows give the linearised developments between 1970 and 1985 for various countries. They show a development that confirms the earlier analysis by Malenbaum: rising levels of throughput in less developed economies and decreasing levels of throughput for the more prosperous countries. This figure suggests that dematerialisation also holds for a more comprehensive set of matter/energy flows and it has been interpreted by some commentators as 'a sign of hope' in resolving current environmental problems (cf. Wieringa et al., 1991; Simonis, 1994; von Weiszäcker and Schmidt-Bleek, 1994).

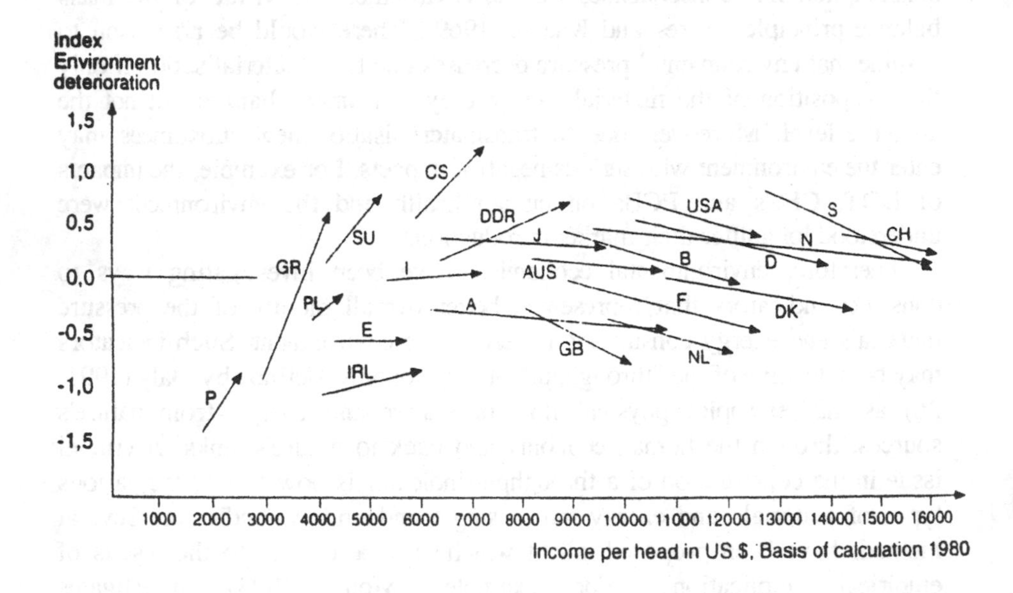

Figure 22. Developments in aggregated throughput. Arrows indicate the linearised development between 1970 and 1985. From Weiszäcker and Schmidt-Bleek (1994) after Jänicke et al. (1989).

The results of Jänicke et al. have been re-examined by De Bruyn and Opschoor (1997) by extending the time-horizon and making some minor improvements in the indicator calculation. Their results suggest that since 1985 there has been an upswing in the levels of throughput for some developed economies. *Figure 23* makes this development explicit for 8 countries between 1966 and 1990. Using the same indicators as Jänicke et al. (1989) we see that the developed economies experienced an increase in their levels of throughput again after 1985. A phase of dematerialisation existed for all countries except Turkey between 1973 and 1985, but this did not continue in the late 1980s. De Bruyn and Opschoor hence conclude that the actual pattern of throughput over time may be more adequately described as N-shaped, similar to the inverted-U shaped curve but with a subsequent phase of 'rematerialisation' that may continue until new technological breakthroughs enable another de-linking phase.[4]

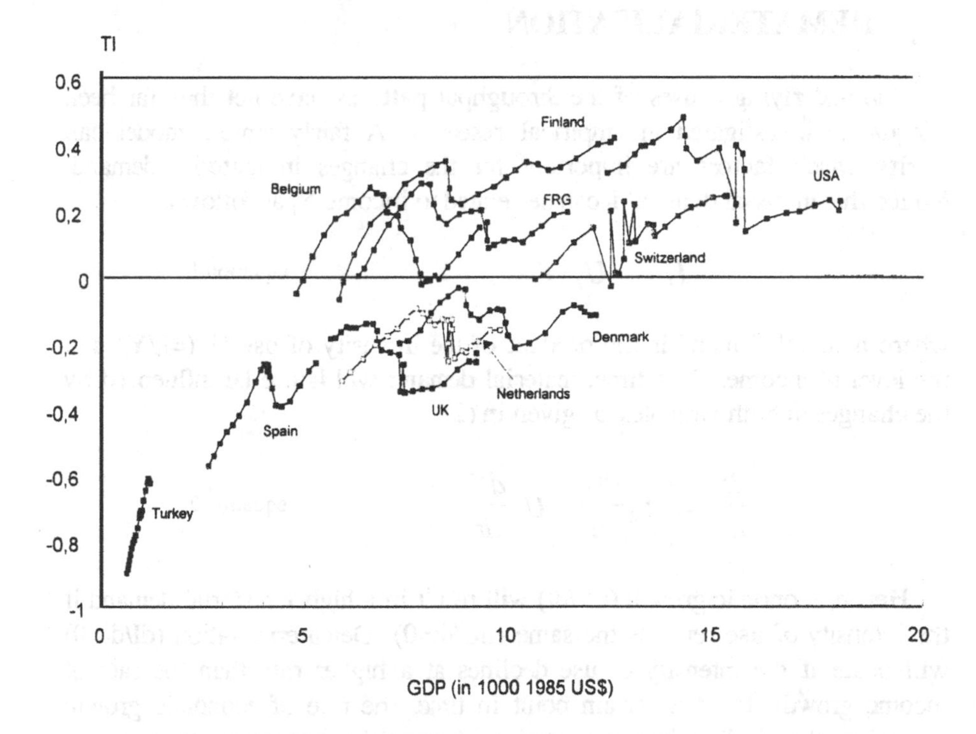

Figure 23. Developments in aggregated throughput. Every dot indicates the one years moving average between 1966-1990. TI= throughput indicator.

These findings may also have consequences for the environment in developed economies. Some empirical work has suggested that there exists an inverted-U, or environmental Kuznets, curve between several pollutants and income (cf. Selden and Song, 1994; Grossman and Krueger, 1995). Given the fact that emissions and wastes originate from the consumption of materials and energy, it can be expected that some pollutants would follow a similar N-curve as described above. Investigations into the patterns of CO_2, NO_x and SO_2 emissions in some developed economies indeed reveals that CO_2 emissions rise after 1985 (De Bruyn et al., 1996). The observation that SO_2 emissions decline may relate to the relatively less expensive costs of abatement and higher local benefits for the communities living in polluted areas (health and forest vitality). Also the international negotiations dealing with the reduction of SO_2 emissions are in a more advanced stage than those dealing with CO_2 emissions.

4. THE IMPORTANCE OF STRUCTURAL AND TECHNOLOGICAL SHIFTS FOR DEMATERIALISATION

The underlying causes of the throughput patterns have not thus far been adequately investigated in empirical research. A fairly simple model can clarify which factors are important for the changes in material demand. Notice that material demand I_t can be related to income Y_t as follows:

$$I_t = U_t \cdot Y_t \qquad \text{equation 1}$$

where material demand is the product of the intensity of use U_t ($=I_t/Y_t$) and the level of income. Over time, material demand will hence be influenced by the changes in both variables, as given in (2):

$$\frac{dI}{dt} = Y_t \frac{dU}{dt} + U_t \frac{dY}{dt} \qquad \text{equation 2}$$

Hence, economic growth (dY/dt) will result in a higher material demand if the intensity of use remains the same (dU/dt=0). Dematerialisation (dI/dt<0) will occur if the intensity of use declines at a higher rate than the rate of income growth. If, at a certain point in time, the rate of economic growth overtakes the decline in the intensity of use, the economy starts to re-materialise again.

A minimum condition for dematerialisation, and the occurrence of an inverted-U shaped curve, is that the intensity of use declines. The intensity of use can be perceived as a kind of efficiency criterion: the amount of materials that are required to 'generate' a certain level of income. Alternatively one may speak of 'material productivity', analogous to the more well-known concept of labour productivity. The total material productivity of an economy will be determined by the efficiency of material use at the level of production processes, products and the consumers. These different levels can be distinguished using the following identity (cf. Roberts, 1990; Tilton, 1986):

$$\frac{I_t}{Y_t} = \frac{I_t}{P_t} \cdot \frac{P_t}{Q_t} \cdot \frac{Q_t}{Y_t} \qquad \text{equation 3}$$

$$IU = EST\ MCP\ PCI$$

where P_t is the mass of materials that is embodied in products and Q_t the number of products. The total material productivity of an economy, the intensity of use, is hence a function of three efficiency ratios. The first ratio, the efficiency state of technology (EST), defines the efficiency of the production technology as the ratio between the mass input in the production process and the mass embodied in the products. If the efficiency of the process technology improves, more materials will be embodied in the products and fewer materials will be wasted during the production process. This will lower the EST and hence decrease the IU. Process innovation is an important driving force of the EST.

The second ratio gives the material composition of products (MCP) and it relates the mass of materials embodied in the products to the number of products produced. The MCP defines the material productivity at the level of the products. Dematerialisation of products implies that less materials are embodied in the same product, which has occurred for example in U.S. manufactured automobiles (Herman et al., 1989) or computers. Dematerialisation of products implies that material is used more efficiently to generate the same product-services to the consumers. Product innovation is an important factor for the decline of the MCP.

The third ratio defines the material productivity at the level of the consumers. It can be perceived as the amount of income spent on (physical) products. Less income spent on material products and more on non-material services implies that the material productivity will increase. This may be achieved due to a change in the structure of final demand (referred to as structural or inter-sectoral changes) or due to increases in the value added of individual products. Adding more knowledge to existing products (such as computers) will increase their value added and lower the product composition of income (PCI).

The changes in the intensity of use over time thus depend on the development of the process technology, product innovation and changes in consumer preferences. How do these factors relate to the inverted-U and N-shaped patterns that have been discussed in the previous section? One could argue, following Malenbaum, that the change in the consumer preferences and the associated change in the structure of production is the main determinant of the inverted-U shaped curve. It does make sense to assume that people in developing countries first show an appetite for material welfare (cars, infrastructure, consumer durables) which increases total material consumption and that only at certain high income levels do services (banking, insurance, recreation) become more important.

The importance of the PCI as the underlying cause of the change in the intensity of use makes it difficult to explain the N-shaped pattern. The idea that consumers, in the course of economic development, should reverse their preferences is untenable from a theoretical perspective as well as from an intuitive point of view. Consumers probably do not start to prefer material consumption goods again after a period in which they preferred more services. Does this imply that the N-shaped pattern is unsupported by economic theory?

5. AN EVOLUTIONARY PERSPECTIVE ON DEMATERIALISATION

Previous empirical work has decomposed the change in energy intensities into structural (intersectoral) and technological factors. Howarth et al. (1991), for example, decomposed the change in energy intensities for eight OECD economies and found, on average, little support for structural changes as an important determinant of the recorded decreases in the energy intensities between 1973 and 1988. The decreases in energy intensities are much better explained by technological improvements in processes and product innovations. These conclusions seem to hold generally for a range of developed economies, a finding confirmed by other studies.

If structural changes in developed economies do not have a marked impact on materials intensities, this implies that the PCI is not the main factor that determines material demand. This need not contradict the inverted-U shaped curve since total aggregate throughput could still decrease if decreases in materials intensities due to technological improvements are faster than the rate of economic growth. If the decrease in the intensity of use has come to a halt, economic growth will simply result in higher levels of throughput, as can be seen from equation 2. This would imply an N-shaped curve. If we accept the fact that structural changes play a minor role in the development of the intensities for developed economies, the main question concerning the explanation for the different patterns deals with the issue of how technology will develop over time. Conflicting views exist on this in economic theory.

In neoclassical economic theory technological change follows a process of Darwinian natural selection at the margin. However, why innovations occur has been poorly understood in neoclassical economics. Whereas technological change was first assumed to be 'autonomous' and 'exogenous' to the neoclassical model, more recently the theory of endogenous growth has incorporated technological change by explicitly investigating the role of human knowledge in generating R&D and welfare. Romer (1990), for example, argues that economic growth can be enhanced by investing in 'human capital' that results in innovations and technological change. Whether innovations are rejected or accepted depends on opportunities for the firm to compete more successfully in the market. Technological change is thus endogenised by making it dependent on a cost-benefit analysis concerning investments. The yields of those investments will gradually improve over time because of the accumulation of knowledge. As a logical result, the economy will gradually become less material- and more knowledge intensive.

Alternatively, it has been suggested that the process of technological change does not follow a smooth process along a path of equilibrium, but is characterised by disequilibrium and an evolutionary path of learning and selection (cf. Dosi and Orsenigo, 1988). Innovations over time may typically come in clusters as the result of a process of creative destruction, first introduced by Schumpeter. Gowdy (1994) has analysed the discussion concerning the process of change in biology and argues that biological evolution as a Darwinistic process of gradual adaption is hard to defend by the fossil record. The intermediate steps between various species that would support gradual change are missing. It is here that the 'punctuated equilibrium' view has been introduced: species remain virtually unchanged for quite a long time, but radical breaks in the equilibrium result in sudden appearance and extinction of species. Evolution takes place not so much on an individual level but on a species- or systems-level, where species co-evolve together in their environment. Gowdy then applies these findings to economics and argues that the economic system may be relatively stable and in an equilibrium during certain times which are followed by a drastic shift in technological paradigms and institutional and organisational structures. Hence the evolutionary path of learning and selection may, given sufficient stability in technological paradigms and institutional structures, move around a certain equilibrium (a so called attractor point), but changes in technological paradigms and institutional structures may shift the attractor point so that in the long run the disequilibrium state is persistent.

It is beyond the scope of the present paper to elaborate the arguments, examples and mathematical treatments that underpin this point of view. But it is interesting to investigate whether the developments in resource use are characterised by a process involving gradual changes, resulting in lowering intensities of use, or by a process of alternating punctuated equilibria. An easy way to present the patterns of the intensity of use over time is to use phase diagrams (cf. Ormerod, 1994, who applied phase diagrams to investigate employment issues). A phase diagram is a scatter diagram where a certain variable is plotted against two dimensions: the value in the current year and the value in the previous year. The various values of the intensity of use over time are then connected with a line. This way of plotting the data has the advantage of illustrating clearly whether there is a gradual improvement in the intensity of use (which would result in a straight negative line) or whether the intensity of use moves around a cycle of punctuated equilibria (or attractor points). As an example we take historical data for steel and energy intensities in the Netherlands.

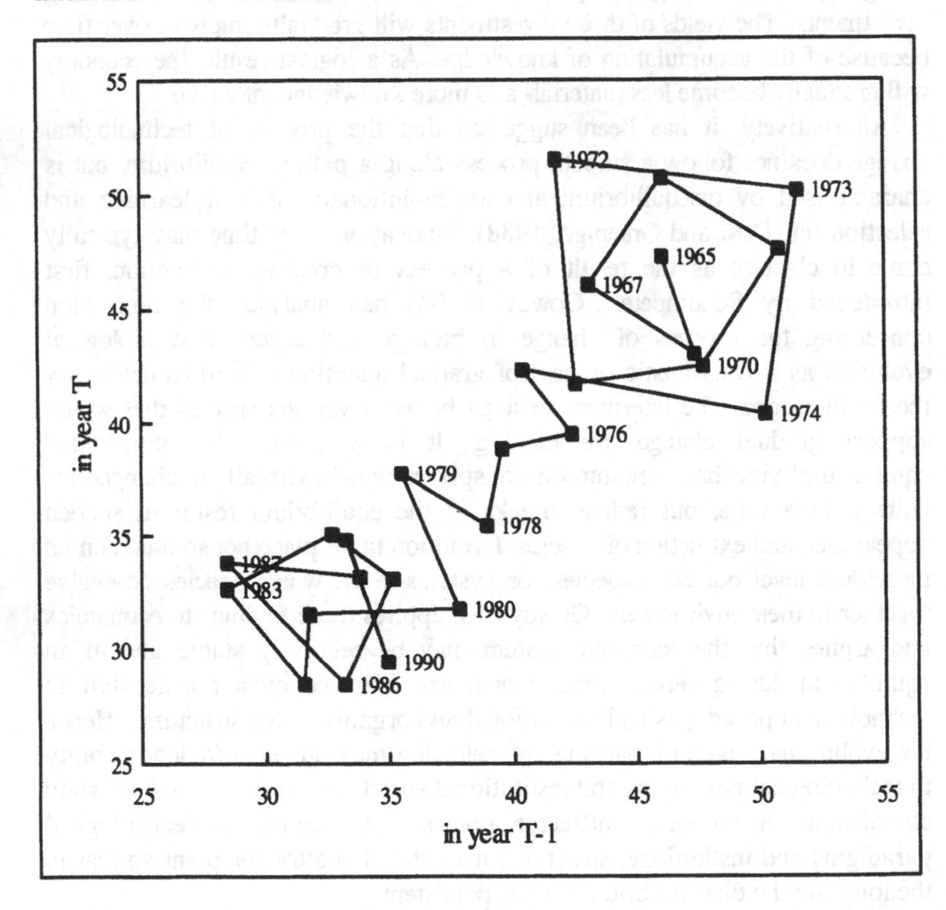

Figure 24. Developments of steel intensities in the Netherlands in (1965-1990): kg/1000 US$ (1985)

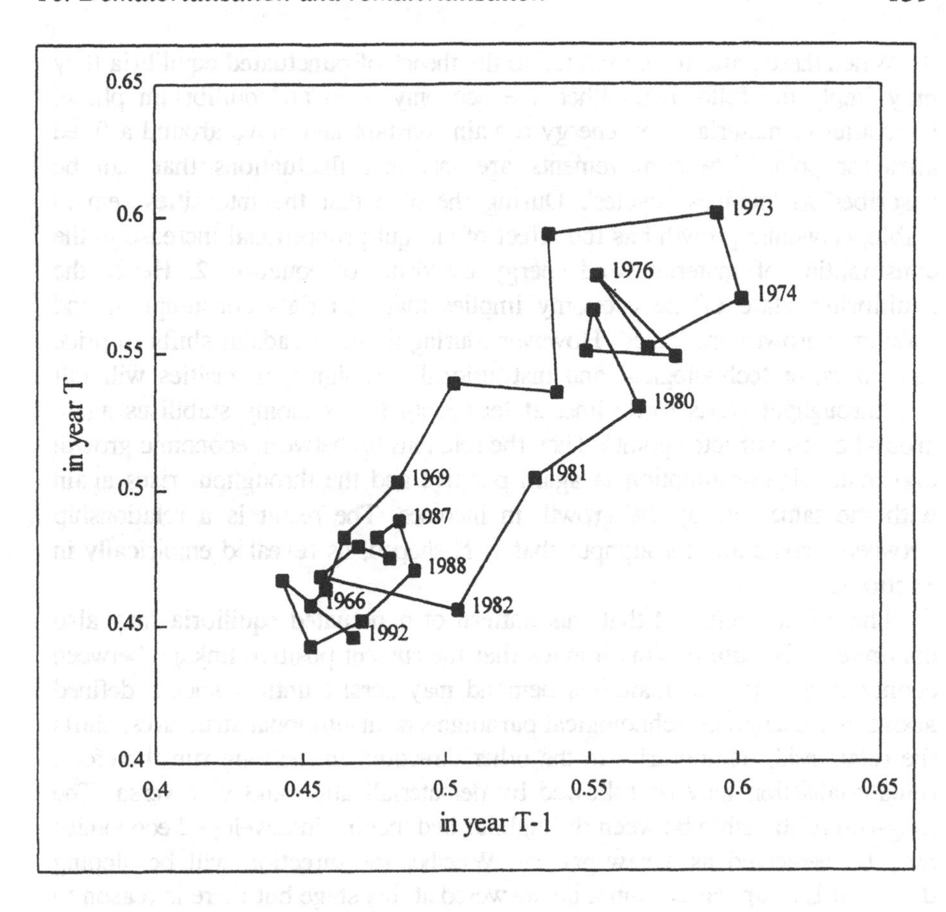

Figure 25. Developments of energy intensities in the Netherlands (1966-1992) :toe/1000$(1985)

Figure 24 and *Figure 25* give the phase diagrams for steel and energy consumption per unit of GDP in the Netherlands, from the mid-1960s until the beginning of the 1990s. Both give evidence of a pattern of punctuated equilibria, although the time period investigated may be too short to reveal the evolutionary forces at work. Steel intensity in the Netherlands remains roughly the same between 1966 and 1974, fluctuating around an attractor point of approximately 45 kg per 1000 US$ (1985). After 1975, however, the steel intensity starts to decline, at least until 1982 when a new equilibrium is reached, fluctuating around the 32 kg per 1000 US$. Energy intensities increase monotonically from 1966 to 1973. From the early 1970s to 1979, energy intensities abruptly stop growing and stabilise around an attractor point. In 1979 intensities start to decline rapidly, at least until 1983 when a new attractor point has been reached at a level about 20% below the attractor point prevalent in the 1970s.

When these patterns are related to the theory of punctuated equilibria they may imply the following. When the economy is in an equilibrium phase, intensities of materials and energy remain constant and move around a fixed attractor point. These movements are marginal fluctuations that can be described as 'business cycles'. During the time that the intensities remain stable, economic growth has the effect of an equi-proportional increase in the consumption of materials and energy by virtue of equation 2. Hence the equilibrium state of the economy implies that materials consumption and economic growth are linked. However, during times of radical shifts in price structures, or technological and institutional paradigms, intensities will fall and throughput starts to decline, at least until the economy stabilises again around a new attractor point[5]. Then the relationship between economic growth and materials consumption is again positive and the throughput rises again with the same rate as the growth in incomes. The result is a relationship between income and throughput that is N-shaped, as revealed empirically in Section 3.

There is a likelihood that this pattern of punctuated equilibria may also continue in the future. This implies that the current positive linkage between economic growth and materials demand may persist until a shock, defined above as a change in technological paradigms or institutional structures, shifts the relationship temporarily in the other direction. In the long run therefore, rematerialisation may be followed by dematerialisation and vice versa. The long-run relationship between throughput and income in developed economies may be perceived as a saw-pattern. Whether the direction will be sloping downwards or upwards cannot be answered at this stage but there is reason to believe that the trend will be towards higher levels of throughput. If we compare the various countries in ***Figure 23***, for example, we still see that higher income countries have higher levels of throughput.

6. DISCUSSION

This chapter has investigated the 'stylised facts' of the consumption of materials and energy during the course of economic development. It has been argued by several authors that the relationship between materials and energy consumption and income would be inverted-U shaped so that after a particular level of income, growing output may be associated with declining (material) input. This phenomenon of dematerialisation has been found in some studies on single materials and energy. However, with a more comprehensive measure of throughput, several developed economies show a phase of rematerialisation since the second half of the 1980s. The resulting N-shaped pattern can be explained by reference to recent advances in economic theory which suggests that the actual economic process may be better explained by dis-equilibrium, non-stability and evolutionary patterns of learning and selection. During times of radical changes in the technological and institutional paradigms, the relationship between throughput and income growth may be altered due to increased substitution possibilities and technological advances in the processing and use of materials and energy. However, such a phase of dematerialisation will not persist indefinitely, and the positive relationship between income growth and throughput growth is likely to be restored, albeit at a lower level of throughput.

Dematerialisation and rematerialisation may hence be two recurring phenomena in the throughput trajectory of developed economies: two sides of the same coin. The environmental implication is that the often-made argument that economic growth can be beneficial to environmental quality is probably invalid. If the N-shaped figure holds for aggregated material input, a similar development should be traced for aggregated environmentally relevant output (emissions and wastes) by virtue of the mass balance principle. For the policymaker, these results imply that there can be less room for optimism that the process of economic growth itself will solve our environmental problems. Instead, institutional and technological breakthroughs may be required to reverse the current trend of rematerialisation into a more environmentally benign trend. We have observed that in the past these radical changes have occurred in the years characterised by high and rising prices of energy and raw materials. These factors appear to have triggered governments and business enterprises to reconsider their use of resources and the associated environmental impacts and to start a process of rationalisation, or restructuring. A new stage of eco-restructuring, possibly based on steep price rises for materials and energy inputs, may be required to prevent environmental disturbances to result in irreversible impacts and to shift the positive relationship between income growth and throughput growth (temporarily) in a new direction.

REFERENCES

Auty, R., (1985). Materials intensity of GDP: Research issues on the measurement and explanation of change, Resources Policy, vol11: p275-283.

Ayres, R.U. and Kneese, A.V. (1969). Production, Consumption and Externalities. Am. Econ. Review, LIX: 282-297.

Ayres, R.U., Schmidt-Bleek, F.B., 1993. Towards a Universal Measure of Environmental Disturbance., Working paper 93/36/EPS, INSEAD Centre for the Management of Environmental Resources, Fontainebleau.

Barnett, H.J., Morse, Ch., (1973). Scarcity and Growth: The Economics of Natural Resource Availability. John Hopkins University Press, Baltimore.

Bossanyi, E., (1979). UK primary energy consumption and the changing structure of final demand. Energy Policy, vol. 7, September: p253-258.
Chesshire, J., (1986). An energy-efficient future: a strategy for the UK, Energy Policy, vol. 14: p395-412
Daly, H.E., (1991). Steady State Economics: Second Edition with New Essays. Island Press, Washington, DC, 297 pp.
De Bruyn, S.M., J.C.J.M. van den Bergh, J.B. Opschoor, (1996). Economic Growth and Patterns of Emissions: Reconsidering the Empirical Basis of Environmental Kuznets Curves, Free University Research Memorandum 1996-48.
De Bruyn, S.M., Opschoor, J.B. (1997). 'Developments in the throughput-income relationship: theoretical and empirical observations', Ecological Economics 20: 255-268.
Dosi, G, Orsenigo, L. (1988). Coordination and transformation: an overview of structure, behaviours and change in evolutionairy environments', In: Dosi, G., Freeman, C., Nelson, R., Silverberg, G. and Soete, L. (Eds.), Technical change and economic theory, Pinter Publishers Ltd., London.
Georgescu-Roegen, N., (1971). The Entropy Law and the Economic Process. Harvard University Press, Cambridge, MA.
Goldemberg, J., (1992). 'Energy, Technology, Development'. Ambio, vol. 21, p14-17.
Gowdy, John M. (1994). Coevolutionary Economics: The Economy, Society and the Environment, Dordrecht: Kluwer Academic Publishers.
Grossman, G.M., Krueger, A.B., (1995). Economic growth and the environment, Quarterly Journal of Economics 112: 353-378.
Herman, R., Ardekani, S.A., Ausubel, J.H., (1989). Dematerialization, in: J.H. Ausubel & H.E. Sladovich (Editors): Technology and Environment, National Acadamy Press, Washington, D.C., p50-69.
Howarth, R.B., Schipper, L., Duerr, P.A., Strnm, S.[1991]. Manufacturing energy use in eight OECD countries: Decomposing the impacts of changes in output, industry structure and energy intensity, Energy Economics, p135-142.
Jänicke M., Monch, H., Ranneberg, T., Simonis, U.E., (1989). 'Economic Structure and Environmental Impacts: East-West Comparisons'. The Environmentalist, vol9, p171-182.
Klaassen, G., (1995). Trading sulphur emission reduction commitments in Europe: A theoretical and empirical analysis, International Institute for Applied System Analysis (IIASA), Laxenburg, Austria.
Labson, B.S. & Crompton, P.L., (1993). Common Trends in Economic Activity and Metals Demand: Cointegration and the Intensity of Use Debate. Journal of Environmental Economics and Management, vol. 25, p147-161.
Labys, W.C. & Wadell, L.M., (1989). Commodity lifecycles in US materials demand. Resources Policy, vol. 15, p238-252.
Malenbaum, W., (1978). World Demand for Raw Materials in 1985 and 2000, McGrawhill, New York, 126pp.
Meadows, D.H., Meadows, D.L., Randers, J., Behrens, W.W., (1972). The Limits to Growth, Universe Books, New York.
Moll, H.C. (1993). Energy counts and materials matter in models for sustainable development: dynamic lifecycle modelling as a tool for design and evaluation of long-term environmental strategies, Styx, Groningen, 396pp.
Nilsson, L.J. 1993. Energy intensity trends in 31 industrial and developing countries 1950-1988, Energy, 18 (4): 309-322.
Ormerod, P.(1994). The Death of Economics. Faber and Faber, London, 230pp.

Radetzki, M., Tilton, J.E. (1990). Conceptual and Methodological Issues, in J.E. Tilton (Ed.) World Metal Demand: Trends and prospects, Resources for the Future, Washington D.C.. pp.13-34.
Roberts, M.C.(1990). Predicting metal consumption: The case of US steel, Resources Policy, vol. 16: p56-73.
Romer, P.M. (1990). Endogenous Technological Change, Journal of Political Economy, vol. 94: S71-S102.
Selden, T.M., Song, D.S., (1994). Environmental Quality and Development: Is There a Kuznets Curve for Air Pollution Emissions?. Journal of Environmental Economics and Management, vol. 27, p147-162.
Simon, J.L., (1981). The Ultimate Resource. Princeton University Press, Princeton, New Jersey.
Simonis U.E. (1989). Industrial restructuring for Sustainable Development: Three points of departure, Science Centre Berlin FS II 89-401, Berlin, 30pp.
Simonis, U.E. (1994). Industrial restructuring in industrial countries, in: R.U. Ayres and U.E. Simonis (Eds.), Industrial Metabolism: Restructuring for Sustainable Development. UNU Press, New York, pp. 3-21.
Tilton, J.E. (1986). Beyond Intensity of Use. Materials and Society, vol. 10, 245-250.
Tilton, J.E. (1990). The OECD Countries: Demand Trend Setters, in: J.E. Tilton (Ed.) World Metal Demand: Trends and prospects, Resources for the Future, Washington D.C., p35-76.
Valdes, R.M. (1990). Modelling Australian steelconsumption, Resources Policy, vol16, p172-183.
Weizsäcker E.U. von, Schmidt-Bleek, F. (1994). Signs of hope for the 21st century?, in Dutch Committee for Long-Term Environmental Policy (Eds.) The Environment: Towards a Sustainable Future, Kluwer Academic Publishers, Dordrecht/Boston/London, p21-45.
Wieringa, K., H.J.M. de Vries, N.J.P. Hoogervorst, (1991). Economie in Rijksinstituut voor Volksgezondheid en Milieuhygiëne, Nationale Milieuverkenning 2, Alphen aan de Rijn, pp. 55-65.
Williams, R.H., Larson, E.D., Ross, M.H. (1987). Materials, Affluence and Industrial Energy Use, Annual Review Energy, 12, p99-144.

---

[1] Such 'optimistic' points of view have in turn been criticized by, for example, Georgescu-Roegen (1971) and Daly (1991) who argue that the ultimate scarce resource is low entropy because of the fixed inflow of solar energy and the finite stocks of concentrated fossil fuels and minerals. Substitution is always from one form of low entropy energy/matter for another. There is no substitute for low entropy itself. Since technology is also bounded by the laws of thermodynamics, scarcity is an inescapable aspect of any society that transforms low entropy sources into dissipative high entropy sources.

[2] Malenbaum has presented his theory not for the absolute consumption of materials but for the relative consumption of materials: the amount of materials per unit of income. This is the 'intensity of use', which would follow a similar inverted-U curve but with a lower turning point (in *Figure 25*, b would be the turning point of this curve). This has no implications for the elaboration of the theory presented here because Malenbaum acknowledged that further movements along the inverted-U curve would eventually result in absolute reductions of materials consumption.

---

[3] See De Bruyn and Opschoor (1997) for an elaboration on how this throughout index has been calculated.

[4] A similar pattern was found in several other countries that are not given in the figure. One may point at the fact that the patterns vary among countries. Most countries show relative strong fluctuations in their index between 1973 and 1980 which may reflect the uncertainty in the resource markets during that period, combined with the nature of the data collected. Since the data exclude changes in stocks, any release or accumulation of stocks of steel, cement or energy will be reflected as consumption in the throughput index. During times of uncertainty both governments and speculators may enter the resource market more actively, which may explain the different patterns over time and across countries given in

[5] Attractor points not only have an implication for the shape of the consumption of resources over time, they also have a clear econometric equivalence in terms of co-integration. The fact thatshocks permanently shift the equilibrium relationship between resources and income to a new attractor point implies that the consumption of resources and income are not co-integrated. The absence of co-integration may imply that most of the estimations proving the intensity of use hypothesis are statistically not supportable (cf. Labson and Crompton, 1993).

Chapter 11

Aggregate resource efficiency

Are radical improvements possible?

FRANS BERKHOUT
Researcher, University of Sussex, Falmer, Brighton, United Kingdom

Key words: **Factor 4, dematerialisation, resource efficiency, materials accounting**

Abstract: **The claims that radical reductions can be made in resource use in industrialised countries are tested using empirical evidence on materials throughput for the United States, Germany, Japan and the Netherlands. While there is evidence of improvements in relative resource productivity, patterns of improvement appear to be long-term, country-specific, and not rapid enough to enable current targets to be met. The chapter concludes with a discussion of policy options for materials management.**

1. INTRODUCTION

Resource productivity has become a major issue in debates about sustainable industrial development. Bold statements about the need for radical improvements in the use of materials and energy resources have achieved recognition in policy circles (Schmidt Bleek, 1994; von Weizsäcker, Lovins and Lovins, 1997). The argument is that productivity improvements are necessary to minimise impacts on the capacity of natural systems to assimilate waste materials and energy. In other words, resource productivity is being promoted not because of fears of resource depletion (as in the *Limits to Growth* debate of the 1970s), but because of concern about the depletion of 'sinks'. These arguments recall a similar point made nearly thirty years ago by Ayres and Kneese (1969):

> "...externalities associated with the disposal of residuals resulting from the consumption and production process...are normal, indeed, inevitable...Their economic significance tends to increase as economic development proceeds, and the ability of the ambient environment to

receive and assimilate them is an important natural resource of increasing value."

This chapter sets out to do three things. First, it attempts to place the current discussion about resource productivity in the wider debate about sustainable development. A curious aspect about the debate about 'Factor 4' and 'Factor 10' is that we live in a period of apparently abundant and cheap raw materials. Scarcity is a consideration in few markets for non-renewable resources. Although there have been periodic concerns about the regenerative capacity of some renewable resources like fish stocks, over-exploitation is not generally seen as a pressing issue. It has also become commonplace to argue that structural change in economies, technological progress and materials substitutions will compensate for the depletion of resource stocks (Tilton, 1986).

On the other hand, those arguing for radically improved resource productivity are generally indifferent about dissipation of residuals back into the environment and with specific environmental impacts. Indeed, the focus on resource productivity is based on the contention that assessment of specific impacts has impeded rather than helped environmental policy (Hinterberger, Luks and Schmidt-Bleek, 1997). The advocates of 'Factor 4' and 'Factor 10' argue that improved resource efficiency is the most effective way of avoiding a depletion of sinks.

Second, the chapter reviews recent data about aggregate resource flows through national economies to assess whether there is evidence for resource productivity improvements. The central question is: is it reasonable on historic trends to expect resource intensities to decline rapidly in future? Dematerialisation has been demonstrated for a range of materials, but does it hold for resource flows as a whole? Are the rates of dematerialisation sufficient to achieve the goals set by those who argue for an 'efficiency revolution'? The Factor 4 argument rests on two claims: that total resource use should be halved while total welfare (or wealth) is doubled over a period of perhaps 50 years. This suggests an average rate of economic growth of about 1.4 percent per year, and mean annual productivity increases of about 2.7 percent.

Lastly, the chapter will assess what role public policy may play in enabling resource productivity improvements to grow more sharply in the future. It will argue that the stress placed on efficiency improvements alone will not always lead to the most effective mitigation of impacts associated with the dissipation of residuals into the environment. Priorities for resource productivity growth need to be set against the background of knowledge about impacts. Eliminating a tonne of aggregates from the national mass throughput may bring fewer benefits than eliminating a gramme of dioxin.

1.1 Resource efficiency and sustainability

However inexact an idea, one shared understanding of sustainability is that current consumption of resources should not be at the expense of future well-being (Solow, 1991). In the long and complex debates about sustainability, broadly two different approaches may be discerned. These could be termed the 'capital' and 'throughput' approaches.

Broadly speaking, environmental economics has sought to understand the capacity of future generations to secure minimum levels of welfare. *Capital* theories of sustainability assume that natural and human capital are to some extent fungible (or substitutable), and that a condition of sustainability is that the aggregate amount of one or both forms of capital is at the least not being reduced (Pearce and Turner, 1990). This account is therefore concerned with the preservation of *stocks* of capital (and of flows between them). The sustainability of current consumption can be judged according to whether capital (the capacity to reproduce the conditions of life) in aggregate is being consumed or accumulated. For this school, the question of resource productivity is important only in so far as it relates to the balance between the rate of depletion of natural resources and sinks, and the rate of capital formation in the economy (this may be the built environment, productive capital, or technological knowledge). One conclusion might be that so long as a requisite proportion of resources are invested in capital formation, there should be indifference about the rate of flow of resources through the economy. This condition would be satisfied, leading to optimal depletion rates and capital formation, if equivalent rates of return on capital were applied to both natural and human capital. The optimal scale of resource use would be tied to the required rate of return, and this in turn to social rates of time preference.

Throughput theories of sustainability are concerned with *flows* of materials and energy between natural and human systems. Human systems consume resources, extract work and value from them, and dissipate energy and materials back into the environment. Understanding these flows generally requires a 'systems' perspective, and typically an appeal to the three conservation laws and the second law of thermodynamics.[1] This tradition originates in the work of Georgescu-Roegen (1971) and has been carried forward in ecological economics. It draws much of its inspiration, language (the idea of 'industrial metabolism' or 'industrial ecology') and method from the thermodynamic view of ecosystems. The flow of materials and energy through human systems is seen as shaping and sustaining organisation of these systems, just as in ecosystems (Proops, 1983; Faber, Niemes and Stephan, 1995). Dissipation of residuals (both energy and materials) out of human systems is taken to represent a loss of availability of these resources, and therefore a loss in the orderliness of the system as a whole (an increase in entropy) (see Söllner, 1997). The appropriate scale of consumption of resources for this school is one in which no net loss in orderliness occurs. For material resources this condition would be met when a balance is established between resource use and dissipation by human systems, and processes of re-ordering (concentration) of mineral resources and regeneration of natural systems. These processes are geological, geomorphological and biological, and operate across many different spatial and time scales. For energy resources, sustainability can, in the long run, only be achieved through a transition towards renewable energy, principally solar energy.

The problem of resource productivity is closely associated with what has been termed the 'throughput' school of sustainability. It is concerned with reducing the flow of materials and energy through human systems by increasing the efficiency with which work and value are extracted from them. Reducing throughput in the human economy will bring less pressure on resources and sinks in the natural environment. However, the advocates of 'Factor 4/Factor 10' make no special claims about the appropriate scale of sustainable materials flows. Typically, they take a normative approach which claims that aggregate flows of materials and energy should not grow in the future, or should be reduced by some fixed amount.[2] For some substances, like carbon, current climate models suggest that a balance between human consumption and dissipation into environmental sinks will be achieved only by radical cuts in total throughputs.

1.2 Resource efficiency studies

Methodologically, energy analysis and mass balance studies have played a dominant role in economic throughput-based analysis. These studies have sought to characterise and measure the physical and energy basis of economic systems. Fischer-Kowalski (1997) has provided an illuminating review of the use of the 'metabolic metaphor' in a range of disciplines, ranging from biology to social anthropology and economics. It is clear that a concern with the physical basis of natural and social systems has waxed and waned since the middle of the last century.

For Fischer-Kowalski and Haberl (1997) one of the main results of this work has been a recognition of the underlying process of 'colonisation' of nature which has accompanied economic and social development. They show that in the transitions from hunter-gatherer to industrialised society, there has been a 20-fold increase in the consumption per capita of energy and materials.[3] This suggests not only that the 'efficiency revolution' is seeking to reverse deep-seated historical trends in resource use, but also provides a basic metric against which to evaluate the feasibility of radical improvements in materials and energy efficiency. A return to levels of materials and energy inputs associated with what Fischer-Kowalski and Haberl term 'agrarian society' (pre-industrial) would achieve an improvement in input intensities of an order of five (Factor 5).[4]

An opposite trend, the 'intensity of use' or 'dematerialisation' hypothesis which suggests that a de-linking occurs between resource intensity and growth at later stages of economic development was developed in the 1970s and apparently confirmed using data on limited sets of mineral and energy resources (Malenbaum, 1978; Jänicke et al, 1989; Bernardini and Galli, 1993). Advocates of the 'efficiency revolution' have drawn strength from this result and sought to apply it in practical ways. More recently there has been a flowering of empirical work on the absolute physical scale and throughput of national, regional and local economies (see for instance, Baccini and Brunner, 1991, Steurer, 1992, Wernick and Ausubel, 1995, Jones, 1995). These have permitted a re-examination of the dematerialisation hypothesis, and a clearer understanding of the relationships between different economic resource flows. A number of critiques of the dematerialisation hypothesis have emerged (Auty, 1985; de Bruyn and Opschoor, 1994).

True comparative analysis of physical stocks and flows has been hampered by the use of different conventions in the treatment of solid, liquid and gaseous inputs to the economy, in the setting of boundaries between the economy and natural systems (as in the treatment of agricultural inputs, and of movements of overburden extracted during mining, potentially a huge quantity of material), and in the presentation of results. Moreover, many studies have been static. Over the past five years much effort has been put into developing standardised approaches for materials accounting by researchers and statistics agencies in Europe, the United States and Japan (Eurostat, 1997). This has promoted further work on 'substance flow' studies for individual substances (mainly heavy metals), and on regional and national 'material flow accounts' and 'physical input-output' tables (Stahmer, Kuhn and Braun, 1997).

The main source of data for analysis in this chapter is *Resource Flows: The Material Basis of Industrial Economies*, a collaborative study published in 1997 (Adriaanse et al, 1997). This provides for the first time comparable, time-series data for materials inputs into national economies (Germany, the United States, Japan, and the Netherlands). Dry-weight inputs either domestically or through imports are presented for about 30 different materials flows. Transformations within the economy, such as the production of semi-manufactured goods, are not taken into account because they are composed of materials which have already entered the economy.[5] The study presents data in two forms: 'total materials requirement' (TMR) which includes so-called 'hidden flows' which do not enter the economy (primarily 'erosion', mining waste and overburden); and 'direct materials inputs' (DMI) that are materials flows which have entered the economy, and for which actual or notional prices existed.

In this analysis the term 'direct resource inputs' (DRI) is used, rather than DMI to denote active and economically-relevant materials in use. These flows will be the main focus of the analysis. This is a different emphasis than in the *Resource Flows* report which highlights the scale of 'hidden flows'.[6] There are three reasons for a focus on direct inputs in this study. First, 'hidden' flows are typically composed of bulky and broadly-inert materials which are transported over relatively short distances. They do not move *through* the economy, but are materials displacements caused on the periphery of it. Strictly speaking they should not be included in an analysis of throughput. Second, the scale of hidden flows is huge, and their inclusion in an analysis of throughput is likely to 'drown out' patterns and dynamics of smaller, economically and environmentally-relevant flows. Third, a focus on hidden flows will add little to the scientific or policy debates about resource use. For instance, if hidden flows were included, the most effective way of achieving improvements in resource efficiency would be to reduce hidden flows, rather than to reduce and manage throughput.

1.3 Resource flows and resource efficiency: four conclusions

Four strong conclusions can be drawn from mass balance studies (Berkhout, 1997). These are: a) the scale of consumption of materials in industrialised economies is convergent, although the composition of throughput varies somewhat between countries; b) industrial economies are 'linear' - most of the material is rapidly consumed and emitted as a waste residual; c) resource flows are growing in absolute terms in most economies; and d) the resource intensity of economic activity is decreasing, but at a relatively slow rate.

Each of these conclusions is discussed below. The analysis presented is deliberately simplified. Averages are calculated for aggregated and disaggregated throughputs, and simple regressions (linear except where stated) have been conducted.[7] No attempt is made here to conduct a more thorough analysis of the underlying causes of patterns of materialisation and dematerialisation through, for instance, decomposition analysis.

No attempt has been made to check the quality of data provided in the *Resource Flows* report. However, similar rules have been used in setting boundaries around resource inputs directly into the economy. This is important, especially in the derivation of agriculturally-based materials flows. In addition, similar categories of materials flows are reported for each of the economies, and there has been an attempt to standardise the derivation of mass numbers from existing national accounts and other data sources. Nevertheless, the issue of data quality remains pressing, and there is a need for further standardisation in the collection of materials flow data. Beyond that, there is clearly a need for national materials accounts to be provided for a greater range of countries. The sample of four countries analysed here may or may not be representative of industrialised economies, but tells us nothing about flows in developing and newly-industrialising countries.

Convergent scale of materials consumption

National materials flow studies show that industrial economies consume materials in similar quantities. Depending on the study, the range appears to be between 10 and 25 tonnes per capita per year (not including water or air, or overburden moved in extracting mineral resources), with most countries falling in the surprisingly narrow range of 19-21 tonnes (Steurer, 1996). The *Resource Flows* study bears this basic result out, identifying a mean per capita range of materials throughputs for the period 1975-1994 of between 15 and 21 tonnes (see *Table 19*). The revealed range of total per capita direct resource inputs for Germany, the United States, and the Netherlands (export adjusted) are extremely small: from 21.09 to 21.93 tonnes.[8] Of this total DRI, 80-90 percent was extracted domestically. This tends to confirm earlier findings that the 'intensity of use' of resources is similar in countries at an equivalent stage of economic development (Malenbaum, 1978).

Japan is revealed to consume per capita about three-quarters the amount of the other economies. There appear to be three possible explanations. First, the dematerialisation hypothesis predicts that learning effects permit late-industrialising economies to follow a less resource-intensive trajectory of economic development. Second, the higher dependence on foreign resource inputs in the Japanese economy (about 50 percent of total throughput) may have created economic and other incentives for efficiency savings and materials substitutions, and thus a relatively less materials-intensive economy. Third, a greater dependence on resource imports, may have enabled Japan to source inputs from higher quality resources and more efficient foreign producers, thus also reducing proportionately the foreign component of its DRI.

Judged on a per capita GDP basis, the range of mean materials intensities is wider, showing a variance of over 40 percent (0.84 to 1.48 kilogrammes per $GDP (1987)). This is primarily because the range of GDP/capita is wider than the range of resource throughputs. This appears to underline the basic result of convergence in resource intensity in economies at a similar stage of development. On both counts, the German economy appears to be the most resource-intensive and Japan the least resource-intensive of the four economies surveyed over the period 1975-94.

Table 19. Mean relative Direct Resource Inputs (1975-1994)

Tonnes	Per capita	Per $1000 GDP (1987)
Germany		
Total	21.93	1.48
Domestic	16.36	
United States		
Total	21.09	1.20
Domestic	18.89	
Japan		
Total	15.58	0.84
Domestic	10.35	
Netherlands		
Total	35.14	
Total (export adjusted)	21.39	1.45
Domestic	18.41	

Source: Adriaanse et al.(1997)

The composition of materials throughputs are also significant, since they may give us a clue about why mean total productivities appear to be convergent. One possible explanation would be that industrial or post-industrial economies, given there similar economic structures, also display similar patterns of materials throughputs. The ranking of materials flows appears to be broadly similar in the four economies. Construction materials, energy carriers, agricultural biomass and industrial minerals are the most significant in terms of scale, usually in that order. However, the relative importance of these flows shows significant variability (see *Table 20*). Construction materials account for between 20 percent (the Netherlands) and 50 percent (in Japan) of materials inputs; and plant biomass between 10 percent (Japan) and 40 percent (in the Netherlands). Industrial minerals account for between 4 to 10 percent of total flows. Large differences are also found in the significance of energy carriers, their importance ranging between 20 percent (Japan)[9] and 39 percent (USA).

Some of this variation may be accounted for by well-known structural characteristics of the economies surveyed, such as the heavy dependence on fossil fuels in the United States, the relatively large scale of agricultural production in the Netherlands, and the greater emphasis placed on infrastructure development in Japan. What is more difficult to explain is how these specific patterns of resource input lead to such tightly convergent aggregate throughputs. The result seems to suggest structurally-based substitutions between broad categories of materials flows at a national level. For instance, large construction minerals flows in Japan seem to substitute for energy carriers, when compared to the United States. Furthermore, these differences in the pattern of materials flows appear to be persistent through time. Energy carriers accounted for 20-21 percent of total throughput in each of the years between 1975 and 1994 in the United States.

No explanation for the variable pattern of materials throughputs can be found in the dematerialisation hypothesis which assumes that dematerialisation will follow a similar pattern in different countries. However, if substitutions exist at a deep level between, for instance energy and construction minerals flows, then dematerialisation in one category of flows may lead to a growth in another category.

Table 20. DRI by materials category (1975-1994)

Percent	Germany	US	Japan	Netherlands (export adjusted)
Energy Carriers	25.26	39.68	20.19	27.14
Metal Ores	3.90	2.09	7.74	2.10
Industrial Minerals	3.44	3.76	10.32	9.03
Construction Minerals	44.23	34.78	49.45	20.04
Plant Biomass	13.81	16.23	9.48	41.13
Animal Biomass	0.01	1.71	0.62	1.76
Semi Manufacturers	7.36	0.90	0.02	-5.70
Final Products	1.99	0.84	0.00	4.49

Source: Adriaanse et al. (1997)

Linearity of economies

Materials flow analysis reveals that industrial economies are highly 'linear', consuming and dissipating materials rapidly over periods of weeks and months (Ayres, 1978). This issue is central to the claims made for a radical improvement in resource efficiency. Linearity is a sign of inefficient use of resources because it signals waste. By contrast 'cyclicity', that is the recovery and reuse of resources, and 'durability', or the immobilisation of materials flows in capital goods, are signs of resource efficiency. In simple terms, cyclicity can be measured as the proportion of materials which are recycled (i.e. used more than once) before being dissipated into the natural environment. Durability can be measured as the proportion of materials flows that is embedded in capital stock. Economies in which a greater proportion of materials throughput is converted into capital stock, or was recycled would be regarded as being more efficient (and sustainable).[10]

There has been considerable dispute over the proportion of materials which become embedded in durable goods and infrastructures. Ayres (1989) estimates that only about 6 percent of throughput accumulated in durables in the US economy (leaving the remaining 94 percent to pass through the economy). More recent estimates put the accumulation of capital stock much higher. Wernick and Ausubel (1995) and Steurer (1996) estimate that about 40 percent and 33 percent of total throughput was accumulated in the US and Austrian economies respectively. The difference between Ayres' and more recent estimates appears to be due to different accounting conventions.

National materials accounts permit the construction of a simple 'linearity index', taking into account the proportion of resource inputs embedded in capital stock and recycled (see *Table 21* for figures for 1990). This appears to show that between one-third and one-half of total DRI throughput is either embedded in capital stock or recycled. Once again, this appears to be a robust result, repeated across different economies. The figures show that, in aggregate, rates of recycling are comparatively small (3-7 percent). As before, Japan appears to be an outlier, with a disproportionate amount of throughput being 'captured' in capital stock. On these figures, and solely by virtue of investing heavily in capital goods and infrastructure, Japan appears to be the least linear of the economies surveyed here.

Table 21. Linearity indicators (1990)[a]

Percent	Capital stock	Recycled	Total
Germany	44.8	3.5	48.3
Netherlands	26.6	6.8	33.4
United States	32.6	4.2	36.8
Austria	33.1	~3.0	36.1
Japan	60.8	n.a.	>60.8

[a]. These figures come from different studies using different conventions for estimating throughput. They should not be viewed as directly comparable. Sources: Adriaanse et al (1997), Wernick (1995), Steurer (1992) and Payer et al. (1995)

The scale of residuals discharged by the economy over short time periods (between 40 percent and 65 percent of total input) appears to place an upper limit on feasible resource productivity improvements. A decrease in linearity by a factor of two would already mean that 80-100 percent of materials inputs were transformed into durable capital, or recycled. A truer limit, given current technology, is set by the nature of the residuals. These can be allocated to three categories of output: atmospheric emissions (20-40% of total output); solid wastes (10-35%); and dissipative wastes (3-6%). As a first approximation we may assume that atmospheric emissions and dissipative wastes are non-recoverable. Efforts to decrease linearity using recycling will therefore focus on the solid waste stream. This suggests that there may be limits to resource productivity improvements at an aggregate level. Under current patterns of materials outputs from economic activity, resource productivity may be restricted to between 30 and 100 percent (Factor 1.3 and Factor 2).

Growth in absolute resource intensities

Improvements in resource productivity will bring environmental benefits primarily by leading to falls in absolute resource flows. Mass balance time-series have previously shown that total throughputs of materials in industrial economies increase through time, as do the absolute amounts of key substances including nitrogen, phosphorous, sulphur and most metals and mineral ores (Ayres, 1989, and Ayres and Ayres, 1996).

This basic result is confirmed in the *Resource Flows* study, although the picture appears more complex than previously thought. *Table* 22 shows data on trends in absolute DRIs from 1975-1994, together with a projection of trends over this twenty year period up to 2050. While the United States and Japan both show clear growth of absolute throughputs over the period 1975-94, a slight decrease in absolute throughputs over the period 1975 to 1990 is observed for Germany. Political unification led to a sharp growth in absolute throughout from 1990 on. The Netherlands followed a growth path up to about 1990, but at a steadily declining rate. Since 1990 there has been an absolute decline in throughput. The poor fits of trendlines to German and Japanese data are due to the oscillating pattern of absolute DRIs through time (this result is similar to that found by de Bruyn, Chapter 10).[11] A consistent upward trend is detectable only in the United States data.

Table 22. Trends in absolute direct resource inputs (DRIs): 1975-1994

	Average rate of growth (1990)[a]	**Total DRI: 1990 (million tonnes)**	**Projected DRI: 2050 (million tonnes)**
Germany[b]	-0.208%	1352	1140(-15.7%)
United States	+1.23%	5402	9370(+73.4%)
Japan	+0.94%	2173	3200(+47.5%)
Netherlands[c]	-0.073%[d]	332	n.a.

a. R^2 values for trendlines are: Germany - 0.0505; United States - 0.9654; Japan - 0.5458; Netherlands - 0.9614

b. DRI trendline for Germany adjusted not to include post-1990 materials inputs.

c. Netherlands DRI has been adjusted to account for exports materials.

d. A better correlation was achieved for Netherlands when a plynomial trendline was fitted.

Source: Adriaanse et al. (1997)

Two main conclusions can be drawn from these results. First, while average data on resource throughputs at the national level reveal a strongly convergent picture, the underlying dynamics of throughputs are divergent.

Some economies show a secular decline in throughputs, while others reveal growing throughputs. Three countries appear to show broadly linear trends in the development of throughputs, while one (the Netherlands) has a downward curving trend. There appears to be no relationship between relative resource intensity and the trend in absolute resource throughputs. In comparing trends, Germany, the most resource intensive, most resembles Japan, the least intensive. There also is no clear relationship between per capita GDP and trends in throughput. The highest income country (the United States) also has the highest and most consistent growth rates. A test on economic growth rates would confirm whether there is a correlation with absolute throughput. The expected result would be for countries with higher growth rates also to have higher rates of throughput growth. The unique pattern and trajectory of absolute resource throughputs appears to raise another challenge to the dematerialisation hypothesis. Far from seeing homogeneity and convergence, we are confronted on the basis of this evidence with diversity and divergence.

Second, on historical evidence, absolute increases in resource throughputs appear as likely as decreases. In projecting current trends forwards 50 years (*Table* 22, final column) throughput in United States and Japan economies is shown to grow by three-quarters and by a half. The German economy, on pre-1990 trends, would have declined by about one-sixth over the same period. However, the forward projection of Dutch trends in *Table* 22 demonstrates why great caution should be taken in interpreting this kind of evidence. On current trends, the Dutch economy will have a negative mass in 2050, so strongly has it been dematerialising during the 1990s.

Decline in relative resource intensities

While absolute flows of many materials are growing, much evidence exists that the relative resource intensity of economic activity is falling. Aggregate environmental indicators for national economies, partially-based on inputs of key bulk materials such as energy, cement and steel, have shown that there may be an 'inverted U-shaped' relationship between per capita income and the intensity of use of materials (Jänicke et al., 1989). There therefore appears to be 'de-linking' between economic growth and the growth in the use of resources. This result builds on the work of Malenbaum (1978) which first suggested that intensity of materials use - the ratio between physical inputs and monetary outputs of an economy - was related to economic development. A very similar relationship between per capita income and quantitative indicators of environmental quality related to materials *outputs* (emissions to the environment) from economic and social activity have also been derived (Grossmann and Kreuger, 1992). This relationship has been called the 'Environmental Kuznets Curve' (Seldon and Song, 1994).

Falling relative resource and pollution intensity of economic activity appears to be a generalised phenomenon. Bernadini and Galli (1993) suggest that the theory of dematerialisation constitutes two basic postulates: a) that the intensity of materials (or energy) use follows the same pattern in all economies, first growing rapidly, then stabilising, before beginning to fall; and b) that the maximum intensity of use declines the later the process of economic development takes place. Late industrialising countries would therefore be expected to follow lower materials intensity trajectories. However, there is also contrary evidence that in periods of relatively higher economic growth, industrialised economies experience a process of 're-linking'. De Bruyn and Opschoor (1994, 1997) argue that this occurs because reductions in materials intensity caused by technological and structural change may be overwhelmed by a 'growth effect'.

Three explanations have been put forward for dematerialisation: a) structural change in which heavy industry and the construction of infrastructures is superseded by more service-intensive economies; b) technological change in which improvements in process efficiency and product design lead to lower materials and energy intensities; and c) regulations which may have an impact on the resource efficiency and environmental impact intensity of production. The interactions between these factors is yet to be properly understood. In the data presented below the factors identified above are not segregated. Only broad trends are plotted. *Table 23*shows trends in per capita DRI intensities for the period 1975-1994, expressed as a percentage at 1990. These figures reflect those for absolute DRI trends, but they give a better representation since they adjust for population. The main result is that trends appear to be in opposite directions, some countries showing per capita declines in relative resource productivities (especially marked in 1990 for the Netherlands) and other showing increases. Paradoxically, Japan, with the lowest per capita throughput also experienced the most rapid per capita growth in throughputs over the period 1975-94.

Table 23. Trends in per capita DRI intensities (1975-1994)

	Growth in per capita DRI[a] (1990)
Germany[b]	-0.28%
United States	+0.37%
Japan	+0.54%
Netherlands	-0.97%[c]

a. R^2 values for trendlines are: Germany - 0.1165; United States - 0.6472; Japan - 0.2558; Netherlands - 0.9706.

b. DRI trendline for Germany adjusted not to include post-1990 materials inputs.

c. As with absolute DRIs, a better correlation is achieved for Netherlands per capita DRI intensities when a polynomial line is fitted with a turning point in around 19985.

Source: Adriaanse et al. (1997).

Normalising for economic growth and for inflation, we find clear evidence of relative dematerialisation in all the countries surveyed. *Table 24* shows the trend in DRI intensities per $1987 for the period 1975-94, expressed as a percentage at 1990. The relative intensities for 1990 are also given. The trends in all cases are downwards, but across quite a wide range. The German and Japanese economies were dematerialising rapidly at rates of 2.6-3.2 percent per year in 1990. Assuming the low economic growth rates implicit in Factor 4 formulation (about 1.4 percent per year), these are the rates of resource productivity gain that would be required to achieve a halving in resource use over a period of 50 years. In the United States' and Dutch economies, on the other hand, resource productivity appears to have improved at about half this rate - about 1.4 percent annually.[12]

These figures confirm the findings of Jänicke et al (1989) of a pattern of relative dematerialisation for higher-income countries. They appear to confirm that trends in resource productivity continued downwards during the 1990s. This contrasts with the finding of de Bruyn and Opschoor (1994; 1997) that there has been evidence of a 're-linking' between economic growth and resource use since the late 1980s in industrialised economies.

Table 24. Trends in DRI intensities with GDP (1975-1994)

	Growth in DRI per $GDP (1990)[a]
Germany[b]	-3.17%
United States	-1.38%
Japan	-2.61%
Netherlands	-1.39%

a. R^2 values for trendlines are: Germany - 0.9604; United States - 0.9215; Japan - 0.9147; Netherlands - 0.9213.

b. DRI trendline for Germany adjusted not to include post-1990 materials inputs.

Source: Adriaanse et al. (1997)

Figure 26 shows that relative declines in DRI per GDP intensities have been fairly consistent. Trajectories of change appear steady for the whole 20 year period surveyed. Clear changes in orientation are observable only for the German economy after 1990 (once again, the unification effect) and in the Dutch economy prior to 1980. The evidence of *Table 24* and *Figure 26* appears to argue for a long-standing incremental improvement in resource efficiency, and casts doubt on the idea of an 'efficiency revolution'. Whether growth in resource efficiency can be expected to continue into the future depends on our understanding of the factors that are bringing it about. The causes of intensity reductions may be explained by returning to two arguments made earlier. First, that dematerialisation is caused by structural change (the relative importance of different types of economic activity over time) and by technical change. Second, that the main opportunities for dematerialisation exist in reducing the fraction of throughput which is neither embedded in capital stock, nor inherently dissipative - that is, non-dissipative mineral resource flows. DRI/GDP intensities shown in *Table 24* may be explained by structural change in economies. The Japanese and German economies were industry-intensive compared to those of the Netherlands and the United States over the period 1975-94.[13] This would suggest that with growing service-intensity of economies, rates of resource productivity improvement slow down. This may be explained by the relatively declining role of resource improvements in industry (process efficiency improvements, product dematerialisation and materials substitutions) and static (or negative) resource efficiency in the provision of services.

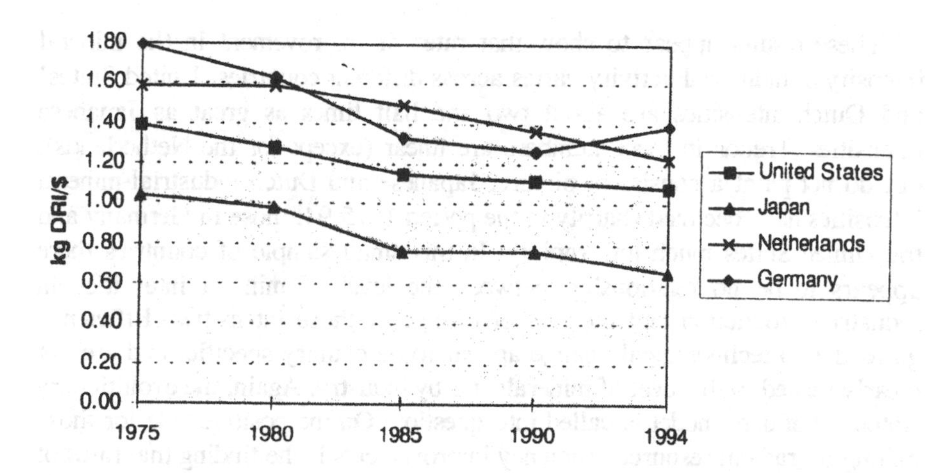

Figure 26. Direct Resource Inputs (DRI) per $ (1987) GDP: USA, Japan, Netherlands, Germany

Von Weiszäcker et al. (1997) place great emphasis on the role of technological innovation in achieving resource productivity gains. Their approach is to argue that technologies already exist which would dramatically improve resource efficiency per unit of service output by the economy. To investigate this claim further, we can investigate the rates of change in resource flows identified most closely with industrial activity. *Table 25* plots the mineral throughput of national economies (metal ores, industrial minerals, semi manufactures and final products) against the value added attributed to industry, expressed for 1990. This is only a partial indicator, since renewable resource inputs to industrial production are not included in the index.

Table 25. Trends in mineral throughputs with industrial value added (1975-1984)

	Growth in minerals DRI per Industrial Value Added ($,1987) (expressed for 1990)[a]	DRI per Industrial Value Added ($,1987) (kilograms)
Germany	-0.71%	1.47
United States	-0.80%	1.88
Japan	-3.09%	0.716
Netherlands[b]	-4.50%	1.73

a. R^2 values for trendlines are: Germany - 0.7529; United States - 0.7443; Japan - 0.9256; Netherlands - 0.9376.

b. Netherlands trendline is polynomial fit.

Source: Adriaanse et al (1997)

These results appear to show that rates of improvement in the mineral intensity of industrial activity varies across different countries. United States' and Dutch intensities are about two and half times as great as Japanese intensities. Trends in these numbers are linear (except for the Netherlands), but do not paint a consistent picture. Japanese and Dutch industrial-mineral intensities have declined sharply in the period 1975-94, those in Germany and the United States much less rapidly. In this small sample of countries there appears to be no relationship between the level of mineral intensities in industrial production and the rate of change of those intensities. Efficiency gains due to technological change appear to be country specific, and are not clearly related to the level of minerals use by industry. Again, the evolutionary 'intensity of use' model is called into question. On the positive side for those calling for radical resource efficiency improvements is the finding that rates of improvement that would be required for a 'Factor 4' improvement in industrial resource intensities (again assuming low rates of economic growth) can be identified as a long-term trend in Japan, and as a tendency during the 1990s in the Netherlands.

The four results which have emerged from an analysis of resource intensities in national economies: the` convergence of mean per capita resource intensities; the linearity of economies; the growth of absolute materials fluxes; and the decline of relative resource intensities, throw up perplexing questions for the analyst and policymaker. Why are per capita intensities in structurally dissimilar economies apparently the same? Why is the composition of throughput variable across different economies? Are less 'linear' economies more sustainable? Can improvements in the relative resource intensity overcome the growth of economic activity? Relative to managing resource throughput, is there an 'optimal' economic growth rate? If resource efficiency is to be a goal of public policy, what are the most effective measures by which to achieve it? Our task here is primarily to deal with the latter question, and identify some policy approaches to managing the throughput of resources in the economy.

2. MANAGING THROUGHPUT THROUGH POLICY

Given that the interaction between human activities and the environment is physically based, environmental policy could, in principle, benefit greatly from material flow accounts and energy budgets. Management of materials and energy inputs and outputs at different levels of aggregation could become a broad goal of environmental policy, helping to define new issues and setting priorities. This approach could also provide integration across a disparate set of environmental policies including pollution control, waste management and prevention, recycling policy, policy on managing toxic substances, as well as providing a link with resources policy.

One of the main conclusions to be drawn is that at an aggregate level resource flows in industrial economies show patterns of stability and persistence. Changes in the scale in resource flows appear, in general, to be slow, while the direction of change may face in different directions, growing in some countries, declining in others. Materials accounting therefore presents a complex and dismaying picture to the policy analyst. The policy issues raised are not self-evident, although some the research community sometimes act as though they are.

Materials accounting evokes fundamental questions to which no simple answers are possible. To give three examples of policy conclusions which have been drawn from materials accounting, some of them uncomfortable:

- if materials and energy intensity is strongly related to the stage of economic development, then one implication is that economic growth which accelerates the transition to a service-based economy is a good way of encouraging dematerialisation.
- if the amount of matter recycled is only a small proportion of the total consumed, and if improvements in recycling rates are slow and costly, then a materials-based environmental policy should devote less effort to seeking to encourage increased cyclicity, and more on reducing materials inputs through a resources policy.
- if what matters is the scale of dissipation of materials and energy flows, rather than the rate of dissipation, then the emphasis some authors place on 'durability' may be misplaced, since rapid 'clean' recycling may be a better alternative.

To caricature, the three messages being presented here are 'encourage more rapid economic growth', 'forget about recycling' and 'durability is a red herring'. We clearly need to understand better the drivers of materials flows before the policy lessons will be clear.

However, a basic framework can be created, and this needs to emphasize the need for a broad-based and integrative approach. In outline, the materials flow approach raises issues in three policy contexts: those concerned with inputs (the front-end); those concerned with flows (throughput); and those concerned with wastes and emissions (the back-end).

The front-end: Resources policy has traditionally been concerned with physical inputs to the economy, and in particular with the over-riding problem of the depletion of resources. Depletion and materials substitution are clearly a significant component of policies to manage throughput, but at a time of abundant and cheap raw materials, resources policy has limited appeal and is therefore unlikely to motivate significant policy intervention.

Resources policy is more likely to be stimulated by concern over 'sinks' for resources once they have been consumed in the economy and dissipated back into the environment. Carbon abatement policies fall in this class of policies in which waste and emissions control policies are linked to resources policies. This *linkage* between the front- and the back-end must be a generic characteristic of integrated materials-based environmental policy. In cases of renewable resources such as forest products the preservation of less tangible resources such as biodiversity may have an important bearing on policies to limit rates of exploitation.

Throughput: Environmental policy traditionally has been not much concerned with flows of materials and energy. The emergence of materials recycling as a policy goal has focused attention somewhat on materials flows. Drawing a parallel with economic policy which is primarily concerned with managing stocks and flows of money in the economy, environmental policy in the future may increasingly be concerned with managing physical and energy stocks and flows.

How should this be done? Materials and energy accounting re-emphasizes the importance of structural and technological change in modifying materials flows. The role of environmental policy in influencing these flows appears to vary across materials types. It appears to be more significant for smaller flows of toxic materials (lead, cadmium), than for bulk flows (oil, timber). Policies to modify materials flows will therefore differ across materials types. Supply-side policies (support for research and development for clean processes and products) are likely to be far more effective than demand-side or waste management policies in modifying bulk materials fluxes.

A materials throughput policy would start by considering industrial flows of materials in the context of natural environmental flows. This is already inherent in any discussion of renewable resources, but can equally well be applied to non-renewable resources, whether natural or man-made. A first step would be to classify materials according to their scale of use; the time-scales over which they are dissipated; the impacts associated with dissipation; and the costs and practicability of arresting or diverting the route of dissipation, or of substitution. Some materials flows are inherently dissipative (i.e. fossil fuel energy carriers), while others could be recycled with high efficiency, and can be treated as a stock within the economy (steel and aluminium). Between these two extremes exist intermediate classes of materials which may flow through the economy rather slowly (building materials), or which are recycled rapidly and then dissipated as quality declines.

A throughput management policy would be concerned with four questions: how do natural flows of given materials compare with industrial flows? do anthropogenic flows cause unacceptable environmental or health impacts? can the rate or nature of dissipation be altered (as with carbon sequestration)? and can higher service value be extracted for materials flows before they are dissipated?

The back-end: The control of wastes and emissions is the traditional arena of environmental policy. Typically control has been exerted through emissions control at industrial production sites applied through locally-negotiated technology-based standards. This site-specific approach to control of air emissions (NOx and SOx, for instance) with regional impacts was integrated within pollution control regimes during the 1970s and 1980s. National emissions control strategies have since also been applied to dissipation of pollutants with global impacts (carbon and CFCs). Note that climate abatement strategies are *materials* control policies. At a purely pragmatic level, materials accounts are important in verifying that international treaty obligations are being met.

More recently, 'priority waste stream' legislation, producer responsibility arrangements and limits on landfilling of solid wastes have imposed more generic limits on the disposal of waste residuals into the general environment. More integrated approaches to industrial emissions control, as in the 1996 EU Integrated Pollution Prevention and Control (IPPC) directive, are also bringing more concerted pressure on the back-end of industrial systems. Through product-oriented environmental policies the problem of dissipative consumption-related emissions is being confronted in a systematic way for the first time. These pressures seem likely to increase in future, but will need to be better coordinated and managed according to environmental goals.

There are essentially two approaches to managing the dissipation of materials into the environment: reducing the aggregate output of the material (typically by reducing its use); and switching the receiving medium. Both approaches have been employed for the full range of wastes, but for those where control regimes have been devised two characteristics stand out. First, they encompass materials used in a few, well-defined processes or products. Second, they have clearly-established and chronic environmental or health impacts where mitigation is widely held to be necessary. Proposals for a generic approach to the consumption and dissipation of materials like chlorine which do not share these characteristics, have not so far succeeded.

Impact or 'sink' analysis is vital to making judgements about what is more and less important and for guiding action in substance control policy. The effectiveness of control will be undermined without an analysis of materials' impacts or sinks. One sink is not equivalent to another. For instance, although it may, in the longer term, prove possible to reach political consensus over the long-term disposal of long-lived radioactive wastes in deep geological repositories, emission of these wastes into the air seems unlikely ever to be regarded as acceptable. To take another example, most lead used in batteries is recycled in Europe and North America, and emissions from this cycling are predominantly in the form of solid wastes. Lead additives to petrol represent by mass a much smaller flow, but one which is dissipated into the air (see Thomas and Socolow, Chapter 12). Clear evidence of health impacts associated with these emissions has been documented, and it is this much smaller flow of material which has been the focus of regulation. Both examples demonstrate that integrated environmental assessment needs to be a core element of a substance control policy.

This conclusion contradicts the assertion of the advocates of the 'efficiency revolution' that environmental impact assessment will condemn environmental policy-making to piecemeal and ineffective approaches. A balance clearly needs to be struck, since impact analysis should not be used as a barrier to policy action. Faster ways of doing sink or impact analysis need to be developed which also allow the comparison of different policy options for modifying materials and energy fluxes as a whole.

3. POLICY OBJECTIVES AND INSTRUMENTS

A materials management policy will be concerned with modifying the scale, flow and dissipation of materials in economies in pursuit of given objectives in resources and environmental policy. Drawing on the work of Bringezu (1994) and others, four broad objectives for a substance flow policy can be put forward[14]:

- *reducing resource inputs to economic activities*: materials and energy efficiency of industrial systems can be greatly improved through incremental and qualitative changes. This objective is sometimes stated in the opposite way as intensifying service outputs.

- *minimising dissipation of residuals into the environment*: emissions from the anthroposphere into the environment would be reduced, either through recovery and reuse, or through the long-term sequestration of toxic or hazardous materials. A key parameter would be the need to match flows associated with human activities with naturally occurring flows.
- *closing substance loops*: recovery and recycling will be effective ways of reducing and managing the flows of some substances. The significance of recycling in meeting environmental objectives is still unclear and should not be overstated.
- *eliminating key materials flows*: some flows of materials can be eliminated either because they are proven to have adverse environmental or health impacts and/or they are economically and sustainably substitutable.

This menu of objectives contains no great surprises. There is nothing here which has not already been attempted in environmental policy. What is being argued for is a more comprehensive approach which would take these objectives as central, and would seek to apply them across the board to all substance flows in the economy.

As argued above, the instruments used to influence substance flows will also vary across materials groups. These would include the standard list of environmental policies ranging from command-and-control regulation, through information policies and voluntary agreements, to market-based instruments. But they would also include a wider set of policies related to science and technology policy, trade and macro-economic management. Structural change in the economy will not be achieved through 'design for the environment' programmes within firms.

Steurer (1996) has made a first attempt to allocate policy instruments to materials flows. He shows that for small volume flows with potentially high impacts (toxic metals, for instance) chemicals policy instruments such as pollution control, bans and substitution, and recycling mandates have already been implemented, often successfully. The key policy task here is to consolidate what has been achieved, incrementally to improve the state-of-the-art in safe waste management and clean recycling, and to extend these direct materials controls more widely geographically, and into new materials groups such as volatile organics.

For medium volume materials like paper and steel which per unit mass have lower impacts, the policy objectives will be to reduce material and energy intensities of production, and to encourage closed loops through optimal recycling. Policy may also be concerned with deeper questions related to the demand for these materials. These materials can be seen as the current focus of environmental policy innovation. In general, the role of direct regulation will be in pollution prevention, and in encouraging recovery, reuse and recycling. Resource intensity and demand will be tackled primarily through Pigovian taxes and support for research and development. Demand for these materials will be influenced by information policies and by macro-economic policy.

For high volume materials like carbon and gravel with very low specific environmental impacts, policy objectives will be concerned with depletion, the energy required to realise their economic value, and their impacts on natural environmental sinks. To date, few of these materials flows have been considered in environmental policy. Substances which exhibit large flows are by definition pervasively used in the economy. Changing the demand for them, and the ways in which they are used will normally require macro-level policies acting over the longer term (i.e. periods on the order of 20-50 years). The role of direct regulation will be very limited, the main policy instruments available being charges and taxes, information instruments, R&D and macro-economic policy.

As a general rule therefore, larger substance flows are likely to be tackled with 'softer' and more front-end oriented policies, whereas smaller flows will be managed through 'harder' policies, some of which have already been implemented (lead-free petrol). Back-end oriented policies will tend to feature in the management of medium-scale substance flows.

4. CONCLUSIONS

This chapter has sought to do three things. The first is to situate the debate about resource productivity in the wider debate about sustainable development. I have argued that there are strong links with what may be termed the 'throughput' tradition in environmental studies, but that unresolved contradictions remain. In particular, present-day advocates are concerned mainly with showing that the throughput of materials in economies is large compared to natural fluxes of many materials. But this type of comparison helps little in defining indicators or objectives for sustainability. Some nutrient cycles can be quite seriously disrupted without causing irreversible environmental change or risks to human health, while others appear to be far more fragile.

The second aim of the chapter has been to reflect on the credibility of the normative goals which have been set by the 'resource productivity' school. I have argued that the theoretical basis of these claims owes much to the 'intensity of use' tradition in resource economics. The empirical analysis presented here gives a mixed picture. While there are examples of declining absolute materials throughputs in national economies (Germany in the 1980s, and the Netherlands in the 1990s) the overall picture is one of absolute growth in direct resource inputs. Relative resource intensities (per capita and per dollar of GDP) have been declining over the past twenty years, but relatively slowly. The rates required to achieve deep reductions in resource throughput, especially given relatively high rates of economic growth, do not appear achievable on historical evidence. Moreover, the trends in relative efficiency gains appear to be steady over periods of decades, suggesting that the factors underlying them are generic and well-established. They relate more to trends in energy efficiency and rates of construction, than to the improved efficiency of industrial minerals use by industry. The fascinating question about whether the rate of relative efficiency gains can be increased (the so-called 'efficiency revolution') by the adoption of smart new materials and energy efficient technologies cannot be answered here.

Lastly, the chapter attempts to set out in broad terms what policy measures exist to manage materials throughput. I argue that many traditional regulatory instruments can be thought of as seeking to manage materials flows, but need to be brought together within a more integrated framework which sets more realistic targets for dematerialisation and decarbonisation. An integrated materials management policy would adopt different approaches to different categories of flows (large, medium and small), with the role of direct regulation diminishing the greater the scale of the flow.

REFERENCES:

A. Adriaanse, S. Bringezu, A. Hammond, Y. Moriguchi, E. Rodenburg, D. Rogich, H. Schütz, 1997, Resource Flows: The Material Basis of Industrial Economies, World Resource Institute, Washington DC.

R.U. Ayres, 1978, Resources, Environment and Economics: Applications of the Materials/Energy Balance Principle, New York: John Wiley & Sons.

R.U. Ayres, 1989, Industrial Metabolism. In: J.H. Ausubel and H.E. Sladovich (eds.): Technology and the Environment, Washington D.C.: National Academy Press.

R.U. Ayres and L.W. Ayres, 1996, Industrial Ecology: Towards Closing the Materials Cycle, Edward Elgar, Cheltenham.

R.U. Ayres and A.V. Kneese, 1969, 'Production, Consumption and Externalities', American Economic Review, 59, 3, pp 282-297.

R. Auty, 1985, Materials intensity of GDP: Research issues on the measurement and explanation of change, Resources Policy, vol II: 275-283.

P. Baccini and P.H. Brunner, 1991, The metabolism of the anthroposphere, Springer, Berlin.

O. Bernadini and R. Galli, 1993, Dematerialisation: Long Term Trends in the Intensity of Use of Materials and Energy, Futures, May, 431-448.

F. Berkhout, 1997, Policy Implications of Substance Flow Analysis, paper to ConAccount on Material Flow Accounting, University of Leiden, January 21-23, 1997.

S. Bringezu, 1994, Strategien einer Stoffpolitik, Paper 14, Wuppertal: Wuppertal Institute, May.
S. de Bruyn, J. vd Bergh and H. Opschoor, 1994, Ecological Restructuring in Industrial Economies: Some Empirical Evidence on Materials Consumption, Research Memorandum 1994-54, Faculteit der Economische Wetenschappen en Econometrie, Vrije Universiteit Amsterdam.
S. de Bruyn and J.B. Opschoor, Developments in the throughput income relationship: theoretical and empirical observations, Ecol. Econ., 20, 3, 255-268.
Eurostat, 1997, Material Flow Accounting: Experience of Statistical Institutes in Europe, Luxembourg.
M. Faber, H. Niemes, G. Stephan, 1995, Entropy, Environment and Resources, 2nd ed, Springer, Berlin.
M. Fischer-Kowalski, 1997, Society's Metabolism: On the Development of Concepts and Methodology of Material Flow Analysis: A Review of the Literature. Presented to: ConAccount Conference on Material Flow Accounting, Univ. of Leiden, January 21-23.
M. Fischer-Kowalski and H.Haberl, 1997, Tons, Joules and Money: Modes of Production and Their Sustainability Problems, Society and Natural Resources, vol 10:61-85.
N. Georgescu-Roegen, 1971, The Entropy Law and the Economic Process, Harvard University Press, Cambridge.
G. Grossmann and A.B. Kreuger, 1992, Environmental Impacts of the North American Free Trade Agreement, CEPR Paper No 644, London: Centre for Economic Policy Research.
F. Hinterberger, F. Luks and F. Schmidt-Bleek, 1997, Materials flows vs. 'natural capital': What makes an economy sustainable?, Ecol. Econ., vol 23, no 1: 1-14.
M. Jänicke, H. Mönch, T. Ranneberg and U.E. Simonis, 1989, Structural Change and Environmental Impact. Empirical Evidence on Thirty-One Countries in East and West, Environmental Monitoring and Assessment, 12, 2, 99-114.
W. Malenbaum, 1978, World Demand for Raw Materials in 1985 and 2000, New York: McGraw-Hill.
W. Moomaw and M. Tullis, 1994, Charting Development Paths: A Multicountry Comparison of Carbon Dioxide Emissions. In R. Socolow, C. Andrews, F. Berkhout and V. Thomas (eds), Industrial Ecology and Global Change, Cambridge: CUP, 157-72.
D.W. Pearce and R.K. Turner, 1990, Economics of natural resources and the environment, Johns Hopkins University Press, Baltimore.
J.L.R. Proops, 1983, Organisation and dissipation in economic systems, J. Soc. Biol. Struct., vol 6: 353-366.
F. Schmidt-Bleek, 1994, Wieviel Umwelt braucht der Mensch? Berlin: Birkhauser.
T.M. Seldon and D. Song, 1995, Environmental quality and development: Is there a Kuznets curve for air pollution emissions?, J. Environ. Econ. Manage., 27: 147-63.
F. Söllner, A reexamination of the role of thermodynamics for environmental economics, Ecol. Econ., vol 22, no 3:175-202.
R.M. Solow, 1991, Sustainability: An Economist's Perspective, in R. Dorfman and N. Dorfman (eds), Economics of the Environment, W.W.Norton, NY, 179-187.
C. Stahmer, M. Kuhn and N. Braun, 1998, Physical Input-Output Tables for Germany, 1990, Eurostat Working Papers, 2/1998/B/1, January, Luxembourg.
A. Steurer, 1996, Material flow accounting and analysis: where we go at a European level, Notes on London Group on Environmental Monitoring meeting, Stockholm, May 28-31.
V. Thomas and T. Spiro, 1994, Emissions and Exposure to Metals: Cadmium and Lead. In R. Socolow, C. Andrews, F. Berkhout and V. Thomas (eds), Industrial Ecology and Global Change, Cambridge: CUP, 297-318.
J.E.Tilton, 1986, Beyond Intensity of Use, Materials and Society, vol 10: 245-250.

I. Wernick and J. Ausubel, 1995, National Materials Flows and the Environment, Annual Review of Energy and Environment, 20, 463-92.

---

[1] The first law of thermodynamics states that energy is neither created nor destroyed during energy transformations, such as the combustion of fossil energy to carry out work. The second law of thermodynamics is complex and far-reaching, but in simple terms relates to the *availability* of energy. With each successive transformation, energy becomes less and less available to do useful work. It is dissipated.

[2] Factor 4 advocates a halving of total materials throughput with doubling welfare. See: Hinterberger et al, 1997 for a justification of these targets.

[3] Hunter-gatherer societies are estimated to have per capita inputs of 10-20 GJ/year of energy and about 1 tonne/year of materials, 'agrarian' societies consume about 65 GJ/capita per year and 4 tonnes of materials per capita/year, while industrialised societies are estimated to have inputs of about 220 GJ/year and 20 tonnes of material. Source: Fischer-Kowalski and Haberl, 1997, p 70.

[4] Physical and energy inputs for 'agrarian society' were derived by investigating the ledgers of a medieval monastry.

[5] Imports of semi-manufactured goods are included.

[6] One objective being to show the extent to which production and consumption in industrialised economies is based upon large materials movements associated with extractive industries in developing economies.

[7] Data for the years 1970, 1975, 1980, 1985, 1990 and 1994 were analysed.

[8] The 'export adjusted' figure is used for the Netherlands throughout this analysis to take account of the 'bulkiness' of Dutch exports, from domestic hydrocarbon production and due to trans-shipment via Rotterdam.

[9] Steurer, 1992 found that energy carriers accounted for only 13 percent of Austrian throughput in 1990.

[10] The distinction between what is durable and what is dissipative is time-dependent. Materials flow through economies on different time scales. Over a longer time horizon materials accumulated in capital stock will also be output as wastes, residuals or corrosion and leachate products. Indeed, wastes associated with durable goods (whether end-of-life vehicles or construction wastes) have come to feature prominently in environmental policy.

[11] The main cause appears to be the cyclical pattern of demand for construction minerals in these economies.

[12] Discounting the period before 1980, the rate of relative DRI intensity improvement has been 1.96% (expressed for 1990) in the Netherlands.

[13] Mean proportion of GDP accounted for by 'industry' for 1975-94: Japan, 37.6%; Germany, 40.8%; Netherlands, 29.9%; United States, 28.9%. Source: OECD, *OECD in Figures: statistics on the member countries, 1996 edition, Paris*, 1996.

[14] Bringezu includes two further aims for a substance flow policy: to increase the use of 'immaterial' inputs such as solar energy; and to impose temporal and spatial 'barriers' to the flow of materials. The role of energy in the management of materials flows, in particular in refining and concentrating materials is clearly a generic issue for substance flow policy. It is not clear that impeding the flow of materials will always be environmentally beneficial. Indeed, it could be argued that the rigidities and inefficiencies built into the economic system through barriers would be environmentally harmful.

Chapter 12

The industrial ecology of lead and electric vehicles

ROBERT H. SOCOLOW AND VALERIE M. THOMAS
Center for Energy and Environmental Studies, Princeton University, USA.

Key words: lead-acid batteries, lead, electric vehicles, lead recycling, industrial ecology

Abstract: **The lead battery has the potential to become one of the first examples of a hazardous product managed in an environmentally acceptable fashion. The tools of industrial ecology are helpful in identifying the key criteria that an ideal lead-battery recycling system must meet to achieve "clean recycling": maximal recovery of batteries after use, minimal export of used batteries to countries where environmental controls are weak, minimal impact on the health of communities near lead-processing facilities, and maximal worker protection from lead exposure in these facilities. The likelihood that the lead battery will provide peaking power for several kinds of hybrid vehicles, a role only recently identified, increases the importance of understanding the levels of performance achieved and achievable in battery recycling. A management system closely approaching clean recycling should be achievable.**

1. THE INDUSTRIAL ECOLOGY OF A HAZARDOUS MATERIAL

This essay is about the two uses of lead in cars -- lead in gasoline and lead in batteries. Our intent is to show how the concepts and tools of industrial ecology (Ayres 1994, Braungart 1994, Socolow 1994) can provide guidance for the management of these contrasting uses of lead.

From the point of view of industrial ecology, which sees the activities of industrialised societies in terms of flows of materials out of and back into the natural environment, these two uses of lead in automobiles could not be more different. The use of lead in a gasoline additive is a dissipative use; the use of lead in a battery is a recyclable use.

The use of lead as a gasoline additive is a dissipative use, because during gasoline combustion the lead is entrained with emissions at the car's tailpipe. After the lead leaves the tailpipe, it disperses in the air and is deposited on soil, crops, and other surfaces. People are exposed not only by inhalation but also through ingestion of food, dust, and soil. Total phaseout of lead additives in automotive gasoline is a goal that is easy to define and to compare with current practice. The goal has already been reached in several countries.

The use of lead in a battery is a recyclable use, because all the lead is confined within the battery as the battery undergoes its cycles of discharge and recharge. Industrial ecology suggests that the goal of lead management can be clean recycling instead of phase-out. Clean recycling requires no lead emissions into the environment except in forms as isolated from human beings and ecosystems as the original lead in the ground. The tools of public policy, including economics, can be combined with industrial ecology to help decide how closely this goal should be approached, at any given state of technology.

Battery-associated lead leaves the industrial system only during lead processing, battery recycling, and battery manufacture. In the United States today, the recycling system for lead batteries is becoming more and more closed: only a small percentage of lead batteries escapes recycling, and only a small percentage of the lead that enters the battery recycling system becomes environmental emissions. Moreover, on a mass basis, these environmental emissions are largely in the form of bulk materials with low concentrations of lead, and these bulk materials are further processed and managed to reduce the potential for lead to be leached. Only a very small percentage of the lead emissions into the environment from the battery recycling system is airborne particulates, the portion of greatest concern from a public health standpoint. Impacts on the health of workers in recycling facilities and in people in communities near these facilities are being reduced. For the lead in batteries, something close to the ideal of clean recycling should be achievable.[1]

Some have concluded that the first-generation electric car powered by lead batteries is "a potential environmental liability," implying that there is greater merit in pursuing the gasoline-powered car with advanced pollution control (Lave et al. 1996, p. 406; see also Lave et al. 1995). In the remainder of this essay, we take issue with this way of framing the discussion. In Section II we argue that it is too early to remove the lead battery from the list of technological options: the lead battery may have several roles in future electric vehicles, not only the role discussed in the LRHM analysis. In Section III we argue, further, that there is reason for optimism that an environmentally responsible management system for lead batteries can be achieved. Quite generally, opportunities for creative policymaking are uncovered when it is widely agreed that choices among technologies can be made in more than one step (Ross and Socolow 1991). With constructive initiatives from several directions, a potential environmental liability does not have to become a real one.

2. TWO ROLES FOR THE LEAD BATTERY IN THE FUTURE VEHICLE

The electric vehicle, in various configurations, is an important technological response to three distinct challenges to the transportation system: urban air quality, oil security, and global warming (Johansson et al. 1996, Sperling 1996). Urban air quality is a critical public health problem throughout the world, exacerbated, in much of the developing world, by fast-growing urban populations. Oil security is an international geopolitical problem rooted in the enduring fact that the Persian Gulf region has a large fraction of the world's lowest-cost oil reserves. Global warming is a deep challenge to a global energy system where roughly three-fourths of primary energy comes from fossil fuels.

In many regions of the world the objective of improving local air pollution is paramount today. California is one of these regions, and its initiatives have driven the development of an electric vehicle of short range, with a battery as the exclusive power source.[2] One version of the battery-powered vehicle uses a large lead battery: this is the only lead battery for advanced vehicles envisioned in the LRHM analysis. A second lead battery has now come into view, however, with potentially greater promise as a commercially effective response to air pollution, oil security, and global warming: the lead battery in the role of peak-power device for hybrid electric vehicles.

2.1 The battery-powered vehicle

The short-range battery-powered electric vehicle has been brought to market in response, especially, to regulatory pressure from the State of California on the major automobile companies to sell a Zero-Emission Vehicle (ZEV). California's ZEV Mandate requires, for each automobile company, that at least 10% of vehicle sales in California be ZEVs by 2003.[3] The ZEV Mandate continues a long tradition of regulatory independence on the part of the State of California, which, beginning in the 1960s, has repeatedly tightened, ahead of the federal government, the limits on emissions of specific combustion products from new vehicles sold there (Calvert et al. 1993).

The battery-powered car confronts several severe challenges today. Its range (the distance it can travel on a fully charged battery before the battery must be recharged) is currently less than 150 km (roughly, 100 miles), far below the range of today's gasoline-powered cars on the road. The time required for battery recharging considerably exceeds the time required for filling a gasoline tank. And the production cost appears to be considerably greater than the production cost of the gasoline-powered vehicle. Principally for these reasons, the battery-powered car is expected to be competitive only in niche markets.

Electric cars running on other kinds of batteries are now entering production. Toyota and Honda are planning to sell, in California and elsewhere, beginning with the 1997 model, a few hundred electric cars using nickel/metal-hydride batteries, and Nissan is entering these same markets with a car using a lithium-ion battery. Both of these batteries promise to give an electric car at least twice the range envisioned for a car with a lead battery. Nonetheless, in these same automotive markets, the higher-volume car General Motors has chosen to introduce, the EV-1, has a lead battery. The cost advantages of the lead battery, apparently, were judged by General Motors to dominate the disadvantages of low energy density.

When a lead battery is removed from an electric car, generally because the battery can no longer go through a cycle of discharge and recharge effectively, the lead in that battery will enter the highly developed recycling system already in place for managing the SLI battery. Because of its size, it is hard to imagine it being abandoned; recovery rates will be close to 100%. Also, it "will be large enough to warrant consideration of disassembly and material segregation (manual or automated) as the first step in recycling," which should simplify handling (Gaines and Singh 1995). By contrast, the recycling system for the nickel (and other metals) in the nickel/metal-hydride battery is just emerging, and the recycling system for the materials in the lithium-ion battery does not yet exist (California Air Resources Board 1995). There may be serious consequences for public health and the environment, if the management of the materials in these newer batteries is done poorly: the risks are relatively unclear. There are few elements in the periodic table whose toxicity is as well studied as the toxicity of lead (National Research Council 1993, Castellino et al. 1995).

The automobile industry has already begun to market range-extending hybrid vehicles that respond both to community concern about urban air quality and to consumer concern for vehicle range. Such vehicles use battery power for in-city driving and a combustion engine for long-distance travel. The lead battery for such a hybrid vehicle would be nearly as large as the lead battery for a vehicle powered only by a battery. Both the all-lead-battery electric car and the range-extending, lead-battery hybrid could be manufactured for decades.

2.2 The Peak-Power Hybrid Vehicle

California's ZEV Mandate was designed to elicit new kinds of vehicles that reduce local air pollution, but it resulted in nearly all engineering initiative being directed towards a single technological response, the short-range battery-powered vehicle. A subsequent 1994 initiative of the US federal government, the Partnership for a New Generation of Vehicles (PNGV) program, is redressing the balance. A collaborative industry-government effort, the PNGV program has focused less on the challenge of air quality and more on the challenges of oil security and global warming. It has set the goal of tripling the fuel economy of the automobile by 2004, relative to a 1994 automobile with comparable performance, size, safety, and emissions. The most promising technological response appears to be the peak-power hybrid car (National Research Council 1996).

A quite general argument provides the rationale for peak-power hybrid cars. The task of driving involves short periods of time, such as when passing another vehicle or climbing a steep hill, when pulses of supplementary power are needed, and long periods of routine driving when base-load power is adequate (Ross 1994). A peak-power hybrid vehicle is any vehicle that uses separate on-board systems to produce peak power and base-load power. (The peak-power hybrid is an entirely different hybrid from the range-extending hybrid discussed above.) A peak-power hybrid vehicle should be able to achieve greater energy efficiency (higher fuel economy) and lower emissions than a vehicle using only one energy system, for three reasons: 1) the base-load system can be designed to operate always at nearly optimal conditions; 2) the peak-power system can be light in weight, because it can be repeatedly recharged by the base-load system; and 3) the peak-power device can improve energy efficiency by doubling as an energy storage device that recovers the energy of braking.

Batteries (as well as flywheels and ultracapacitors) are well matched to the task of providing peak power for hybrids, because they can be repeatedly recharged on board (MacKenzie 1994, Illman 1994, Sperling 1995, Office of Technology Assessment 1995, National Research Council 1996). Only recently has battery research addressed the objective of peak power, after decades of concentration on the objective of range. Of special relevance to the argument in this essay, the first commercially interesting peak-power batteries emerging from this research are lead-acid batteries (Juergens and Nelson 1995, Nelson 1996, Keating et al. 1996). As in the competition among batteries for battery-only electric vehicles, the cost advantage enjoyed by the lead battery will improve its prospects in the competition for providing peak power. A peak-power battery would have much less lead than a battery providing range.

The base-load partner in a peak-power hybrid vehicle must give the vehicle good range and convenient refuelling. Generally, these objectives are met by storing the fuel on board in a tank. Two distinct peak-power hybrid vehicles involving a battery for peak power are in view: a combustion-engine/battery hybrid and a fuel-cell/battery hybrid (National Research Council 1996).[4]

The peak-power hybrid using a combustion engine for base-load power and a battery for peak power will use on-board fossil fuel both to power the car and to recharge the battery, thereby not requiring electricity generated at a separate power plant. Candidate base-load partners in the combustion-engine/battery hybrid include the gasoline engine, the diesel engine, the stirling engine, and the gas-turbine engine. These hybrids do not qualify as zero-emission vehicles, but they may be the best candidates for meeting the goals of the PNGV program (National Research Council 1996), and they may be impressive overall, when judged against the joint objectives of local air quality, oil security, and global warming (Ross and Wu 1995). Prototype combustion-engine/battery peak-power hybrids are now being designed.

Although the peak-power hybrid using a fuel-cell for base-load power and a battery for peak power lies somewhat further in the future, it may be even more impressive. It promises zero or near-zero emissions. It promises high energy efficiency. It promises advantages in addressing oil security, because the fuel for the fuel cell can be a product derived from a fossil fuel other than petroleum or from a non-fossil energy source such as wood chips (Williams, 1994).

The fuel cell derives its power from the electrochemical oxidation, instead of the combustion, of a fuel on board. In the prototype fuel-cell vehicles being tested today, the fuel is hydrogen, and the hydrogen combines electrochemically with the oxygen in air to make water. The underlying chemical reaction is not an obstacle to achieving long range. And the end of the life cycle of a fuel cell may not be complicated either. Today's most promising contender for transportation applications, the proton-exchange-membrane (PEM) fuel cell (Prater 1994), uses carbon electrodes and a platinum catalyst, as well as casing and tubing that do not require exotic materials. Safe management seems possible.

Electric and hybrid vehicles are being encouraged by strong industries. The electric utilities see large new markets with potentially attractive load characteristics, and aerospace and other high-technology companies hope to manufacture advanced batteries, fuel cells, and related components. However, the industrial system delivering transportation today, based on petroleum and on vehicles where both base-load and peak power are provided by a single large internal combustion engine, will not be easily abandoned. Indeed, it has already evolved considerably to improve fuel economy and to reduce emissions, and it has the potential to evolve further. The same approaches that brought past improvements will bring future improvements; these include streamlining and lowering the weight of the vehicle, improving the control of fuel injection, incorporating better catalytic treatment of the exhaust, and modifying the fuel (Office of Technology Assessment 1991, DeCicco and Ross 1994, Ross 1994). Time will tell whether dramatically different vehicles can capture a large market share from vehicles that continue to be improved incrementally.

3. ENVIRONMENTAL AND HEALTH IMPACTS OF LEAD IN GASOLINE AND BATTERIES

3.1 Health consequences of lead exposure

The dangers of exposure to high levels of lead have been understood for centuries. Beginning in the 1970s, however, epidemiological studies reported adverse effects on child development at what had previously been considered low and normal levels of lead exposure, down to a blood lead concentration of 10 μg/dL. (The standard measure of human exposure to lead is the concentration of lead in blood, expressed in the US as micrograms per deciliter, or μg/dL, a deciliter being one-tenth of a liter.[5]) A recent US National Research Council review concluded:

The weight of evidence gathered during the 1980s clearly supports the conclusion that the central and peripheral nervous systems of both children and adults are demonstrably affected by lead at exposures formerly thought to be well within the safe range. In children, blood lead concentrations around 10 μg/dL are associated with disturbances in early physical and mental growth and in later intellectual functioning and academic achievement" (National Research Council 1993, p. 93).

As a result of these studies, in 1991, the US Center for Disease Control reduced its threshold definition of dangerous levels of lead in children's blood from 25 μg/dL to 10 μg/dL (US Dept. of Health and Human Services 1991a).

The concentration of lead in blood is generally considered to be a measure only of recent lead exposure, because over a period of 20 to 30 days lead is deposited to bone, where it accumulates throughout life, at least until the age of 50 to 60. But lead in bone is not inert. Bone lead remains in equilibrium with blood lead and, especially in people with high past exposures, lead released from bone can make a significant contribution to blood lead concentrations. Moreover, the release of lead from bone to the bloodstream increases during menopause, pregnancy, and old age. Maternal blood lead is passed to the foetus, and maternal blood lead levels in the 10 μg/dL range have been associated with adverse effects on foetal and infant development (Silbergeld et al. 1988, National Research Council 1993).

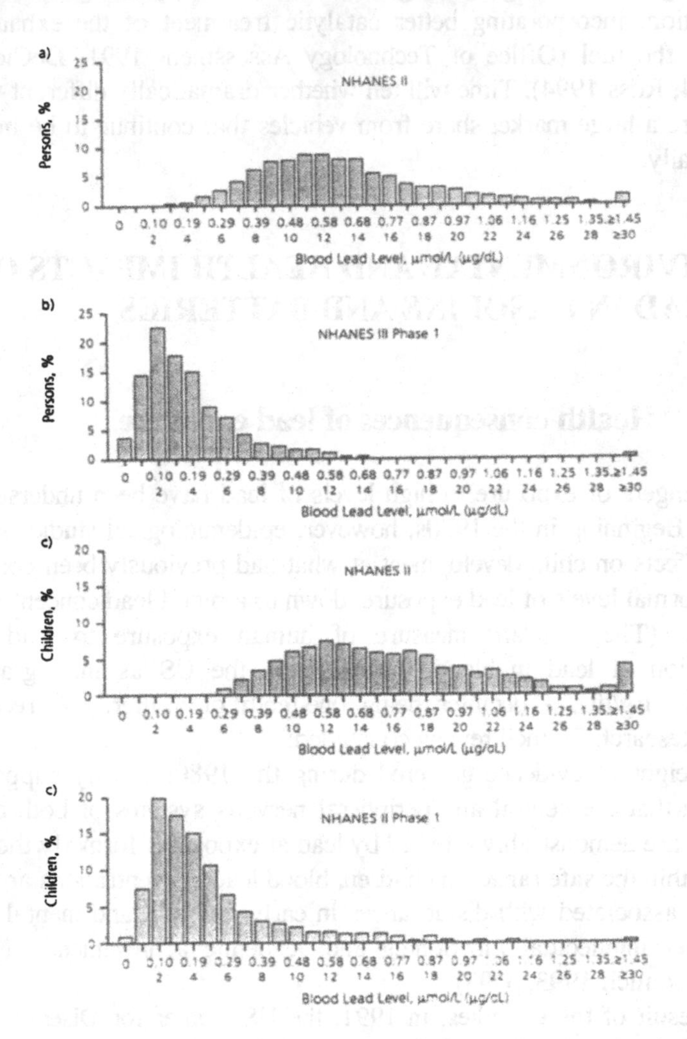

Figure 27. Histograms of blood lead levels in two national samples. The samples are from the National Health and Nutrition Examination Surveys known as NHANES II and NHANES III, roughly a decade apart. The data in NHANES II are from 1976-80 and the data in NHANES III are from 1988-91. Both the full samples (A,B) and the subsamples of children from age 1 to 5 (C,D) are shown. The units of blood lead level are given in both micromoles per liter and micrograms per deciliter, two units which differ by a factor of 20.7 (one-tenth the atomic weight of lead). Source: Pirkle et al 1994

The average blood lead concentrations in the US population have been remarkably well documented. In 1976-1980 and 1988-1991, the National Center for Health Statistics and the Centers for Disease Control conducted

National Health and Nutrition Examination Surveys, which included measurements of lead in blood. The two studies, known, respectively, as NHANES II (Annest 1983) and NHANES III (Brody 1994), allow close comparison (Pirkle 1994).

Figure 27 compares the distributions of blood lead levels in two representative samples from the NHANES studies, each consisting of, roughly, ten thousand people. *Figure 27* also compares the distributions just for children between 1 and 5 years old, from subsamples of roughly two thousand children. In NHANES II the mean blood lead level for the whole sample was 12.8 μg/dL, and the mean for children of age 1 to 5 was 15.0 μg/dL. (These are arithmetic means.) But in NHANES III these means had plummeted, in each case by more than a factor of four, to 2.8 μg/dL and 3.6 μg/dL, respectively. Focusing just on the children aged 1 to 5 with blood lead levels below 10 μg/dL, the change is particularly striking: at the time of NHANES II, only 12% of the children in this age group were in this category, whereas, at the time of NHANES III, the percentage had grown to 91% (Pirkle, 1994).

3.2 Lead in Gasoline

The NHANES II study of US population blood lead levels was carried out between 1976 and 1980, a time of major reduction in gasoline lead additives in the United States. The NHANES II data made possible an analysis of the relation between US population blood lead levels and total lead used in US gasoline. Comparing data on US consumption of lead gasoline additives for nine six-month intervals between 1976 and 1980 with NHANES II blood lead data revealed that blood lead concentration was highly correlated with gasoline lead use (0.93 correlation coefficient). During those years the average blood lead level dropped by about one-third, and the rate of use of lead in gasoline production dropped by about one-half (Annest 1983).

Air lead levels also fell sharply when gasoline lead use was reduced. The emissions inventories of the US Environmental Protection Agency show that leaded gasoline was by far the largest source of lead air emissions in the US during the 1970s and 1980s (US Environmental Protection Agency 1993). As lead in gasoline was reduced, US air lead levels fell correspondingly. In 1979 the average concentration of lead in urban air was 0.8 μg/m3, and it was less than 0.1 μg/m3 in 1988 (US Environmental Protection Agency 1990).[6]

Studies of the effect of reduced use of lead in gasoline on population blood lead levels have been conducted in several other countries. In all cases, decreased use of lead in gasoline results in decreased blood lead levels. In three separate European studies, the average lead concentration in blood fell by 6 to 9 μg/dL when the lead concentration in gasoline was reduced from about 0.40 to 0.15 g/L. These studies -- in Turin, Italy (Bono et al. 1995), in Belgium (Ducoffre et al. 1990), and in Tarragona Province, Spain (Schuhmacher et al. 1996) -- were conducted over intervals of four to eight years. They also measured substantial reductions in lead concentrations in air.

The evidence of these and other studies has led to widespread recognition of leaded gasoline as a significant contributor to human lead exposure. Where lead has been eliminated from gasoline, it is properly viewed as a major achievement.

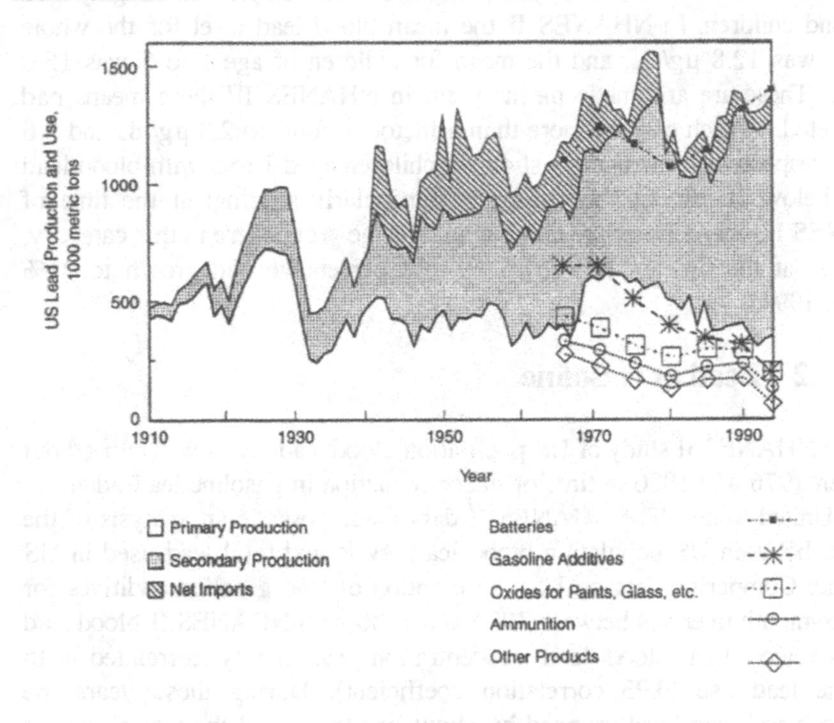

Figure 28. History of US lead production and consumption

Production is in three categories, shown additively: primary, secondary (recycled), and net imported lead. Consumption is in five categories, also shown additively: batteries; gasoline additives; oxides for paints and glass; ammunition, and other products. Sources: US Bureau of Mines 1993, US Bureau of Mines 1995, US Department of Commerce 1975 (p. 603), Woodbury, 1993.

3.3 Lead in Batteries

The ratio of secondary (recycling) lead production to primary lead production has increased nearly continuously in the United States for the past 80 years. *Figure 28* shows the details of production during that period, as well as lead consumption, by category of lead product, for the past 30 years.

Primary lead production has fallen steadily since 1970, along with the total use of dissipative lead products (gasoline additives, oxides, and ammunition). Secondary lead production in the same period closely tracks production of batteries.

The dominance of batteries and battery recycling in the lead flows through the US economy is also seen in *Figure 29*, which shows sources, products, and fates. The total annual US lead consumption (averaged over 1993 and 1994), about 1400 thousand metric tons, is dominated by 910 thousand metric tons of secondary (recycled) production, the rest being primary production and imports. Batteries are the main product, consuming 1180 thousand metric tons.

Although recycling rates are not as high as in the United States, the global pattern of lead use and recycling is similar. In 1990, global lead production, 5.9 million metric tons, is the sum of about 3.3 million metric tons of primary production and about 2.6 million metric tons of secondary production. The total production for batteries, 3.7 million metric tons, is more than sixty percent of production for all products (Thomas and Spiro 1994).

The US battery industry estimates that, annually from 1990 to 1994, from 93% to 98% of lead available for recycling was actually recycled. "Available lead" is estimated from the production of batteries for transportation in previous years, taking into account exports and imports of batteries as well as battery lead in scrap (Battery Council International 1995). The Battery Council International's methodology can be illustrated for 1994, when the "recycling rate" was 98%. The numerator, 834 thousand metric tons, is simply the mass of lead arriving in batteries at secondary smelters in 1994, derived from questionnaires sent to the 11 companies that own all the US secondary smelters. The denominator, 849 thousand metric tons, is an estimate of the domestic supply of lead in batteries that should have become available as previously produced batteries reached the end of their useful life, taking into account exports and imports. Exports and imports are in three categories: lead in new batteries (both batteries as separate products and batteries in vehicles of the appropriate year), lead in worn-out batteries, and lead in waste and scrap derived from batteries.[7]

Figure 29 shows an expanding stock of lead within the industrial system. Entering the system from mineral extraction (bottom left) is more lead than leaves the system via disposal and export (bottom right) and recycling waste (top right). The same expansion is displayed in the leftward tapering (trapezoidal) shape at top center, which reflects the time lag between production and eventual recycling of batteries (on average, about three years).

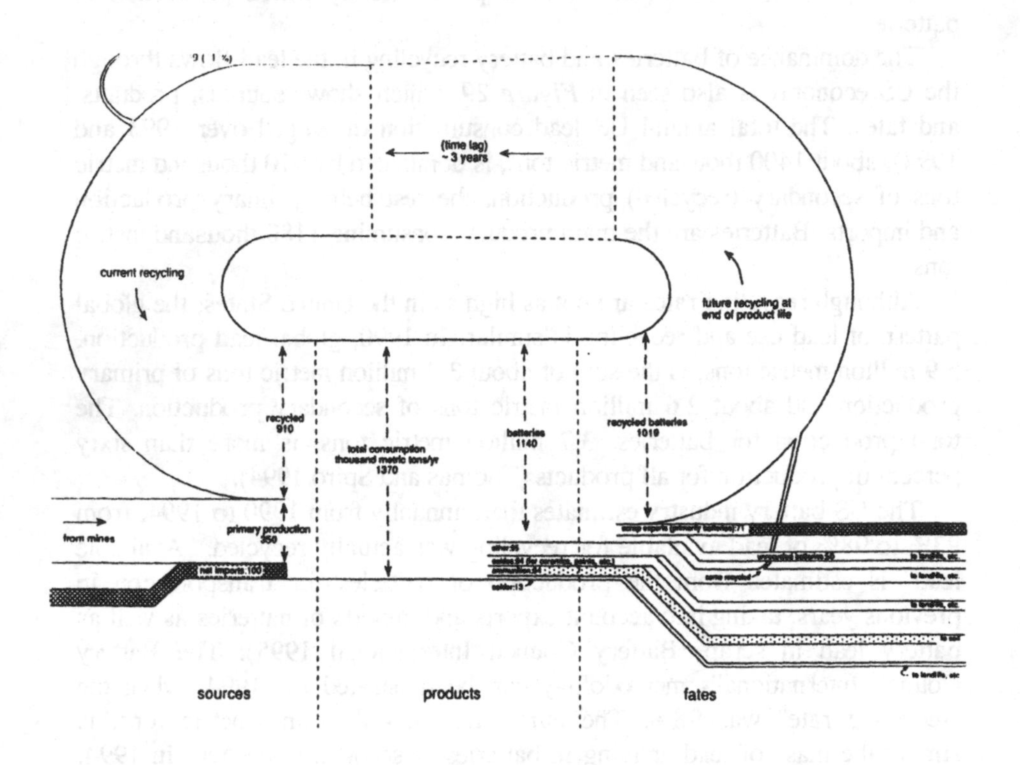

Figure 29. Flows of lead through the US economy in 1993-94

Data displayed are averages of data for 1993 and 1994, in thousands of metric tons of lead per year.) "Other products" are bearings, brass, bronze, cable covering, caulking lead, casting materials, pipes, and sheet lead. Emissions to water or air -- the primary health concerns -- are not shown because they are too small to be visible at this scale. Also not shown is the roughly 0.6 thousand metric tons of lead used in aviation gasoline (about a billion litres with a lead concentration of 0.4 - 0.8 grams per litre). Sources: Battery Council International 1995, US Bureau of Mines 1995, US Department of Commerce 1994, US Environmental Protection Agency 1994a.

Almost all recycling waste is solid waste. Past practices of the lead battery manufacturing and recycling industry have left the United States with over 50 lead-contaminated Superfund sites, characterised by high concentrations of lead in the soil and lead in groundwater or surface water (Roque 1995). Today, waste disposal from secondary lead recycling and battery manufacture is regulated by the Resource Conservation and Recovery Act (RCRA). To conform to RCRA regulations, the slag that is the solid aggregate remaining from the smelting process is treated to reduce the ability of lead to be leached. In a common method of treatment, "stabilisation," leachability is reduced by mixing the waste with portland cement (Lebo 1996).

Figure 29 (top left) shows 7 thousand metric tons of lead leaving the US industrial system as recycling wastes. These recycling wastes are reported in the Toxic Release Inventory as land disposal and off-site transfer of lead from 15 secondary lead smelters (US Environmental Protection Agency 1994a). Since the total production in 1994 at all smelters was about 910 thousand metric tons (US Bureau of Mines 1994), the recycling wastes are about 0.8% of secondary production.

Figure 29 also shows that about 50 thousand metric tons of lead are in batteries that are not recycled but remain in the US. Some of these batteries end up in landfills, where leaching can result in lead moving into groundwater or surface water. As of the mid-1980s, battery lead was estimated to represent 65% of the total lead in municipal waste landfills (Franklin Associates 1988). The concentration of lead in leachate from municipal waste landfills has been measured, and found to be as high as 1600 micrograms of lead per liter of leachate (μg/L) for pre-1980 landfills, and as high as 150 μg/L for post-1980 landfills. The US Environmental Protection Agency presents a worst case scenario, where this leachate is diluted by a factor of only 100 by groundwater before reaching well-water (US Environmental Protection Agency 1991). The resulting maximum lead concentration in drinking water from a pre-1980 landfill, 16 μg/L, can be compared with the current US Safe Drinking Water Act goal to reduce lead concentrations in drinking water to below 15 μg/L.

Some unrecycled batteries do not go to landfills, but are discarded directly on the land (in backyards, for example). Although lead is quite immobile in soil and tends to be adsorbed by soil particles, the lead in a battery discarded on the land is relatively mobile; this is because up to half of the lead in a battery is in the form of a paste of lead oxide and lead sulphate, and because the acid electrolyte can enhance lead migration in soil (US Environmental Protection Agency 1991).

Figure 30 summarises the record from 1975 to 1990 for the United States, showing the consumption of lead in gasoline additives and batteries, along with the average blood lead levels in the population. The use of lead in batteries has increased, although not steadily, and lead in automotive gasoline has disappeared, while population blood lead levels have plummeted.

Hazards to the communities near secondary smelters and the potential for improvement Human health and the environment are adversely affected in communities near many secondary smelters today. Some of the smelters are located in developing countries, and many of these smelters receive a

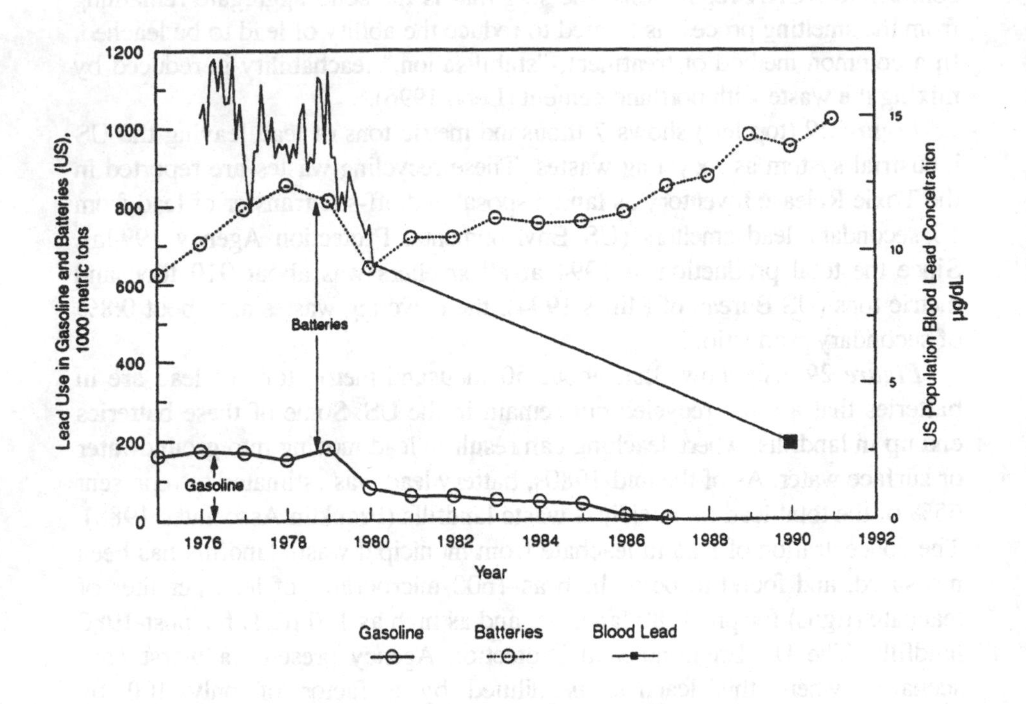

Figure 30. **Annual US lead use in gasoline and batteries and US population blood lead levels from 1975 to 1990.**

Battery data are from US Bureau of Mines 1993 and US Bureau of Mines 1995. Gasoline data are from Nriagu 1990. Both the battery and the gasoline data are referenced to the x-axis: they are not shown additively. Blood lead data are from NHANES-II (Annest 1983) and NHANES-III (Pirkle 1994). NHANES-II data are 28-day averages from February 1976 to February 1980 (Annest 1983, Figure 1). The single NHANES-III blood-lead data point at 1990 is an average for 1988-91. The dotted line connecting the NHANES II time series with the single NHANES III data point is a only a guide to the eye: there are no intermediate data points

significant fraction of their lead from industrialised countries, largely as scrap. Many of the smelters in developing countries are characterised by very large environmental lead emissions, documented by high lead concentrations measured in river and marsh sediments and in soils (Greenpeace 1994).

Results of numerous studies document that people sustain marked increases in blood lead levels when they live near secondary lead smelters and other stationary lead sources (ATSDR 1988).

However, as a result of investments in pollution control and materials management, the adverse environmental impacts of some secondary smelters on nearby communities are becoming much less serious. A recent study at a secondary lead smelter in California retrofitted with state-of-the-art pollution control systems found no statistically significant impact on the blood-lead levels of the children in a nearby community. The study examined blood lead levels and lead in the household environment of children between the ages of 1 and 5; data were obtained from two communities, the nearby community and a control, with a sample size of 122 children in each community. The average blood lead level in the exposed and control group were nearly the same, 3.8 and 3.5 μg/dL, respectively (Wohl et al. 1996).[8]

This improvement is reflected in air emissions factors for secondary smelters (total mass of air emissions divided by total lead production) that have decreased substantially in recent years in the United States. Previous analyses have assumed that the air emissions factor is at least 0.02%, a value that was derived by the US Environmental Protection Agency from data reported for its 1987 Toxic Release Inventory (Lave et al., 1996; Roque 1995). But current reports from secondary smelters indicate air lead emissions that are an order of magnitude lower. The California Air Emissions Board (CARB), for example, routinely uses an air emissions factor of 0.002%.

The CARB air emissions factor turns out to be based only on a single three-day "source test" conducted at CARB's behest in November, 1990, at a large secondary smelter owned by GNB in Vernon, California (McLaughlin 1996).[9] There is considerable evidence to confirm CARB's air emissions factor as characteristic of many secondary smelters today. We have calculated air emissions factors for 17 of the 18 secondary lead smelters in the United States, by combining Toxic Release Inventory data from 1994 with information on the capacities of US secondary lead smelters. The data are presented as a histogram in ***Figure 31***. The production-weighted average air emissions factor (including both fugitive and point emissions) is 0.006% (a total of 57 metric tons of air emissions from the secondary production of 910 thousand metric tons). The distribution of air emissions factors in ***Figure 31*** is consistent with the air emissions factors for two California smelters (Steele and Allen 1996). It is also consistent with the 0.001% air emissions factor reported by Boliden Bergsoe, a firm that owns a secondary smelter in Landskrona, Sweden (Karlsson 1996). Even lower air emissions factors are reported in a summary of a confidential industry survey (Electric Power Research Institute 1993). The average air emissions factor for secondary smelters has fallen in recent years because some smelters have invested in pollution control to meet tighter regulations and others have gone out of business. In 1995, the US Environmental Protection Agency established limits on air emissions from secondary lead smelters; as these limits come into effect, air emissions factors can be expected to fall further.

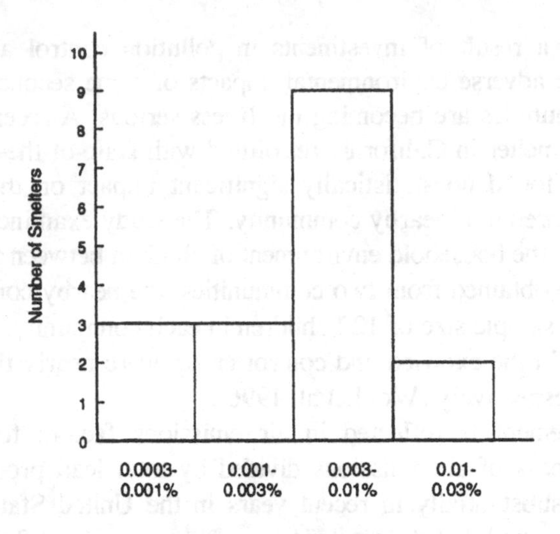

Figure 31. Histogram of the values of the total air emissions factors for 17 US secondary lead smelters in 1994

The emission factor for each smelter is its total (stack plus fugitive) air emissions, as reported in the US Environmental Protection Agency's 1994 Toxic Release Inventory, divided by the midpoint of its estimated capacity. Each bin of the histogram is an approximately factor-of-three interval. Sources: US Environmental Protection Agency 1994a, US Environmental Protection Agency 1994b.

Childhood blood lead screening is now routine throughout the US. Given the continuing progress in reducing lead emissions reported by the companies owning secondary lead smelters, these companies should be challenged to demonstrate that blood lead levels in nearby communities are indistinguishable from those in similar communities far from smelters. The collection and publication of blood lead levels in communities near all secondary lead smelters would go a long way toward clarifying the extent of lead exposure from lead battery recycling.

Occupational exposure is the hardest challenge to clean recycling. A complete assessment of the level of worker exposure associated with US battery recycling must take into account that some of the lead in US batteries is exported as a component of lead scrap, and that some of this lead is processed in smelters in the less developed countries where worker exposures are very high (Greenpeace 1994).

In the US the 1978 lead standards of the Occupational Safety and Health Administration (OSHA) have had an important role in controlling occupational lead exposures (Silbergeld et al. 1991). The standards require removal of workers to a lead-free area if the blood lead level exceeds 50 μg/dL for six months or more, and immediate removal if the blood lead level exceeds 60 μg/dL, until such time as the blood lead level is reduced to 40 μg/dL or less. There is increasing pressure to revise these standards (Spurgeon 1994). In 1991 the US Public Health Service declared a goal of eliminating occupational lead exposures exceeding 25 μg/dL by the year 2000 (US Dept. of Health and Human Services 1991b). Even at 25 μg/dL, pregnant female workers will be at risk for passing lead directly to developing foetuses; blood lead levels above 10 μg/dL in women can result in fetal exposure in excess of the recommendations of the Centers for Disease Control.

Instances of high worker exposure continue to surface in the United States. A 1991 investigation of an Alabama battery reclamation facility (where batteries are cut open and the lead and plastic are sent to other plants for further processing) found that 14 of 15 workers had blood lead levels exceeding 50 μg/dL. The neighbouring community was also affected: 12 of 16 employee children had blood lead levels exceeding 10 μg/dL, with three children having levels greater than 40 μg/dL. A survey of 11 adults and 5 children (aged 6 to 17) living within one block of the plant found average blood lead levels of 11.6 μg/dL and 9.8 μg/dL, respectively. In response to this information, OSHA declared the facility an "Imminent Danger," and all workers were removed from their jobs until their blood lead levels declined to 40 μg/dL. The magnitude of the health and safety fines ($1.2 million) contributed to the decision by the owner to close the facility (Gittleman et al. 1994).

Reducing worker blood lead levels to those of the general population is a task of unknown difficulty. The extent of worker exposure today is revealed in data collected by Battery Council International, a trade organisation, which collects data on worker blood lead levels in US secondary smelters and battery manufacturing plants. In the first three months of 1996 about 6% of workers in secondary smelters had levels exceeding 40 μg/dL and 56% had levels exceeding 20 μg/dL; the percentages for battery plants were 2.5% and 44%, respectively (Battery Council International 1996).[10]

4. CONCLUSIONS

This is a period of great fluidity in automotive engineering. A multiplicity of options for providing the energy for vehicles are being explored, and each new capability has implications for several options at once. The battery-powered car could not penetrate even niche markets at this time, if it did not incorporate innovations related to fuel economy and the management of electricity on board -- innovations that have broad implications affecting the rates of progress for other vehicle concepts. The research and development process would be proceeding more slowly, if state and national governments were not actively promoting electric and hybrid vehicles through the mandating of prototype development and pilot-scale marketing.

An important argument against thwarting the development of the lead battery for electric vehicles is that the lead battery may have a second, entirely different, and potentially more important assignment than its assignment in today's EVs. In addition to providing stand-alone power for short-range vehicles, the lead battery may provide peak power in several kinds of hybrid vehicles. These peak-power hybrids may be particularly suited to address, simultaneously, urban air quality, oil security, and global warming. The fuel-cell/battery peak-power hybrid, in particular, may expand options for the global transportation system, by facilitating the introduction of renewable fuels.

When a social experiment is dangerous to the public even at a small scale, it is important that it be deterred. But there is another class of experiments, of which the pilot programs for electric vehicles are an example, where the experiment itself presents negligible risks, but full-scale implementation has the potential for serious harm. For this class of experiments, there may be a large societal and environmental cost in suppressing the experiment. Freeman Dyson captures what is at stake in an essay entitled, "The Cost of Saying No":

It is not enough to count the hidden costs of saying yes to new enterprises. We must also learn to count the hidden costs of saying no...It often happens in technological development that one design turns out to be not merely better but enormously better than its competitors, for reasons that could not have been predicted in advance. There is no way to find the best design except to try out as many designs as possible and discard the failures (Dyson 1975).

Comparison of lead in gasoline with lead in batteries shows the limitations of mass balance analysis in identifying the most important environmental and public health consequences of a product system. Often, on a mass basis, the principal emissions are in the form of bulk solids, such as slags; yet, several much smaller mass transfers have far greater health impacts. Meriting detailed and sustained attention are exposures to workers and to people in neighbouring communities.

A mass balance can be indispensable, however, in accounting for consumer products that escape the recycling system; in the case of unrecycled batteries, the health and environmental impacts are still largely unexplored. And a mass balance can provide detailed information about imports and exports; here an important issue is the export of scrap containing lead from used batteries to countries where environmental controls are weak.

Generalising, a proper risk assessment must combine mass-flow analysis with hazard analysis. It must fully explore the impacts on human health and the environment that result when a hazardous material leaves an industrial system and undergoes subsequent transformations. (Relative to lead, such transformations are even more complex in the case of several other toxic metals, such as mercury, cadmium, and chromium.) Defining system boundaries that exclude these transformations from consideration will impoverish any analysis.

Little can be done to prevent the lead in gasoline, after combustion, from finding its way to people in the dangerous form of fine particles. But much can be done to affect the fate of lead in batteries. From a societal perspective, the appropriate goal for the lead battery industry should be clean recycling, where almost all used batteries are retrieved and where, during recycling, lead releases to the environment have almost no environmental and public health impacts.

The small and shrinking share of primary lead in today's lead batteries has two complementary implications for clean recycling: 1) Unless primary production is much dirtier than secondary production, secondary production should be the principal focus of clean recycling. 2) A clean recycling system can be achieved only if both primary and secondary smelters meet stringent environmental standards.

Achieving a battery management system that approaches the industrial ecology ideal of clean recycling will require much work world wide. The environmental performance of many of the world's primary and secondary lead smelters is far below the environmental performance of the world's best smelters. In the poorly performing facilities, worker lead exposure is very high, and people in the nearby communities have elevated blood levels. Some of the worst environmental performance is found in smelters in the developing world (Greenpeace 1994, Woolf 1994). No nation can be said to have a clean recycling system unless all its batteries at end of life go to secondary smelters and battery manufacturing plants that meet demanding environmental and occupational health standards.

It will not be enough for all facilities to reach the standards of today's best managed facilities. Even in today's best managed facilities, workers have blood lead levels exceeding the levels of well-documented adverse health effects. Better documentation of the recycling system is needed, including better measurements of air emissions, community exposures, worker exposures, and lead transfers across countries. And data need to be acquired or independently confirmed by government agencies, environmentalists, and others who can be expected to have an independent view.

The lead industry as a whole should be challenged to approach the industrial ecology ideal. Some intrinsically dissipative uses of lead, including leaded gasoline, are already being phased out, in favour of more benign substitutes. The lead battery, by far the most important lead-containing consumer product that does not dissipate lead with use, can become one of the first examples of a hazardous product managed in an environmentally acceptable fashion. The tools of industrial ecology are well matched to the charting of progress.

ACKNOWLEDGEMENTS

Reprinted from Journal of Industrial Ecology, 1:1 (Winter 1997), pp 13-36 with permission from MIT Press Journals, Cambridge, MA.

We wish to thank Mark Baumgartner, L. Pasha Dritt, James J. Fanelli, and Margaret Steinbugler for research assistance. We have benefited from many reviews of an earlier draft of this manuscript and related discussions. Particularly helpful in increasing our understanding of advanced automobile technologies and associated policies were Andrew Burke, John DeCicco, Frederick Dryer, Patrick Grimes, Joseph Keating, Thomas Kreutz, Jameson McJunkin, Paul Miller, Joan Ogden, Marc Ross, William Schank, Daniel Sperling, and Robert Williams. We have been informed about battery management and associated emissions by David Allen, Marijke Bekken, Douglas Bice, Leland Deck, Katie Chiampou, Robert Fletcher, Linda Gaines, Lester Lave, Neal Lebo, Leonard Levin, William McKusky, Carol McLaughlin, Francis McMichael, Saskia Mooney, Joseph Norbeck, Chris Whipple, and Linda Wing. We have been alerted to public health issues by Robert Elias, Philip Landrigan, Ellen Silbergeld, and Joel Schwartz. We have enjoyed many exchanges about the emerging principles of industrial ecology with Brad Allenby, Jesse Ausubel, Robert Frosch, Thomas Graedel, John Harte, Reid Lifset, David Rejeski, Thomas Spiro, Iddo Wernick, and Stefan Wirsenius. Comments by John DeCicco, Harold Feiveson, Francis McMichael, Ellen Silbergeld, and Robert Williams have enabled us to sharpen our argument. We are grateful for the assistance of Kathy Shargo and Laura Schneider in the preparation of Figures and Elaine Kozinsky in the preparation of the manuscript.

REFERENCES

Annest, J.L., et al. 1983. Chronological trend in blood lead levels between 1976 and 1980. The New England Journal of Medicine, 308(25):1373-1377.

ATSDR. 1988. The Nature and Extent of Lead Poisoning in Children in the United States: A Report to Congress. PB89-100184. US Dept. of Health and Human Services, Atlanta, GA.

Ayres, R.U., 1994. Industrial metabolism: Theory and policy, in The Greening of Industrial Ecosystems, B.R. Allenby and D.J. Richards, eds. Washington, DC: National Academy Press, pp. 23-37.

Battery Council International 1995. 1994 Recycling Rate Study. Chicago, IL.

Battery Council International 1996 Survey of Industry Blood Lead Levels, January 1 - March 31, 1996. Final Report. Unpublished.

Bono, R., C. Pignata, et al., 1995. Updating about reductions of air and blood lead concentrations in Turin, Italy, following reductions in the lead content of gasoline. Environmental Research 70:30-34.

Brody D.J., et al. 1994. Blood lead levels in the US population. Journal of the American Medical Association 272(4):277-283.

Braungart, M. 1994. Product life-cycle management to replace waste management. Industrial Ecology and Global Change, R. Socolow, C. Andrews, F. Berkhout, and V. Thomas, eds. Cambridge University Press, pp. 335-338.

California Air Resources Board 1995. Reclamation of Automotive Batteries: Assessment of Health Impacts and Recycling Technology.

Calvert, J.G., J.B. Heywood, R.F. Sawyer, J.H. Seinfeld 1993. Achieving acceptable air quality: Some reflections on controlling vehicle emissions. Science 261:37-45

Castellino, N., P. Castellino, N. Sannolo 1995. Inorganic Lead Exposure: Metabolism and Intoxication. Boca Raton: Lewis Publishers.

DeCicco, J. and M. Ross 1994. Improving automotive efficiency. Scientific American, December 1994, 52-57.

Ducoffre, G., F. Claeys, P. Bruaux, 1990. Lowering time trend of blood lead levels in Belgium since 1978. Environmental Research 51:25-34.

Dyson, F.J. 1975. "The hidden cost of saying No." Bulletin of the Atomic Scientists, June 1975, pp. 23-27.

Electric Power Research Institute (EPRI), 1993. Lead-Acid Batteries for Electric Vehicles -- Life-Cycle Environmental and Safety Issues. EPRI Project 2415-38. Final Report, December 1993.

Franklin Associates 1988. Characterization of Products Containing Lead and Cadmium in Municipal Solid Waste in the United States, 1970 to 2000. Prairie Village, KS.

Gaines, L. and M. Singh 1995. Energy and environmental impacts of electric vehicle battery production and recycling. Proceedings of the 1995 Total Life Cycle Conference: Land, Sea, and Air Mobility. Society of Automotive Engineers Publication P-293. Warrendale, PA: Society of Automotive Engineers.

Gittleman J.L., M.M. Engelgau, J. Shaw, K.K. Wille, P.J. Seligman 1994. Lead poisoning among battery reclamation workers in Alabama. Journal of Occupational Medicine 36(5):526-532.

Greenpeace 1994. Lead Astray: The Poisonous Lead Battery Waste Trade. Washington, DC: Greenpeace USA.

Hwang, R., M. Miller, A.B. Thorpe, D. Lew 1994. Driving Out Pollution: The Benefits of Electric Vehicles. Berkeley, CA: Union of Concerned Scientists.

Illman, D.L., 1994. Automakers move toward new generation of "greener" vehicles. Chemical and Engineering News, August 1, 1994. pp. 8-16.

Juergens, T. and R.F. Nelson 1995. A new high-rate, fast-charge lead/acid battery. Journal of Power Sources 53: 201-205.

Johansson, T.B., R.H. Williams, H. Ishitani, J.A. Edmonds 1996. Options for reducing carbon dioxide emissions from the energy supply sector. To be published in Energy Policy.

Karlsson, S. 1996. "Can Metals be Used Sustainably? The Example of Lead." Annual Report, S. Karlsson, G. Berndes, S. Wirsenius, eds. Institute of Physical Resource Theory, Chalmers University of Technology and Goteborg University, AFR-Report 121, Swedish Environmental Protection Agency, Stockholm.

Keating, J., B. Schroeder, R. Nelson 1996. Development of a valve-regulated lead-acid battery for power-assist hybrid electric vehicle use. Proceedings of the ISATA Conference, Florence, Italy, June 1996. Available from Bolder Technologies Corporation, Wheat Ridge, Colorado.

Lave L.B., C.T. Hendrickson, F.C. McMichael 1995. Environmental implications of electric cars. Science 268:993-995.

Lave L.B., A.G. Russell, C.T. Hendrickson, F.C. McMichael 1996. Battery-powered vehicles: Ozone reduction versus lead discharges. Environmental Policy Analysis 30(9):402-407.

Lebo, N. 1996. Exide Corporation, Reading, PA. Private communication.

MacKenzie, J.J. 1994 The Keys to the Car: Electric and Hydrogen Vehicles for the 21st Century. Washington, DC: World Resources Institute

McLaughlin, C. 1996. Air Resources Board, Sacramento, CA. Private communication.

National Research Council 1993. Measuring Lead Exposure in Infants, Children, and Other Sensitive Populations. Washington, DC. National Academy Press.

National Research Council 1996. Review of the Research Program of the Partnership for a New Generation of Vehicles: Second Report. Washington, DC: National Academy Press.

Nelson, R.F. 1996. Bolder TMF technology applied to hybrid electric vehicle power-assist operation. Eleventh Annual Battery Conference of the Institute of Electrical and Electronics Engineering (IEEE): Applications and Advancements, Piscataway NJ. Available from Bolder Technologies Corporation, Wheat Ridge, Colorado.

Nriagu, J.O. 1990. The rise and fall of lead in gasoline. Science of the Total Environment 921:13-28.

Office of Technology Assessment of the US Congress 1991. Improving Automobile Fuel Economy: New Standards, New Approaches. OTA-E-504. Washington DC: US Government Printing Office.

Office of Technology Assessment of the US Congress 1995. Advanced Automotive Technology: Visions of a Super-Efficient Family Car. OTA-ETI-638. Washington DC: US Government Printing Office.

Pirkle J.K., D.J. Brody, E.W. Gunter, R.A. Kramer, D.C. Paschal, et al. 1994. The Decline of Blood Lead Levels in the United States: the National Health and Nutrition Examination Surveys (NHANES). Journal of the American Medical Association 272(4):284-91.

Prater, K.B. 1994. Polymer electrolyte fuel cells: A review of recent developments. Journal of Power Sources 51:129-144.

Roque J. 1995. "Pollution prevention for emerging industries: The case of electric vehicles." Reducing Toxics. R. Gottlieb, ed. Island Press.

Ross, M. and R. Socolow 1991. Fulfilling the promise of environmental technology. Issues in Science and Technology: VII(3) 61-66.

Ross, M. 1994. Automobile fuel consumption and emissions: Effects of vehicle and driving characteristics. Annual Review of Energy and the Environment 1994, 19:75-112.

Ross, M. and W. Wu 1995. Fuel economy analysis for a hybrid concept car based on a buffered fuel-engine operating at an optimal point. Society of Automotive Engineers Technical Paper Series 950958. Warrenton, PA, Society of Automotive Engineers.

Schuhmacher, M., M. Belles, et al. 1996. Impact of reduction of lead in gasoline on the blood and hair lead levels in the population of Tarragona Province, Spain, 1990-1995. Science of the Total Environment 184: 203-209.

Silbergeld E.K., P.J. Landrigan, J.R. Froines, R.M. Pfeffer 1991. The occupational lead standard: A goal unachieved, a process in need of repair. New Solutions, Spring: pp. 20-30.

Silbergeld E.K., J. Schwartz, K. Mahaffey. 1988. Lead and osteoporosis: Mobilization of lead from bone in post-menopausal women. Environmental Research 47:79-94.

Socolow, R. 1994. Six perspectives from industrial ecology. Industrial Ecology and Global Change, R. Socolow, C. Andrews, F. Berkhout, V. Thomas, eds. Cambridge University Press, pp. 3-16.

Sperling, D. 1995. Future Drive: Electric Vehicles and Sustainable Transportation. Island Press.
Sperling, D. 1996. The case for electric vehicles. Scientific American, November 1996, 54-59.
Spurgeon A. 1994. Occupational lead exposure: Do current exposure standards protect workers from harm? Indoor Environment 3:112-118.
Steele N, and D.T. Allen 1996. Evaluation of recycling and disposal options for batteries. Proceedings of the Institute for Electrical and Electronics Engineering (IEEE) Symposium on Electronics and the Environment, Dallas, Texas, May 1996, 135-140.
Thomas, V. and T. Spiro 1994. Emissions and exposures to metals: Cadmium and lead. Industrial Ecology and Global Change, R. Socolow, C. Andrews, F. Berkhout, V. Thomas, eds. Cambridge University Press, pp. 297-318.
US Bureau of Mines 1993. Lead. Annual Report 1991. authored by W.D. Woodbury.
US Bureau of Mines 1995. Lead. Annual Review 1994 . Mineral Industry Surveys.
US Dept. of Commerce 1975. Historical Statistics of the United States: Colonial Times to 1970. Series M242-247. Washington, US Government Printing Office.
US Dept. of Commerce 1994. Export Data: Lead Waste and Scrap from Lead Acid Batteries (HTS #7802.000030); Lead Waste and Scrap Not from Lead Acid Batteries (HTS # 7802.000060); Lead Acid Batteries Used for the Recovery of Metal (HTS #8507.100020); Lead Acid Batteries NESOI Used for the Recovery of Metal (HTS # 8507.200020).
US Dept. of Health and Human Services 1991a. Preventing Lead Poisoning in Young Children: A Statement by the Centers for Disease Control. Public Health Service.
US Dept. of Health and Human Services 1991b. Public Health Service. "Healthy People 2000: National Health Promotion and Disease Prevention Objectives." DHHS Pub No. (PHS)91-50212. Washington, DC. US Gov. Printing Office, pp. 303-304. As cited in J.D. Kaufman, J. Burt, B. Silverstein 1994. Occupational lead poisoning: Can it be eliminated? American Journal of Industrial Medicine. 26:703-712.
US Environmental Protection Agency 1990. National Air Quality and Emission Trends Report 1988. EPA 450/4-90-002.
US Environmental Protection Agency 1991. Lead Battery Risk Assessment. SCSP-00144.D. Sept. 6 Draft/Final 1.
US Environmental Protection Agency 1993. National Air Pollutant Emission Trends, 1900-1992. Office of Air Quality Planning and Standards. EPA-454/R-93-032.
US Environmental Protection Agency 1994a. Toxic Release Inventory. Access to on-line data base via Toxnet: Toxicology Data Network, National Library of Medicine, National Institutes of Health, US Department of Health and Human Services.
US Environmental Protection Agency 1994b. Secondary Lead Smelting. Background Information for Proposed Standards. 2 volumes. EPA 453/R-94-024a.
Williams, R.H. 1994. The clean machine. Technology Review, April 1994, 198-207.
Wohl A.R., A. Dominquez, P. Flessel 1996. Evaluation of lead levels in children living near a Los Angeles County battery recycling facility. Environmental Health Perspectives 104(3):314-317.
Woodbury W.D., D. Edelstein, S.M. Jasinski 1993. Lead Materials Flow in the United States 1940-1988. Draft. US Bureau of Mines, Washington, DC.
Woolf, A. 1994. Lead astray: The poisonous lead battery waste trade. United Nations Environment Program (UNEP) Industry and Environment 17(3): 16.

---

[1] Clean recycling has a dynamic dimension as well, since product demand varies over time. For several decades the lead recycling system has managed the starting-lighting-ignition

(SLI) battery that provides starting power for gasoline-powered vehicles. The system has grown to a global scale of more than a hundred million recycled batteries per year. A clean recycling system must assure that the mining and primary processing of the lead required to provide for an increasing stock of lead products is conducted in environmentally responisible ways. It must also assure environmentally responsible contraction. Although no successor to the SLI battery is in view, some day a competitive alternative might challenge its domination. Battery lead in cars on the road would no longer have a predictable destination in a next genereation of batteries, and change would have to occur carefully to avoid large accumulations of lead stocks and large releases of lead to the environment. The scaling back of succesful recycling systems is a general issue that industrial ecology can illuminate.

[2] When a battery-powered electric vehicle substitutes for a gasoline-powered vehicle, local air pollution will be reduced, provided that the generation of the electricity to recharge the battery does not produce compensating air pollution in the same airshed. There will be no compensating increases in local air pollution if the electricity comes from hydropower, nuclear power, geothermal energy, wind, photovoltaics or solar thermal power. Even if the electricity is produced nearby from fossil fuels, there will be little compensating local air pollution from two of the significant air pollutants, volatile organic compounds and carbon monoxide, because it is much easier to reduce these emissions at a central station power plant than onboard a vehicle (Hwang 1994). But if the electricity is produced nearby from fossil fuels, there will be substantial compensating local air pollution from nitrogen oxides (which are particularly important in smog formation), even if pollution controls are in place, because central station power plants are not decisively better than vehicles in controlling nitrogen oxide emissions. There will also be local pollution from sulfur oxides and particulates, especially if the energy source for the electricity is coal. Of course, when the production of electricity that recharges a battery occurs at a power plant a considerable distance grom where the battery is substituting for a gasoline engine on the road, most of the local environmental impacts associated with the production of the electricity will be displaced to the vicinity of the power plant; they do not disappear.

[3] The original (1990) ZEV Mandate also required that 2% of sales be ZEVs by 1998 and5% by 2001. In 1996 these two deadlines were rescinded as a result of negotiations between the State of California and the automobile industry, but the 10% sales requirement for 2003 was reaffirmed.

[4] A variant on the fuel-cell/battery all-elctric hybrid vehicle might use an advanced battery for base-load power.

[5] The international unit is micromoles per liter. One microgram per deciliter is 20.7 micromoles per liter. The factor of 20.7 is one-tenth the atomic weight of lead.

[6] These values are averages over readings at 139 urban sites where 24-hour measurements are performed everey six days. The value that is reported for averaging across sites is the highest of the four quarter-year averages of these 24-hour readings.

[7] The denominator is obtained by adding 840 thousans metric tons of "available" lead from batteries for transportation, plus 125 thousand metric tons of "available" lead from industrial batteries for motive and stationary power, and subtracting 116 thousand metric tons of net exports of lead associated with these batteries. The four largest contributions to the total available lead from batteries for transportation are 1) the 591 thousand tons of lead in the 66.1 million batteries in 1990 passengers cars and light commercial vehicles (derived by assuming that 8.94 kg is the average mass of lead in these batteries); 2) the 103 thousand tons of lead in the 6.3 million batteries in 1191 trucks and heavy-duty

commercial vehicles (16.3 kg average mass of lead); 3) the 47 thousand metric tons of lead in the 3.6 million 1991 marine batteries (13.0 kg average mass); and 4) the 36 thousand metric tons of lead in the 1.9 million 1991 golf carts (18.4 kg average mass of lead). The net exports of lead associated with these batteries is the difference between 127 thousand metric tons of exports and 11 thousand metric tons of imports (Battery Council International 1995).

[8] This experiment had the unfortunate complication that levels below 5 μg/dL were reported as below the detection limit and were not distinguished from one another. In computing the average exposures, all readings below 5 μg/dL were set equal to 2.5 μg/dL.

[9] Air emissions were sampled from six stacks and two vents (the vents were considered to account for the fugitive emissions), and the total air emissions rate was found to be 210 grams per hour. Assuming that this rate was characteristic of the average emission rate for the plant during its 8500 hours of operation in 1990, the air emissions from the Vernon plant would have totaled 1.8 metric tons that year. By CARB's estimate, 93.000 metric tons of lead were processed that year, and thus the air emissions factor (1.8/93,000) is 0.002% (McLaughlin 1996(.

[10] The smelter data were reported by five companies owning smelters, with 2,111 tested employees; the full distribution of blood lead levels reported is 44.3% below 20 μg/dL, 33.2% between 20 μg/dL and 30 μg/dL; 16.7% between 30 μg/dL and 40 μg/dL; 5.3% between 40 μg/dL and 50 μg/dL; and 0.5% above 50 μg/dL. The battery-plant data were reported by 20 companies owning battery plants, with 17,343 tested employees; the complete distribution of levels reported is 56.3% below 20 μg/dL, 28.7% between 20 μg/dL and 30 μg/dL; 12.5% between 30 μg/dL and 40 μg/dL; 2.3% between 40 μg/dL and 50 μg/dL; and 0.2% above 50 μg/dL (Battery Council International 1996).

Chapter 13

Dissipative emissions

Manageable or inevitable?

ROB VAN DER VEEREN
Researcher, Institute for Environmental Studies, Vrije Universiteit, Amsterdam

Key words: nutrients, agriculture, sustainable environmental management

Abstract: The highly productive present day agriculture in western European countries uses large amounts of fertilisers and pesticides. In certain regions, intensive husbandry has resulted in manure production far exceeding the absorbtion capacity of the crops grown. Due to run-off and leaching, nutrients - once important and valuable inputs - have now become a major threat to ecosystem functioning. In order to protect biodiversity, and to prevent ecosystems from adverse effects of excess amounts of nutrients, e.g. eutrophication of rivers and lakes, flows of these substances shoud be managed. However, due to the highly dissipative character of this substance flow, their emissions are largely inevitable. This has implications for sustainable nutrient management.

1. INTRODUCTION

Non-dissipative emissions can be recycled, in principle, as pointed out by van Beukering and Curlee. This is different in case of dissipative emissions, where resources more or less disappear into the environment. Nature can handle large amounts of those wastes, but when certain "threshold levels" are exceeded, dissipative emissions become disruptive. One group of dissipative substances for which maximum emission levels have been established to prevent environmental degradation is nutrients. Nutrients are essential for the realisation of production potentials in agriculture (Langeveld, 1994), but when applied in excessive amounts, leaching and surface run-off can cause algae blooms in semi-stagnant waters and lakes, and enhance the "breaming"-process (a shift in the type of fish in rivers and lakes from a wide variety to a situation dominated by white fish such as bream). If we want to have both, high levels of agricultural production and biodiversity, nutrient emissions need to be managed. Whether this is possible is the main focus of this chapter.

2. A SHORT INTRODUCTION TO NUTRIENT FLOWS

The output of agriculture in most countries has increased considerably. This has mainly been the result of the widespread use of fertilisers, and the international transports of animal feed stuffs. This released farmers from the constraints of the farm level nitrogen cycle (Lanyon, 1995). The increase in use of artificial fertilisers enabled specialisation of former mixed farms, resulting in concentration and intensification of both animal husbandry and arable farming. Dutch intensive livestock farming has also greatly benefited from being closely located to the Rotterdam harbour, which provided the Netherlands a comparative advantage, due to the relatively cheap imports of animal feed stuffs like tapioca from overseas countries. The spatial separation of production of feed stuffs and animals is causing environmental problems in both areas with intensive animal husbandry, facing nutrient excesses, like most European countries (e.g. the before mentioned problems with respect to eutrophication, but also possible threats on ground water quality; cf.*Figure 32*) and in feed stuffs producing regions facing nutrient depletion (e.g. problems related to soil degradation, desertification, and erosion (Stoorvogel and Smaling, 1990). One of the ways to overcome these environmental problems is by closing nutrient cycles. This could take place at different levels; for example on a world wide scale by transporting excess amounts of manure from the excess areas to areas with shortages, for example from the Netherlands to India, as suggested by Prins (1995)[1]. However, closing cycles at this scale involves transports of manure over long distances. Some people argue that this is not sustainable and nutrient cycles should be closed at regional or farm level. This is done at farms applying "integrated farming" or "ecological farming" practices[2]. Eco-labeling can stimulate consumer demand for products produced through such practices (Meulenberg, 1996). Increased vegetarianism would also reduce demand for animal products and therefore contribute to reducing manure production. However, none of these possibilities are likely to be put in practice on a large scale in the short term, therefore, policy should, at least, be aimed at end-of-pipe type of measures. According to the Netherlands Scientific Council for Government Policy (1992), the same agricultural production can be realised using significantly less inputs and causing less environmental impacts. Their report "Ground for choices" has however been criticised for being too optimistic in the way application of technological development has been analysed; "..divergence between maximality and optimality" (Strijker, 1992).

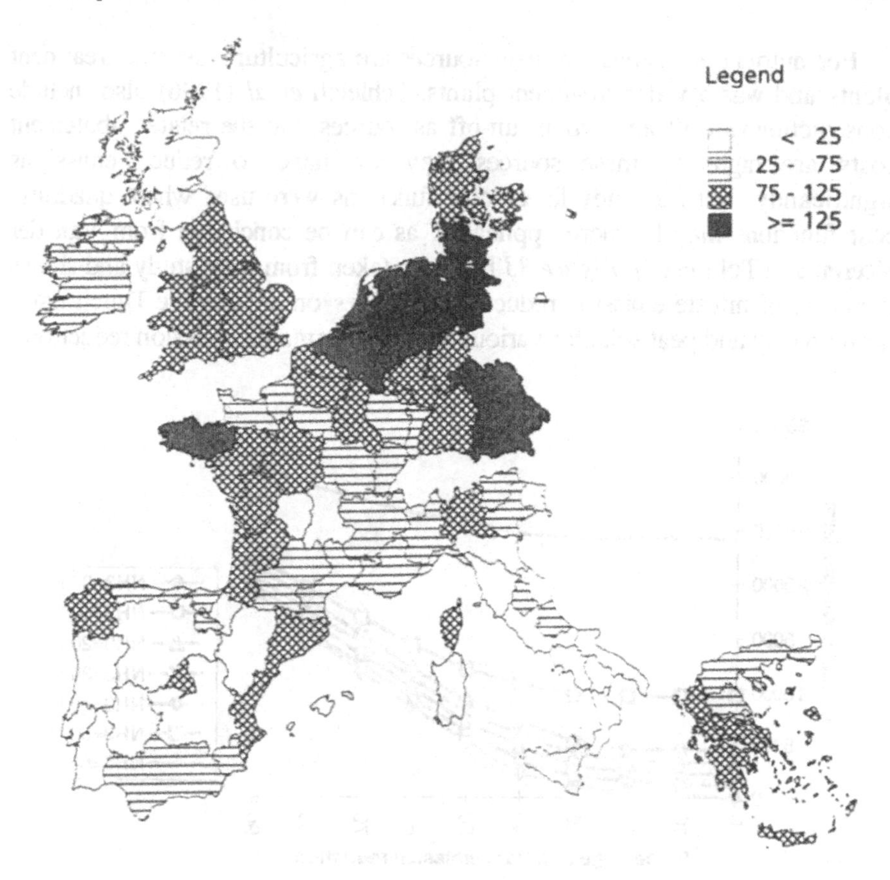

Figure 32. Net nitrogen surplus in the European Union by region in kg N per hectare in 1990/1991

Source: (Brouwer *et al.*, 1995)

3. REDUCING DISSIPATIVE NUTRIENT EMISSIONS IN RIVER BASINS

In economic studies on nutrient flows in watersheds, the focus is often on cost-effective reductions of loads to particular seas or lakes (cf. Van der Veeren, 1997; Schleich *et al.*, 1996; Gren, *et al*, 1995). These models combine relative abatement costs of sources and regions with transport or transmissions to sinks and impact areas so as to calculate (and equate) marginal costs.

Sources with the lowest relative abatement costs should reduce more than sources with relatively high costs, when achieving this reduction in loads at least costs is the objective.

For nutrient emissions the main sources are agriculture, sewage treatment plants and waste water treatment plants. Schleich *et al* (1996) also include construction run-off and urban run-off as sources, but the related abatement costs are high, so these sources may not have to reduce emissions significantly. In their study linear cost functions were used where quadratic cost functions may be more applicable, as can be concluded from Van der Veeren and Tol (1997). *Figure 33* has been taken from their study and shows the costs of nitrate emission reduction strategies on an average Dutch dairy farm on clay and peat soils for various levels of ammonia emission reduction.

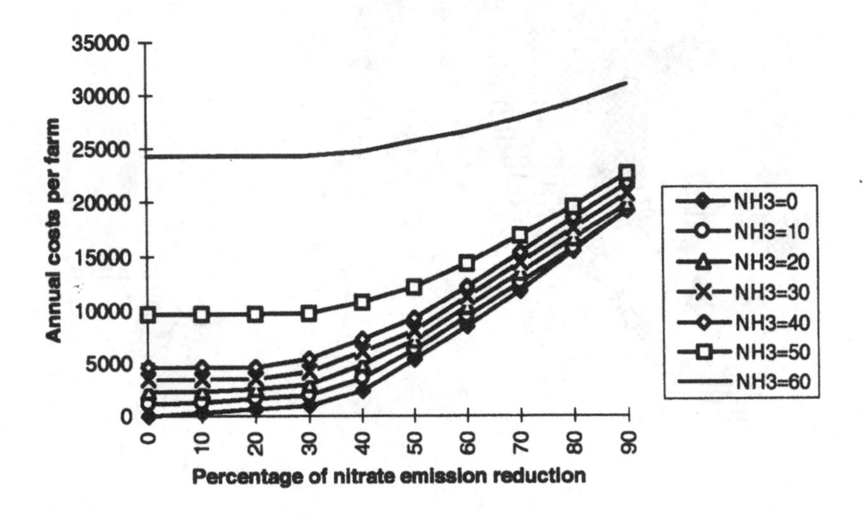

Figure 33. Costs of nitrate emission reduction strategies on an average Dutch dairy farm on clay and peat soils in DFl/year (Van der Veeren and Tol, 1997)

Van der Veeren and Tol (1997) use a linear programming model to determine from a large number of possible nutrient emission reduction measures on farm level the cheapest possible options to reduce emissions by various percentages. For the dairy farms on clay/peat soils, first nutrient application rates will be reduced, since this is the cheapest option, but when emission restrictions become increasingly tight, more expensive options need to be applied, such as changes in manure application techniques.

Van der Veeren (1997) uses quadratic cost functions for a number of agricultural activities (e.g. arable farming on clay and peat soils, arable farming on sandy soils, dairy farming on clay and peat soils, dairy farming on sandy soils, pig breeding farms, pig feeding farms) and for sewage treatment plants (including waste water treatment plants). His analysis, using preliminary data and simplifications, shows that differences between agricultural sectors may be very important when emission reduction targets are to be met at least costs. He finds that costs of a 50% reduction of nitrate loads to the North Sea can be brought down by half by a clever allocation of nitrate emission strategies, compared to the situation of a flat rate. *Table 26* has been taken from his study and shows percentages of nitrate emission reductions by the various agricultural activities in Switzerland when a cost-effective allocation is established in the Rhine river basin.

Table 26. Percentages nitrate emissions reduction for Switzerland when a cost-effective allocation of nitrate abatement strategies is applied in the Rhine river basin (Van der Veeren, 1997).

Source	Percentage nitrate emission reduction
arable clay	21
arable sand	69
dairy clay	5
dairy sandy	17
pig breeding	9
pig feeding	34
sewage treatment	16

From *Table 26*, it appears that agricultural activities on sandy soils should reduce their emissions more than their counterparts on clay/peat soils. This can be explained by the fact that nutrients get washed out more easily on sandy soils, and farmers will have to apply more nutrients for the same crop yield. This overapplication can be seen as a risk premium. It is this risk premium that will be reduced at low costs when nutrient emissions have to be reduced (Van der Veeren, 1997). Arable farms have to reduce their emissions to a larger extent than animal husbandry, because for the first type of agriculture, it is cheaper to shift from manure to fertiliser application, than it is for the latter to change manure production and handling.

Such a cost-effective allocation of nutrient abatement strategies can save up to US $250 million/ year, compared to a flat rate emission reduction policy (Van der Veeren, 1997). The latter policy is what the International Rhine Committee has advocated in its Rhine Action Programme (RIZA, 1990). This might have been done because it *sounds* fair, that everybody should reduce by the same percentage. However, the same objective can be achieved at lower costs, when emission reductions are allocated in a cost effective way. Such a cost effective allocation implies a more tailor-made emission reduction policy and is therefore more difficult to implement. The difference in costs between a flat rate emission reduction and a cost effective allocation with the same water quality, can be seen as an estimation of the costs for easy implementation of an emission reduction policy. The extra costs compared to a cost effective allocation puts the argument of fairness in another dimension. Not many people will agree that asking more money from society than necessary to achieve the same objective(s) *is* fair.

4. REDUCING DISSIPATIVE NUTRIENT EMISSIONS AT FARM LEVEL

From the previous section it is clear that differences in cost-effectiveness, between types of nutrient emitting activities, should be taken into account when dissipative emissions of nutrients have to be reduced.

For a proper analysis of nutrient flows, nutrient balances are used on soil level, animal level and farm level (Berentsen and Giesen, 1995). An example of a nitrogen balance on dairy farms is given in *Table 27.* From this nitrogen balance it appears that only about 15% of the total available nitrogen on an average dairy farm on sandy soils in the Dutch province of Gelderland in 1988 is found in final products. This percentage, the so-called nutrient efficiency, is low for The Netherlands with its intensive livestock farming, as compared to Spain or Poland, where agriculture is more extensive and nutrient efficiency for nitrogen is 26% and 54% respectively (Langeveld, 1993). The low efficiency rate for Dutch agriculture may be a reason for environmental organisations to encourage national governments to pose restrictions on the intensity of agricultural activities, because lower agricultural production means lower nutrient production per hectare, and therefore, if fertilisation rates remain the same, reduces nutrient surpluses per hectare. However, reducing the intensity of agricultural activities, might also reduce the production per farm. This means, that, in order to produce the same amount of agricultural output, more farms will be needed. Only when the relative number of farms increases less than the relative difference in nutrient efficiencies between extensive and intensive farming practices, can extensification be expected to reduce total nutrient emissions. Therefore, one should not only look at nutrient efficiency, but relate this to the realised production. For example, the nitrogen surplus per tonne of milk produced in the Netherlands is about 35 kilograms, as it is in Poland. However, in Spain, where the number of cows per hectare is about 14% of the number of cows per hectare in the Netherlands, the nitrogen surplus per ton of milk produced is about 190 kilogram (Langeveld, 1993).

Different measures can be taken to reduce dissipative emissions of nitrogen[3], including changes in animal diets, changing the level of nitrogen use, changing land use, stable adjustments and changing the number of animals on the farm. These measures are often analysed in agricultural economic models (Van der Veeren and Tol, 1997; Van Eck, 1995; Berentsen and Giesen, 1994; Leneman *et al.*, 1992).

According to those models, nitrogen emissions can be reduced significantly. However, only when very drastic measures are implemented, nitrogen emissions can be reduced to levels below which adverse effects on the environment are prevented (Van Eck, 1995).

Table 27. Nitrogen balance for an average dairy farm on sandy soils in the Dutch province of Gelderland in 1988 (Van der Veen *et al.*, 1993)[4]

Source of nitrogen bought	Amount (kg/year)	Source of nitrogen sold	Amount (kg/year)
Concentrates :		Dairy cattle/milk	1440
dairy cattle	1818	Intensive livestock	213
intensive lifestock	756	Manure	50
silage	374	Roughage in stock	238
other roughage	544	Total nitrogen in products	1941
fertiliser	5534		
deposition	888		
mineralisation	3171		
		denitrification	1015
		nitrate leaching	2577
		ammonia volatilisation	2158
		nitrogen storage in soils	5394
Total nitrogen bought	13085		

With respect to phosphates, the situation is even less hopeful. Since soils have a limited capacity to absorb phosphates, every kilogram applied but not taken up by crops will add to the saturation level of the soil and eventually result in increased phosphate loads to surface waters, where they, together with nitrates, stimulate eutrophication. According to Oenema and van Dijk (1994) the maximum excess amount of phosphate per hectare that is likely not to harm environment, is about 1 kilogram phosphate per hectare per year. Present excess amounts vary between 55 kilogram phosphate per hectare per year on Dutch dairy farms and 65 kilogram phosphate per hectare per year on Dutch arable farms (Oenema and van Dijk, 1994). On experimental farms, i.e. farms applying good agricultural practice, inevitable phosphate losses between 45-55 kilograms per hectare per year on Dutch dairy farms and around 70 kilogram phosphate per hectare per year on Dutch arable farms have been accomplished (Oenema and van Dijk, 1994)[5]. Experimental fields show what is technically feasible, and results on those fields show inevitable losses between 25 and 50 kilogram per hectare per year (Oenema and van

Dijk, 1994). This still is significantly more than the "environmental constraint" of 1 kilogram per hectare per year.

5. CONCLUDING STATEMENTS

Everybody should do whatever he or she is best equipped for. This is an economic principle already stated by Adam Smith in 1776 (Smith, 1976). He showed that specialisation of labour could increase productivity compared to a situation in which everybody is doing everything. The heavily concentrated and highly productive Dutch agriculture exploits its comparative advantage of being able to use cheap food stuffs, coming in from the Rotterdam harbour (Bieleman, 1996). The concentration of agricultural activities in certain parts of the Netherlands has lead to the development of and continuous improvements in the necessary infrastructure. When society wants this sector to transform from the present concentrated production to a more wide spread, less productive, ecological agriculture, and, at the same time, wants to maintain present production levels, this will increase the number of nutrient emitting farms, with per farm smaller inevitable emissions. This will enhance the diffuse source character of agricultural nutrient emissions. Smaller emissions taking place at a larger number of farms, will result in less pressure on the environment around farms, compared to intensive agricultural production practices. The environment might be able to cope with these emissions. However, as was described in the previous section, even small emissions can ultimately result in unwanted environmental pressures. When, on the other hand, high emissions take place at a limited number of locations, agricultural nutrient emissions will get a point source character. Agricultural point sources will than have characteristics similar to point sources in other industries in that they are generally addressed by technology-based approaches (Lanyon, 1994). Acceptable technologies can be readily identified and installed at the site. Thus, concentration of polluting activities enables direct abatement of emissions. This leaves opportunities for nature restoration projects in undisturbed regions outside the agricultural areas, which would enhance diversity in landscape. Whereas intensification might be preferred from the point of view of nutrient emissions management, it requires intensive pest management and raises questions related to the welfare of animals. These negative consequences should not be neglected, but are not the focus of this chapter. As described before, intensification could allow for nutrient recycling on a larger scale. This would partly resolve one of the most remarkable features with nutrient flows; the ever increasing nutrient depletion in the areas that supply the western European countries with animal feed stuffs, while at the same time, the latter regions suffer from problems related to nutrient excesses in soils and waters. A solution would be to transport manure surpluses to regions in need of this valuable resource. For this, a couple of problems will have to be tackled. First, the costs related to collection, shipping and handling might be high[6]. Therefore it is important to share the costs among all parties involved. This means not only agriculture, but also those sectors suffering from high nutrient loads in ground and surface waters, such as drinking water purification plants, fisheries and recreation. But also the farmers in the receiving regions, who benefit from decreased soil degradation. May be, helping poor countries to improve their production opportunities can also be subsidised by development assistance. Second, the receiving regions have to accept the manure not as a waste product, but as a valuable resource, which can help to restore their degraded soils. Both problems might be solved when parties are more informed.

The last section shows that not only nutrients, but also nutrient emissions are inevitable for agricultural production. Therefore, we can not have both; high levels of agricultural production and an undisturbed environment. The partly inevitable character of nutrient emissions urges society to make clear trade-offs between these two partly exclusive needs. However, when demand for reductions in environmental impacts results in monetary compensations to those who will have to reduce emissions, it will appear to be possible to manage at least part of those emissions. Therefore, the answer to the question raised in the title is both; dissipative emissions of nutrients are both manageable and inevitable.

ACKNOWLEDGEMENTS

The author wishes to thank Bert Lyklema, Mariet Hefting, Ben Kuiper, Paul Berentsen (Wageningen Agricultural University), Marlies Schuttelaar (Delft University of Technology), Richard Tol (Institute for Environmental Studies, Vrije Universiteit, Amsterdam) and all participants for their valuable contributions to the workshop "From farm level to river basins: The fate of nutrients", and Richard Tol, Hans Langeveld (Stichting Onderzoek Wereldvoedselvoorziening, Vrije Universiteit, Amsterdam), and Marlies Schuttelaar, (Delft University of Technology) Annemarie Willig for their valuable comments.

REFERENCES

Berentsen, P.B.M. and G.W.J. Giesen (1995). An environmental - economic model at farm level to analyse institutional and technical change in dairy farming, in *Agricultural Systems*, 49: 153-175.

Berentsen, P.B.M. and G.W.J. Giesen (1994). Economic and environmental consequences of different environmental policies to reduce N-losses on dairy farms, in *Netherlands Journal of Agricultural Science*, 42(1): 111-19.

Bieleman, J. (1996) Van traditionele naar technologische vruchtbaarheid en verder ..., in *Frisse kijk op mest; een vitale econologie voor het landelijk gebied*, Verslag studiedag, KLV, 7-18.

Bodeke, R. and P. Hagel (1994). *BEON-studie naar de effecten van de teruglopende nutriëntenblasting van de Nederlandse kustzone*, BEON rapport nr. 94-17, BEON project RIVO 93 E 03, RIVO rapport C016/94, RIVO-DLO.

Brouwer, F.M., F.E. Godeschalk, P.J.G.J. Hellegers, H.J. Kelholt (1995). *Mineral balances at farm level in the European Union*, Onderzoeksverslag 137, Agricultural Economics Research Institute (LEI-DLO).

Eck, G. van (ed.) (1995). *Stikstofverliezen en stikstofoverschotten in de Nederlandse landbouw, Project Verliesnormen, Deelrapport 3*, Ministerie van Landbouw, Natuurbeheer en Visserij.

Gren, I-M., K. Elofsson, P. Jannke (1995). *Costs of nutrient reductions to the Baltic Sea*, Beyer Discussion Paper Series No. 70, Beyer International Institute of Ecological Economics, The Royal Swedish Academy of Science, Stockholm, Sweden.

Langeveld, J.W.A. (1994). *Concepts in nutrient balance calculations for modelling applications; Examples for nitrogen and phosphorous in the Dutch dairy sector*, Staff Working Paper, WP-94-07, Centre for World Food Studies (Stichting Onderzoek Wereldvoedselvoorziening van de Vrije Universiteit), Amsterdam.

Langeveld, J.W.A. (1993). *Nutrient efficiency in dairy farming; Nitrogen, phosphorous and potassium use in the Netherlands, Spain and Poland*, Staff Working Paper, WP-93-05, Centre for World Food Studies (Stichting Onderzoek Wereldvoedselvoorziening van de Vrije Universiteit), Amsterdam.

Lanyon, L.E. (1995). Does nitrogen cycle?: Changes in the spatial dynamics of nitrogen with industrial nitrogen fixation. In *Journal of Production Agriculture*, Vol. 8, no. 1: 70-78.

Lanyon, L.E. (1994). Symposium: Dairy manure and waste management. Dairy manure and plant nutrient management issues affecting water quality and the dairy industry. In *Journal of Dairy Science*, 77:1999-2007.

Leneman, H., G.W.J. Giesen, P.B.M. Berentsen (1992). *Kosten van reduktie van stikstof- en fosforemissie op landbouwbedrijven*, Vakgroep Agrarische Bedrijfseconomie, Landbouwuniversiteit Wageningen.

Meulenberg, M.T.G. (1996). Consumentengedrag en mestproblematiek: van lijden naar leiden?, in *Frisse kijk op mest; een vitale econologie voor het landelijk gebied*, Verslag studiedag, KLV, 47-55.

Netherlands Scientific Council for Government Policy (1992). *Ground for choices; four perspectives for the rural areas in the European Community*, reports to the Government 42, SDU The Hague.

Oenema, O. and T.A. van Dijk (eds.) (1994), *Fosfaatverliezen en fosfaatoverschotten in de Nederlandse landbouw, Rapport van de technische projectgroep 'P-deskstudy'*. LNV, VROM, V&W, Landbuwschap and Centrale Landbouworganisaties, Den Haag.

Prins, (1995), pers. comm.

RIZA (1990). *Verwachte reductie van lozingen van prioritaire stoffen in Nederland tussen 1985 en 1995; Rijn Aktie Plan en Noordzee Aktie Plan (RAP/NAP)*, DBW/RIZA nota 90.067, Rijkswaterstaat Dienst Binnen Wateren/RIZA, Lelystad, november 1990.

Schleich, J., D. White, K. Stephenson (1996). Cost implications in achieving alternative water quality targets, in *Water Resources Research*, Vol. 32, No 9, 2879-2884.

Smith, A. (1976). *An inquiry into the Nature and Causes of the Wealth of Nations*, Vol.1 and 2, edited by R.H. Campbell and A.S. skinner, Clarendon Press, Oxford, 1976.

Stoorvogel, J.J. and E.M.A. Smaling (1990). *Assessment of soil nutrient depletion in sub-Saharan Africa 1983 - 2000*, Vol. 1 Main report, report nr. 28, DLO-Staring Centre, Wageningen.

Strijker, D. (1992). Grond voor keuzen - visie van een econoom; een paradigma ter discussie, *Spil* 5, 17-20.

Veen, M.Q. van der, H.F.M. Arts, J. Dijk, N. Middelkoop, C.S. van der Werf (1993). *Stofstromen in de Nederlandse landbouw, deel 1; nutrientenstromen op melkveebedrijven in Gelderland*, Landbouw-Economisch Instituut (LEI-DLO), Centrum voor Agrobiologisch Onderzoek (CABO-DLO), The Hague.

Veeren van der, R.H.J.M. (1997). *From farm level to river basins: Cost-effective strategies to reduce nutrient emissions to the North Sea*, Paper presented at the 2nd seminar on environmental and resource economics, University of Girona, department of economics, May 19-20.

Veeren van der, R.H.J.M. and R.S.J. Tol (1997). *Cost-effective nutrient emission reduction strategies: building the data set*, W97/2, Institute for Environmental Studies.

Verreijken, P., H. Kloen, R. Visser (1994). *Innovatieproject Ecologische Akkerbouw en Groenteteelt, Eerste voortgangsrapport*, rapport 28, Dienst Landbouwkundig Onderzoek, Instituut voor Agrobiologisch en Bodemvruchtbaarheidsonderzoek (AB-DLO), Wageningen.

---

[1] Other provocative ideas include manure disposal at sea. The rationale behind this idea is that adding nutrients to the sea might be beneficial, because reductions in phosphorous loads to the North Sea taking place especially since the early eighties, might have caused decreased algae development and declining fish populations. Thus, manure disposal at sea would help both agriculture and fisheries. However, nutrient cycles in the North Sea are not understood well enough to be able to predict the possible consequences of such measures (cf. Bodeke and Hagel (1994) describe conflicting impacts of phosphorous loads on the development of shrimps in different areas in the North Sea). In addition the public opinion does not want to use the North Sea to dispose agricultural waste.

[2] "Integrated" and "ecological" farming are two types of biological farming practices. Their main difference is in the degree to which environmentally harmful substances are allowed to be used in production; in integrated farming it should be as little as possible, whereas ecological farming does not allow for any use of environmentally harmful substances at all. This implies that in the latter case achievement of environmental objectives is guaranteed, while it is uncertain in the first. Integrated farming can be regarded as a transition from conventional to ecological farming practices (Vereijken, *et al.*, 1994).

[3] i.e. denitrification, nitrate leaching, ammonia volatilisation, and nitrogen storage.

[4] In most nitrogen balances, losses are not described in detail as has been done in this table, but simply generalised and called nitrogen loss, or at best divided into ammonia volatilisation and nitrate leaching.

[5] According to Oenema and van Dijk (1994) the inevitable phosphate losses on arable farms applying good agricultural practice has a high variance and a low explanatory power. This may be one of the reasons why this number is higher for those farms than found in practice.

[6] According to Prins (1995), both environmental pressure and costs related to manure transport can be low when shipped using oil tankers, which return empty from Rotterdam to oil exporting regions.

Chapter 14

Recycling of materials

Local or global?

PIETER VAN BEUKERING AND RANDALL CURLEE

Pieter Van Beukering is researcher, Institute for Environmental Studies, Vrije Universiteit, Amsterdam. Randall Curlee is head,Global Change Analysis Section, Oak Ridge National Laboratory, Tenessee.

Key words: International trade, recycling, waste, north-south

Abstract: The development of the international trade in secondary materials is discussed in this chapter. Trade in recovered materials is growing rapidly, in particular flows from developed to developing countries. The economic and environmental costs and benefits of this trade are assessed. It is concluded that the information base on which to make these evaluations is poor, but that comparative advantages, properly exploited, could lead to economically and environmentally beneficial global markets for recyclables which would supplement primary materials markets.

1. INTRODUCTION

Recycling is generally considered an important strategy for alleviating the pressures of society on the environment. Natural resources can be saved, emissions can be decreased, and the burden of solid waste can be reduced. Likewise, recycling in the cases of some materials is an important economic activity that creates employment and attracts investments.

The term "recycling" has two dimensions—recovery and utilisation. Recovery refers to the diversion and collection of waste materials from consumers and producers. Utilisation refers to the processing of diverted waste into new and useful materials and products. In recent years the industrialised countries of the North have observed significant increases in the quantity of waste recovered and utilised. These trends have resulted from higher disposal costs, increased public concern about the health and environmental impacts of waste disposal, and a general perception that recycling can result in resource conservation. In many countries of the North, policies have been adopted to encourage or mandate the recovery of waste materials. To a lesser extend, policies have also been adopted to mandate the utilisation of wastes—for example, only modest steps have been taken with regard to mandating recycled material content in selected products and government procurement practices that favour recycled materials.

Another trend is the increasing trade of secondary materials between the North and the South. Waste materials recovered in the North increasingly are being exported to the South for utilisation. As a result, the North has developed into a net supplier of recyclable waste while the South has developed into a net importer. As is the case with any commodity, international trade of secondary materials allows countries with different comparative advantages to exercise those advantages to bring about a more efficient allocation of resources. In the absence of market failures, international trade in secondary materials produces gains in both the North and the South. However, when market failures occur — such as when health and environmental aspects are externalised — international trade may lead to an increase rather than a decrease in total environmental damage. Further, international trade in secondary materials may lead to development patterns in the South that are in contrast to the preferences of both South and North.

The increased trade of secondary materials between the North and the South raises the question of whether recycling in the South is different from recycling in the North, and whether international trade in secondary materials has positive economic, environmental and social impacts. Not only are these issues relevant for national policymakers who must decide about legislation concerning this type of trade; these issues are also important in relation to the negotiations on the Transboundary Movements of Hazardous Wastes (Basel Convention) and those within the World Trade Organisation (WTO).

This chapter highlights the specific factors that drive recycling in both regions and addresses the various impacts of international trade of recyclable waste. Section 2 presents empirical evidence to illustrate a number of global trends in recycling and trade. Section 3 discusses the impacts of these developments for OECD and non-OECD countries. Finally, conclusions are drawn with respect to potential social and economic costs and benefits. In addition, suggestions are made for future research.

2. EMPIRICAL FINDINGS

Empirical evidence on recycling on a global scale is scant. Often data on recycling and international trade of secondary materials are only provided on national levels. Still, based on the scarce information available, several tendencies in global recycling can be identified and are presented below. It should be realised that these specific examples are no justification for generalising these tendencies across all secondary flows. Generalisation would require additional empirical analysis. Special emphasis is placed on differences between industrialised and developing countries.

2.1 Recycling is increasing on a global scale

The value of material residues is increasingly recognised by entrepreneurs and municipal organisations. As a result, the recycling rates for various materials have grown rapidly on a global scale. Several examples support this trend. The global average recovery rate of waste paper increased from 29 to 40 percent between 1973 and 1991 (Beukering and Duraiappah 1996). The global recycling rate of aluminium increased from 20 to 30 percent between 1972 and 1988 while glass recycling in Europe increased rapidly from 20 to 39 percent between 1981 and 1989 (UNEP 1991). Recycling of plastics waste, a relatively new process, has increased by 22 percent between 1993 and 1994 (APME 1996).

Forces driving recycling differ between the North and the South. In industrialised countries, initial incentives were provided primarily by national governments. Regulation, the development of recovery schemes and government procurement schemes for recycled products have created the necessary conditions for increased recycling. Although developing countries are in the process of developing policies and regulations for recycling, alternatives to waste disposal have not yet received much attention. Recycling in the South is generally a purely market-driven practice.

2.2 International trade of secondary materials is growing asymmetrically

The ratio between internationally recovered and traded secondary materials is increasing gradually. This implies that relatively more secondary materials are exported for utilisation in foreign markets. For example, in 1975, only 5.8 percent of globally recovered waste paper crossed borders. This rate increased to 14.2 percent in 1994. The trade in waste plastic increased from 1.5 million tonnes in 1993 to 2.4 million tonnes in 1994. Cheaper freight transport and overall trade liberalisation may have been factors which contributed to this development (Anderson at al. 1992). The overall specialisation of the recovery sector and the recycling industry created supply and demand structures for a larger range of secondary materials, which can only be matched through international trade.

A comparison between secondary and primary material movements shows how these trade flows are structurally different (see *Table 28*). While industrialised countries are the major importers of primary flows, secondary flows generally find their way to developing countries. The North is characterised as a net-exporter of secondary materials, while the South is characterised as a net-importer. This is not a static situation. In the period 1990-1994, imports of various secondary materials to the South grew much faster, almost 10 times as fast as those to the North[1].

Table 28. Direction of primary and secondary materials (1991-1992; tonnage in percentages)

from exporter to importer	Secondary plastic	Primary plastic	Secondary metal	Primary metal	Secondary paper	Primary paper
Export North	78	76	90	75	91	89
Export South	22	24	10	25	9	11
Import North	31	64	51	65	51	85
Import South	69	36	49	35	49	15
North-North	27	53	46	51	51	78
North-South	51	23	44	24	40	11
South-North	4	11	5	14	0	7
South-South	18	13	5	11	9	4

Source: Compiled from TRAINS, UNCTAD 1994

This typical trade pattern raises the question whether developing countries have a comparative advantage in the utilisation of secondary materials, as compared to the production with virgin materials. A number of factors may, support the hypothesis that the South does have a comparative advantage. Compared to primary production, many of the less sophisticated recycling processes have a higher labour intensity (mainly caused by the required manual sorting before processing). Many current recycling methods use relatively simple and less capital-intensive technologies. Since financial, physical, and technological capital are relatively scarce in the developing countries, while there is an abundance of relatively cheap labour, the South may have a comparative advantage in the utilisation of secondary materials. The same can be said for energy and other raw materials. Recycling processes generally consume less energy for production. Since most developing countries are energy and materials importers (which requires the use of their limited hard currency) recycling is a comparatively attractive option for production in the South.

In view of the data limitations and uncertainties, great caution should be taken in interpreting these empirical results. However, it may be concluded that the mutual interdependencies between consumption and production patterns of trading partners' in North and South have increased. Particularly in developing countries, this increase has been more prominent for secondary material flows than primary material flows.

3. IMPACT OF TRADE ON RECYCLING

A better understanding of trends in the international trade of secondary materials is important for the firms involved in such trades. However, it is more important to study (1) the health and environmental impacts of these international trades, (2) the impacts of these trades in promoting or hindering the development of countries in the South, and (3) implications of these trades for technology development in both the North and the South. It is not likely that a unique answer to these questions can be provided. The outcome will differ from material to material, from country to country, and, given asymmetric technological development in both secondary and primary production processes, will also not be constant in time. Nevertheless, various positive and negative impacts of recycling and trade in secondary materials can be identified, and the pros and cons of alternative policies in both the North and South can be evaluated.

As stated earlier, international trade in secondary materials can take gain of comparative advantages in both the North and the South. Comparative advantages may be derived from differences in wage rates, the availability of physical and financial capital, differences in the skills of the labour force, and variations in allocations of natural resources. By utilizing these comparative advantages, the total quantity of materials recycled in the North and the South may increase. Several cases, not only between the North and the South but also between Southern countries have been reported in the literature which support this positive relationship between trade and recycling:

Namibia does not posses proper infrastructure to recycle wastes safely and cleanly. It is currently not feasible to set up recycling plants in Namibia due to the small amounts of recyclable materials available, and because of the lack of water. The existence of trade channels is an appropriate incentive to recover and store secondary materials until the quantity is sufficient to export it to neighbouring countries, such as South Africa where a market for recyclates exists. Recycling of glass, cans and used oil waste from Namibia takes place in South Africa (Kohrs 1996).

It has been estimated that the waste recycling industry in Colombia provides employment for 1 to 2 percent of the labour force. International trade in recyclable materials takes advantage of the existing differences in technical capabilities and the need for raw materials. For many years, Colombia has imported scrap iron from the Netherlands to serve as an important input in the steel industry (Pacheco 1992).

About 90% of the waste paper collected in Hong Kong is exported. Unlike waste paper, all aluminium cans are exported since no company producing aluminium cans operates in Hong Kong (Yeung & Ness 1993). If the export possibility did not exist, this waste would have to be landfilled or incinerated. "In Phnom Penh, garbage collection trucks are rarely seen but the city's mostly-female waste pickers are a common site. These pickers go from door to door to pick-up reusable or recyclable items. The materials are then sold to middlemen who export most of the items to Vietnam and Thailand" (Lapid 1997).

Several waste materials are collected by itinerant waste buyers and waste pickers in Kathmandu. After sorting and cleaning, these materials are exported to the neighbouring country, India, where these materials are recycled. Again it is doubtful whether these recovery activities would be performed without the demand from the Indian recycling industry (Beukering and Badrinath, 1995).

The cases described above appear to show two distinct patterns of recovery and recycling: the first, conducted by poor urban dwellers, mainly as a way of supplementing very low incomes and leading to re-use in traditional economic activities; the second, more industrialised resulting in large-scale industrial reuse of materials, and sometimes being linked to the global trade in waste. In the cases where the overall level of recycling increases both in the North and the South, and the increased recycling activities substitute for virgin production processes, the positive impacts are likely to outweigh the negative effects. International trade of secondary materials may promote economic growth in both the North and the South while it may also result in fewer health and environmental damages, and in the conservation of total resources. *Table 29* shows the potential gains of increased recycling.

Table 29. Reduction in environmental impact by substituting virgin for secondary production

	Potential Savings (percent)			
	Aluminium	Steel	Paper	Glass
Energy use	90-97	47-74	23-74	4-23
Air pollution	95	85	74	20
Water pollution	97	76	35	-
Mining wastes	-	97	-	80
Water use	-	40	58	50

Source: Reprinted with permission from Bartone, 1990

On the other hand, international trade of recyclables may lead to unacceptable outcomes due to market distortions, political and cultural differences, and institutional barriers. For example, since the distinction between hazardous waste and secondary materials is not always easy to make, developing countries run the risk of importing unwanted materials which cause additional burdens for the waste management sector. A similar problem, or concern, may arise if countries in the North and countries in the South have different standards of processing secondary waste. While one environmental standard may be acceptable to the South (i.e., its people, firms, institutions, and government) because of the existing economic and institutional bottlenecks, that same standard may not be acceptable to the North. Because of a potentially high degree of contamination, working with waste can be an unhealthy activity which may increase morbidity. Is it acceptable to have higher morbidity in the South than in the North? Who should decide what are acceptable differences in the environmental performance of recycling activities?

Further, international trade in secondary materials may lead to development in the South that is in contrast to the preferences of both the South and the North. Relatively cheap imported materials may damage local markets for recyclable waste. In Europe this effect was clearly demonstrated when the recovery of waste paper in Germany increased rapidly in the late-1980s as a result of the introduction of new laws which forced producers to take back their packaging materials. Since the capacity of European recyclates markets at that time was insufficient to absorb this growing supply, a large amount was exported to neighbouring countries. The price of waste paper even reached negative levels, which in turn almost caused the collapse of the waste paper recovery sector in these countries. Subsidies by for example the Dutch government prevented the recoverers from bankruptcy. In most cases, governments in developing countries will not be in the position to act as a buffer to these external shocks. Therefore, some waste experts claim that policies to promote re-use, recycling and minimisation of waste generation should include measures to protect the local recycling market against the importation of cheap waste materials from the industrialised countries (Klundert 1997).

Other closely related problems may also arise. The "leakage" of the export of recovered materials may reduce the incentive to set up recycling facilities domestically. Trade in secondary materials may have a significant impact on technology development in both the South and the North. In the South, the availability of secondary materials may lead the developing countries to invest in low-tech recycling processes as an alternative to investments in high-tech virgin material production. This may place the developing country on a different development path. Technologies in the North may also be impacted. For example, restrictions on the export of plastic waste may lead the developed countries to develop and refine new and sophisticated chemical recycling methods as substitutes for the low-tech secondary processes now available. If chemical recycling can be made economically viable, these chemical processes may be applicable to a much larger range of plastic wastes, and, as a result, the total quantity of plastics recovered and utilised in the North could increase dramatically. The availability of international markets may mean there is little incentive to develop these more sophisticated and robust technologies.

These potential positive and negative effects indicate the complexity of evaluating the impacts of the international trade of secondary materials and potential impediments to that trade. Various unknown parameters play a role in this respect. Crucial are the substitution effects between secondary versus primary materials on the one hand, and local versus internationally traded secondary materials on the other hand. Moreover, different degrees of social and environmental externalities accrue with both secondary and primary production which may differ between the North and the South and which need to be incorporated in decisions on trade in recyclable waste. Life cycle analysis (LCA) provides a useful tool to capture these financial and external effects of trade of secondary materials.

An issue which becomes particularly important for implementation of the findings of these analyses is the present condition of national and international markets to trade secondary materials. Do barriers exits that may prevent the optimal allocation of resources, and what are the potential impacts of those distortions ?

Market distortions may include the following. First, as the international market of secondary materials is not as well established as the primary materials market, information constraints may exist. Excessive uncertainties about supplies and prices may create unfavourable trade conditions for recyclates. Second, national governments may intervene in international transactions by creating trade barriers through the establishment of import and export tariffs or quotas on secondary materials, partly as a way of supporting the development of national recycling infrastructures. Export controls may yield short-term economic gains, but generally only at significant long-term costs. In addition, government intervention on the national level, such as subsidies for recovery schemes, may create artificial markets for the recovery and utilisation of secondary materials. The continuation of these markets depends on the continuation of government support, which leads to increased risk for firms that invest in infrastructures to provide recycling. Third, market distortions may arise from increased market concentration. For many recyclable materials a trend towards oligopolisation or even monopolisation may take place. Such concentrations in market power may have undesirable economic effects on the recycling sector.

4. FURTHER RESEARCH NEEDS

Further research is needed to address these fundamental issues. The following specific research topics are identified as being most important:

- an empirical analysis of international flows of secondary and virgin materials between and within OECD and non-OECD countries, and a cross-country analysis of the driving forces of these material flows;
- the development of research tools for the analysis of financial and external costs and benefits related to recycling processes in both the North and the South, such as a combined methodology of life cycle analysis and economic valuation (see Chapter 9);
- the impact of trade in secondary materials on the technological progress in the secondary and primary industries of the importing countries, and vice versa;
- identification and assessment of market distortions that may be present in national and international markets for secondary materials;
- the political and institutional incentives and barriers to the international trade in secondary materials; and
- the design of policy instruments that can be implemented to counter market distortions that may lead to unacceptable conditions for countries in the North and/or the South.

5. CONCLUSIONS

For quite different reasons, recycling of secondary materials has become important for countries in both the North and the South. International trade in secondary materials allows both regions to exercise their comparative advantages; and in the absence of market distortions, international trade will lead to higher recycling rates, possible improvements in health and environmental conditions, and opportunities for economic growth in both regions. However, international trade of secondary materials occurs within markets that are distorted, and as a result there are no guarantees that the theoretical benefits of international trade will be realised. On the contrary, the overall health and environmental conditions may be eroded; development in countries of the South may be hampered, and innovation that may lead to improved and lower-cost recycling technologies may be hindered.

ACKNOWLEDGEMENTS

Table 29 has been reprinted from Resource Conservation and Recycing, 4, C. Bartone, Economic and policy issues in resource recovery from municipal wastes, pp. 7-23, 1990, with kind permission from Elsevier Science - NL, Sara Burgerhartstraat 25, 1055 KV Amsterdam, The Netherlands.

REFERENCES

Andersson, T., Folke, C. and Nyström, S. (1995). Trading with the environment; ecology, economics, institutions and policy. Earthscan Publications Ltd., London.

APME (1996). Information system on plastic waste management in Western Europe: European Overview 1994 data. Association of Plastics Manufacturers in Europe. Montrouge.

Bartone, C. (1990). Economic and policy issues in resource recovery from municipal wastes, Resources Conservation and Recycling 4, 7-23.

Beukering, P.J.H. van, and Badrinath, G.D. (1995). Recycling in Nepal: The Indian Connection. Warner Bulletin. No. 46, p. 6-7.

Beukering P.J.H. van, and A. Duraiappah (1996).The Conomic and Environmental Impact of Waste Paper Trade and Recycling in India: a Material Balance Approach. CREED Working Paper No. 10. International Institute for Economic Development. London. p. 1-26.

Beukering P.J.H. van, and A. Duraiappah (1997). International Trade and Recycling in Developing Countries: Waste Paper in India. Warmer Bulletin. No. 52, p. 8-9.

Klundert (1997). Policy aspects of urban waste management. Issue Paper for the Programme Policy Meeting Urban Waste Expertise Programme. p. 1-8.

Kohrs, B. (1996). Waste management in Namibia. Warmer Bulletin. No. 50, p. 10-11.

Lapid, D. (1997). Southeast Asia regional consultation report: summary of findings. Center for Advanced Philippine Studies. p. 1-4.

Pacheco, M. (1992). Recycling in Bogotá: developing a culture for urban sustainability. Environment and Urbanization. Vol. 4, No. 2. p. 74-79.

UNEP (1991). Environmental Data Report 1991/1992. Third Edition. United National Environmental Programme. Oxford. p. 336-359.

Yeung & Ness (1993). Three recycling industries in Hong Kong: market structures, vulnerabilities and environmental benefits. Asian Journal of Environmental Management. Vol. 1, No. 1, p. 51-59.

---

[1] This figure represents the average growth rate of imports of metal slag, tyres, paper and non-ferrous metals, based on trade statistics from the UNCTAD.

Chapter 15

Philips Sound and Vision

Environment and strategic product planning

JACQUELINE CRAMER[1]
Professor in Environmental Management, Tilburg University, Tilburg.

Key words: **Strategic environmental product planning, life cycle analysis of future products.**

Abstract: **This article examines the experiences gained within Philips Sound & Vision with the development and implementation of strategic environmental product planning. For this purpose the company developed a methodology called 'STRETCH' (Selection of sTRategic EnvironmenTal CHallenges). From initial experiences with the application of STRETCH within Philips it was learned that environmental objectives can be attuned very well to the business strategy. In order to assess the environmental benefits gained through the application of STRETCH, it is recommended to elaborate methodologies that can describe the complex product systems of the future. The scope of the present LCA approach and related methodologies is not sufficient for this purpose.**

1. INTRODUCTION

In recent years experience with the use of life cycle analysis (LCA) within industry has been gained. Most efforts have focused on assessments of the environmental effects of *present* products throughout the whole life cycle. On the basis of these assessments, proposals have been formulated on how to improve the environmental performance of products. Most attention has been paid to incremental changes in existing products within a time scale of about one to three years. Incremental improvements provide significant progress in the early stages by capitalising on 'low hanging fruit' (the easy improvements). After that first period, incremental changes become less profitable in terms of both economic and ecological efficiency. Then more far-reaching environmental improvements begin to deliver a higher reduction in environmental impact at relatively lower costs (Arthur D. Little, 1996). These latter improvements usually require more fundamental, strategic choices.

Contrary to incremental improvements, relatively little experience has been gained with the implementation of such more far-reaching product improvements. Within Philips Sound & Vision growing attention is being paid to the latter type of improvements under the heading of 'strategic environmental product planning'. This endeavour clearly differs from the incremental approach. First of all, the objective is not to focus on products currently produced by the company, but on those to be sold in three to fifteen years. Moreover, the orientation is not so much the environmental issues currently at stake, but the major environmental bottlenecks of tomorrow. As a result, the way in which life cycle analysis is used in such activities, also differs.

This article discusses the experiences gained within Philips Sound & Vision with the development and implementation of strategic environmental product planning. In doing so, attention is paid to the role of life cycle analysis.

2. THE 'STRETCH' METHODOLOGY

In order to be able to implement strategic environmental product planning within Philips Sound & Vision a methodology is needed. This methodology should address the following questions: What opportunities and threats does the environmental issue present for a company such as Philips, particularly for Sound & Vision? What technological options are available for dealing as adequately as possible with environmental problems? And finally, the most crucial question: Which environmental opportunities should be selected to enhance the business and improve the environmental performance of its products?

To cope with these questions a methodology called 'STRETCH' (**S**election of s**TR**ategic Environmen**T**al **CH**allenges) has been designed and tested within Philips Sound & Vision (Cramer, 1997). To underpin the methodology, data is needed on the key drivers that will determine future business strategy in general. For instance, in the case of Philips Sound & Vision the collection of data consisted of information about economic factors (i.e. future market perspectives of the consumer electronics sector in general and of the company itself) and the technological innovations to be expected. Moreover, some information was needed about cultural trends and the possible set of environmental issues at stake in the future. On the basis of this information, a limited number of plausible scenarios related to possible future product market strategies can be formulated. These scenarios are used to help prioritise, select and finally implement the most promising environmental challenges to be adopted by the company. The 'STRETCH' methodology consists of five activities (see *Table 30*).

Table 30. Steps in the STRETCH methodology

Steps	Explanation
1	the identification of the crucial driving forces that will influence the business strategy in general;
2	the design of a limited number of plausible scenarios that the company can adopt on the basis of step 1, leading to a list of potential product market strategies;
3	the specification of potential environmental opportunities and threats for each scenario on the basis of a checklist of environmental design options;
4	the selection of environmental challenges per product leading to a substantial improvement of its environmental performance; and
5	the implementation of the environmental challenges ultimately selected.

To illustrate the plan sketched above, a description of how the methodology has been applied within Philips Sound & Vision is given below:

2.1 Step 1: Identification of crucial driving factors

Data collection about the market perspectives of consumer electronics products and their technological perspectives is no easy job. Due to rapid multimedia development, the consumer electronics products' business environment is changing quickly. The sector is moving from analog to digital signalling and to an integration of modes (such as text, sound, and visuals) that used to be completely separate. Moreover, a variety of once separated businesses are now starting to converge and compete against each other. This holds in particular for the following sectors: communications, entertainment and information and business/consumer electronics. Finally, the multimedia development gives end consumers the opportunity to be less passive than before and have more control over the choice of selection and/or interactivity.

It is very hard to predict the future trends in multimedia. Roughly speaking, four aspects can be distinguished in the realisation of multimedia (Den Hertog, 1994):

- hardware: the development and production of multimedia technology or multimedia hardware (including standardised software tools);
- software: the development and the provision of multimedia services and applications;
- distribution: the transportation or distribution of multimedia services and applications to the users; and
- application/use: the real use of multi media applications and services. This can be professional use within or between organisations, and consumer use.

2.2 Step 2: Design of plausible scenarios

Today the producers of multimedia hardware (especially the computer hardware and consumer electronics companies) are active in new growth markets due to stagnating turnovers and price erosion in their traditional markets. At the same time, the development and distribution sectors are growing. The application/use of multimedia seems to be the greatest bottleneck in the takeoff of multimedia on a large scale. All kinds of questions are still unanswered. For instance, what the future home will look like in relation to multimedia developments and how multimedia will penetrate professional circles.

On the basis of a literature study and interviews the following three plausible scenarios can be formulated for thinking about future developments within the consumer electronics sector:

Scenario 1: The consumer electronics companies continue to focus on making the hardware;

Scenario 2: The consumer electronics companies shift their emphasis towards the development of software/multimedia; and

Scenario 3: The consumer electronics companies concentrate on providing services.

Philips Sound & Vision has recently chosen a particular mix of the three scenarios as a starting point for its strategic planning for the coming 5-7 years. On the basis of this particular scenario, concrete product market strategies have been formulated.

Information on cultural trends has been provided by Philips Corporate Design (PCD), a department that works closely with trend labs to spot future cultural trends. Experience has shown that people can no longer be so easily classed as having a particular lifestyle. The consumer combines various lifestyles. However, according to these trend lab studies, positive drivers for environmental awareness are in particular: 1. time; 2. quality instead of quantity; 3. health; 4. growing consciousness of waste and how a product is made; and 5. homeliness. The trend seems to be that the consumer will become more critical, ethical, spiritual, emancipated, demanding and creative. In order to gain a more detailed indication of the influence of cultural trends, focused marketing research at the level of individual products is necessary.

As explained above, one particular scenario had already been chosen by Philips Sound & Vision and elaborated in detail to distinguish between activities which are already mainstream for Philips, new to Philips and new to the outside world. For each of these three categories a list of products has already been defined which represents the particular category of activities. This list of products formed a good starting point to analyse the environmental opportunities and threats for the coming 5 - 7 years with the support of various Philips' experts. The most interesting in this respect were products in the categories: new to Philips and new to the outside world.

2.3 Step 3: Specification of potential environmental opportunities

After collecting and integrating available data in steps 1 and 2, the Environmental Competence Centre (ECC) of Philips Sound & Vision/Business Electronics identified a number of promising environmental opportunities (step 3). The particular environmental issues which will be headline news in the coming 5 years, or even beyond that, cannot be predicted with great precision. The Environmental Competence Centre (ECC) of Philips Sound & Vision has therefore developed *a general checklist of environmental product design options* that serves as a guideline for prioritisation (see *Table 31*).

Table 31. Checklist of environmental design options

Classification	Items
Minimisation of production impact	Minimisation of waste, emissions and energy use
	Respect for biodiversity
Minimisation of product impact	Reduction of toxic substances
	Minimisation of materials consumption (e.g. through miniaturisation, weight reduction; systems integration)
	Minimisation of use of non-renewable resources
	Minimisation of fossil energy consumption (e.g. through energy efficiency and durable energy use)
Efficient distribution and logistics	Produce where you consume
	Direct distribution to consumer
Intensity of use	Lease vs sell, Collective use
Durability of products	Reuse, Technical upgrading, Longer lifetime
	Reparability, Refurbishing, Ageing with quality
Recyclability of materials	Reduction of materials diversity
	Materials cascading
	Design for disassembly
	Selected, safe disposal

This checklist has been compiled on the basis of various sources (e.g. Business Council for Sustainable Development, 1993; De Boer, 1995; Fussler and James, 1996). The list of environmental design options has served as a tool to assess the environmental challenges at stake when a company implements the product market strategies formulated in step 2. To prioritise these potential challenges, the ECC organised two sequential meetings. The meetings were held at division level with various key persons, namely representatives of strategy development within Sound & Vision, representatives of Philips Corporate Design and environmental experts from Sound & Vision. This group of people then formulated a number of criteria to guide the process of prioritisation. These criteria were:

1. environmental improvements should provide a business opportunity or competitive advantage;
2. projects should have clear environmental relevance;
3. environmental improvements should preferably be quantifiable;
4. environmental problems directly related to health and safety issues require more attention; and
5. implementation should not be hampered because of difficulties in co-operation with third parties or because of lack of expertise within the company.

With the help of the criteria mentioned above, the expert group made an initial selection of nine promising projects. These projects were related to the items mentioned in *Table 31* (e.g. energy reduction, material reduction, recycling and enhancing the durability of products). Based on expert judgement of the brainstorming team it was assessed that these projects would lead to environmental improvements. At this stage the potential environmental improvements could not be specified yet.

2.4 Step 4: Selection of environmental challenges per product

In step 4 the preliminary choices of the most promising environmental design options made at the level of Sound & Vision were discussed in brainstorming sessions with the various Business Groups (BG's). One of the first companies to try structuring such a brainstorming approach is Dow (Fussler and James, 1996). The way in which this company designed the brainstorming process has been an inspiring example in developing the methodology within Philips Sound & Vision and is now being more widely used within other Philips business units.

In principle, each BG could choose two ways to proceed: a quick or a more time-consuming brainstorming approach. For both brainstorming approaches at BG level key data were collected in advance about the product and its market and technological perspectives. Depending on the issue at stake, specific representatives from the relevant organisation were asked to attend the brainstorming session. These were marketing people, product managers and technical people. However, other stakeholders, key suppliers or customers were also involved in specific issues.

The quickest, usually preferred brainstorming approach was to review the environmental design options initially selected at Sound & Vision level and pick out those design options that are most relevant for that particular BG. In this case focused brainstorming sessions were organised with relevant specialists to elaborate each individual environmental design option in great detail.

Each participant in a brainstorming session had to generate ideas about environmental improvements that could lead to eco-efficiencies in the order of magnitude of a factor of 4 (this means a reduction of the burden on the environment to one quarter of the current levels; see Chapter 4 for details). This high eco-efficiency target was set as a guiding principle to elicit creativity, and not as a fixed goal to be reached. As a starting point, data was provided about the environmental performance of the present product with respect to the particular environmental design option (e.g. the present energy consumption during use or the amount and type of material used).

Then the brainstorming team picked the most promising ideas for each design option in terms of both environmental and economic benefits. No detailed life cycle assessment or life cycle cost study was made at this stage. The selection was based on a qualitative expert judgement of those participating in the brainstorming session. The same holds for the assessment of the economic value of the ideas. In fact, a similar qualitative assessment of the business value and competitive advantage was adopted as described in Fussler and James (1996). The ideas ultimately selected have been translated into R&D and/or concrete product development plans. It is only after results become available from the development of products in the coming years that concrete life cycle analyses and/or life cycle cost studies can be performed to assess the environmental and economic benefits in a quantified way.

The second, more time-consuming approach, was to organise an intensive brainstorming session for each product. In this case the initial selection of specific promising design options at Sound & Vision level was not taken as a starting point for further elaboration but just as an initial input for the brainstorming session at product level. In advance, the environmental performance of the present product was assessed on the basis of eco-indicator studies (PRé/Duijf Consultancy, 1994). During brainstorming sessions, which took two days, all the environmental design options mentioned in *Table 31* were taken into account one after the other. The result was a well-grounded list of creative ideas to enhance the business through specific environmental challenges. Here, again, the increase in eco-efficiency could only be evaluated qualitatively on the basis of the expert judgement of the brainstorming team. In order to set priorities across different categories of design options, we used the Eco-compass tool, developed by Fussler and his colleagues (Fussler with James, 1996; see Chapter 18). Following further investigation of the most promising ideas, more precise quantitative assessments can be made in the future on the basis of life cycle analysis and a life cycle cost study.

The elaboration of the various environmental challenges was tailor-made to each BG and each priority item. To illustrate this point the following three examples will be given:

1. the reduction in the energy intensity of Consumer Electronics products;
2. the reduction of the material intensity of Consumer Electronics products; and
3. the development of potential strategies to enhance the durability of products.

With respect to the item 'reduction in the energy intensity', an intensive brainstorming session was held in the BGs TV, Audio and VCR in order to generate and select more far-reaching environmental improvements in the energy consumption during use and standby. As improvements could be made in various parts of the product (e.g. in the components or in the printed circuit board), experts from various backgrounds were present at these workshops. The ideas that these experts proposed are being elaborated in a technical, economic and marketing sense.

Secondly, the 'reduction of the material intensity of Consumer Electronics products' was elaborated in a specific tailor-made way. In order to generate ideas for the reduction of material intensity, close co-operation was established between Philips and one of its main suppliers of materials. Various brainstorming sessions were held to identify promising alternative materials that are lighter, but at the same time have the appropriate functionality for fulfilling the demands on the product. The results of these brainstorming sessions are currently being elaborated in R&D projects.

The project related to 'the development of potential strategies to enhance the durability of products' was elaborated in a slightly different way. First, a summary of the potential options for optimising the life of products was made on the basis of a literature survey. Next, the capability of Philips Sound & Vision in meeting these options as a way to achieve further optimisation of the life of its products was assessed. At this stage it was found important to gauge the view of the outside world on this matter.

To this end, the Environmental Competence Centre organised a brainstorming session with external stakeholders in the Netherlands which was attended by 15 representatives from environmental, consumer and women's groups, from the Dutch Ministry of Housing, Spatial Planning and the Environment and the Dutch Ministry of Economic Affairs, from relevant research institutes and from Philips. The participants at this session were asked which five (not more) activities they thought Philips Sound & Vision should give the highest priority in the context of the theme of 'optimising product life'.

The reactions of the participants suggested a clear prioritisation (Cramer, 1996). Particular attention was given to the following topics: (a) making more robust constructions, (b) designing modular constructions, and (c) selling the use of products/product leasing. These results were presented in brainstorming sessions with the BGs Audio and VCR. Establishing which additional methods stand a good chance of success in the future of Philips Sound & Vision is currently part of further internal consultation and investigation.

Initial results show that products usually break down due to thermal problems (too high temperature) or defective components or joints. Only after more information has been gathered on the various advantages and disadvantages of improving the durability of the products will Philips take concrete action. A PhD student from Delft University of Technology was appointed to the ECC/S&V in 1996 to elaborate concrete designs related to the durability issue.

The three examples clearly show that it usually takes a number of brainstorming sessions and specific R&D initiatives before a final assessment is made of the most promising environmental opportunities to be implemented. Through these sessions and specific projects, learning experiences are built up that are used to reduce the present uncertainties about environmental opportunities and market perspectives. When the company has learned more about these more far-reaching environmental improvements, it becomes easier to integrate these endeavours into the regular product development process.

2.5 Step 5: Implementation of the environmental challenges ultimately selected

The promising environmental challenges are presently being elaborated in particular R&D or product development projects. Before final decisions can be made, data on market perspectives and consumer trends should be collected. If necessary, results of specific consumer tests assessing the interest in the new product should also be available. Finally, a more precise assessment can be made of the environmental and economic benefits gained on the basis of a life cycle analysis and life cycle cost study. Because the further elaboration of promising environmental challenges takes time, the final implementation of the results has not taken place yet.

3. CONCLUSIONS

Until now no structured methodology is in use for attuning environmental considerations to the business strategy of companies. The Philips Sound & Vision Environmental Competence Centre has developed *a methodology* for this purpose. This methodology is called 'STRETCH' (**S**election of s**TR**ategic

EnvironmenTal CHallenges). The objective of STRETCH is to incorporate environmental considerations into the business strategy and to select strategic environmental challenges in an early phase of business development.

The application of STRETCH provides the possibility of meeting *three main objectives.* First, focusing on long-term environmental product design strategies can elicit innovations that may enhance the competitive position of the company. Through the integration of eco-efficiency goals into product innovation in general, a company does not aim to beat the competitors purely on environmental grounds, but on its innovative product strategy in general. In this way economy and ecology can go hand in hand. By taking environmental aspects into account at an early stage of product development, more far-reaching improvements can be made in future consumer electronics products compared with the current range of products. Strategic environmental management, like those taken by Philips Sound & Vision, are still more the exception than the rule (IVA, 1995). This approach, however, could provide a way forward to substantial improvements in eco-efficiency.

Second, the environmental opportunities and threats to be expected in the future can be anticipated in an earlier phase. Through this early warning system an attempt can be made to diminish the negative consequences in an early stage and a response is not required when it is actually too late. In this way actions are more pro-active rather than defensive. The company can even be one step ahead of all kinds of government demands and public pressure by redirecting product development in the context of sustainability in a more fundamental way. By pro-actively integrating environmental aspects into the earlier phases of the product creation process, external criticism can be avoided and the lead taken in environmental priority setting.

Third, as a result of more far-reaching environmental improvements even higher eco-efficiencies are expected to be reached than through incremental improvements. At this stage, the exact data on eco-efficiency gains to be realised within the strategic projects currently being carried out within Philips Sound & Vision cannot yet be provided. These will be collected during the execution of the projects. It is, however, questionable whether the present life cycle analysis (LCA) approach and related methodologies can be used for this purpose. It requires an environmental assessment of the complex product systems of the future. The existing methodologies focus on environmental assessments of present products and their related emissions in specific countries or regions. The performance of a LCA of the present product is only a small part of a much broader assessment related to the strategy and planning of future products.

On the basis of the STRETCH methodology, Philips Sound & Vision has prioritised several projects for further investigation. This prioritisation has been based on expert judgement without much support from analytical tools.

In order to improve the quality of such judgements, the development of additional analytical tools is needed. Moreover, until now little experience has been gained with life cycle assessments of *future* instead of *present* products. The research challenge for the coming years therefore is to devote more attention to methodologies that can describe the complex product systems of the future.

REFERENCES

Arthur D. Little (1996). *Sustainable Industrial Development; Sharing Responsibilities in a Competitive World,* Conference Paper, Arthur D. Little, The Hague.

Boer, M. de, *Milieu* (1995). *Ruimte en Wonen (Environment, Space and Housing),* Ministry of Housing, Physical Planning and Environment, The Hague.

Business Council for Sustainable Development (BCSD), (1993). *Getting Eco-efficient,* Report of the First Antwerp Eco-Efficiency Workshop, BCSD November 1993, Geneva.

Cramer, J. (1996). Pros and Cons of Optimising the Life of Consumer Electronics Products, in: *Proceedings of the First International Working Seminar on Reuse,* organised by the Eindhoven University of Technology, Eindhoven, November, p. 73-84.

Cramer, J. (1997). *Development and Implementation of STRETCH (Selection of sTRategic EnvironmenTal CHallenges), External Report,* Environmental Competence Centre, Philips Sound & Vision/Business Electronics, Eindhoven, The Netherlands, January.

Fussler, C. with P. James (1996). *Driving Eco-innovation; A Breakthrough Discipline for Innovation and Sustainability,* Pitman Publishing, London.

Hertog, P. den (1994). *Multimedia als Innovatie: Fake or Future? (Multimedia as Innovation: Fake or Future?),* STB-TNO, nr. 94/002, Apeldoorn.

IVA (Royal Swedish Academy of Engineering Sciences) (1995). *Environmental Management; From Regulatory Demands to Strategic Business Opportunities,* Stockholm, Sweden.

PRé/Duijf Consultancy (1994). *Project Eco-indicator,* PRé/Duijf Consultancy, Vught, November.

---

[1] The author worked at Philips Consumer Electronics from April 1995-February 1997, in secondment from the Centre for Technology and Policy Studies (TNO). Since then she works at AKZO Nobel.

Chapter 16

Unilever

Chain Management in the Food Industry

CHRIS DUTILH
Environmental Manager, Unilever Nederland, The Netherlands

Key words: eco-design, food products, LCA

Abstract: Unilever has been involved in environmental studies for its food products for eight years. During that period life cycle assessment (LCA) has proven to be a handy tool. A number of useful observations have been made in attempting to use the outcome of such studies for environmental product improvement. These include i) in product development trade-offs between environmental improvements and other product characteristics need to be made; ii) functional units cannot always be unambiguously defined; iii) it is advantageous when LCA-studies are conducted by the industries involved; iv) sometimes there is only a limited choice for alternatives; and v) direct interactions between customer and producer is most effective in product improvement processes.

1. INTRODUCTION

Unilever, as a major producer of food products, has stated in its environmental policy that: "Unilever is committed to meeting the needs of customers and consumers in an environmentally sound and sustainable manner, through continuous improvement in environmental performance in all our activities" (Unilever, 1996). This paper will provide a perspective for that policy-statement, as it describes experiences from eight years of life cycle studies for food products. Within Unilever there are also other centres of LCA-expertise, particularly in Unilever Research (UR), where studies are being conducted for detergents and personal products (UR Port Sunlight, UK) and on packaging (UR Vlaardingen, NL).

2. THE FOOD PROCESS TREE

Like for all products, the food process tree starts with the extraction of raw materials from the eco-system, and it ends with emissions into the eco-system during waste processing. Food raw materials are generated in the agricultural stage where biomass is created from CO_2 and sunlight, with the input of nutrients and water. Therefore the process tree for food

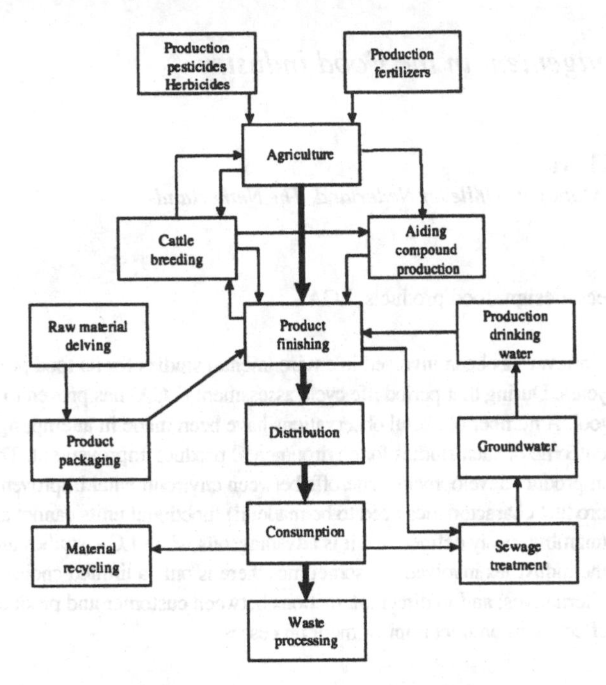

Figure 34. Process Tree for food products

products (*Figure 34*) usually starts with the production of chemical inputs for agriculture from primary raw materials and finishes with the provision of nutrition followed by waste processing steps. All individual processing stages consume energy and lead to emissions of waste.

Each operation in the process tree can be positioned in the scheme depicted in *Figure 35*. This scheme distinguishes between the environmental impact generated within the firm, and all the rest which are generated outside the firm. For instance, the production of fertilisers may be located towards the left of the x-axis, while 'product finishing' may be located further to the right of this axis. Improving the internally-generated impacts usually leads directly to cost reduction ("pollution prevention pays"). Improving the impact elsewhere usually only succeeds if there is a clear win-win situation for all parties involved. The scope for firms to influence externally-generated impacts may be very limited.

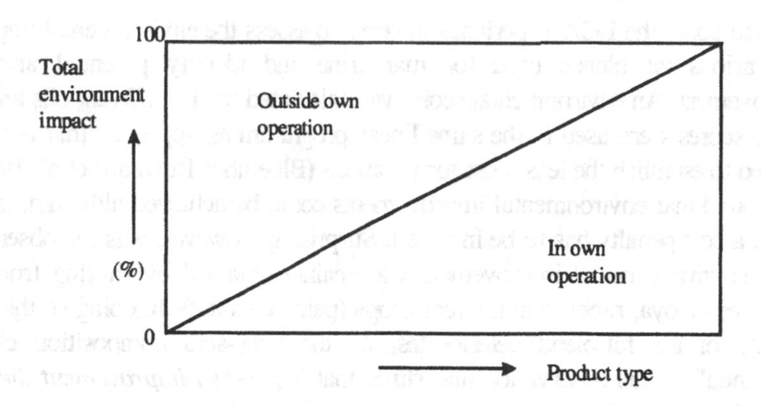

Figure 35. Split of direct and indirect environmental impact for each operation involved in the creation of food

3. UNILEVER'S EARLY LCA-EXPERIENCE FOR FOOD

Unilever recognised at an early stage that Life Cycle Assessment (LCA) is an useful tool to explore the scope for continuous improvement. That is particularly the case as Unilever not only manufactures the products it sells, but it also determines the specifications and selects the suppliers.

In 1989 the first comparative LCA-study was conducted for two different margarines. It took about one year to collect sufficient data to make a sensible comparison. At that time the evaluation was made on economic grounds, using the DESC-model which had been developed by TME at the request of Unilever (Krozer, 1990, 1992; see also Chapter 8). It was very rewarding to discover that data could be found for all production stages in the process tree. At the same time, however, it became clear that the value of such data should not be overestimated. The environmental impact caused by the production of sunflower seeds, for instance, varies a lot between countries, and even within a country between individual farmers. That raised the question: *what are the relevant data to use in LCA-studies?* This first exercise revealed the need for a simple manual on LCA, which could be used to inform suppliers on the background of requests for specific information. Such a manual has been produced with the help of the Centre for Environmental Science (CML) at Leiden University (Van den Berg, 1994). In that manual the experiences of the margarine study illustrate the LCA process.

In 1990 a software tool for conducting LCA-studies was developed in conjunction with TME (PIA, 1993). This software, called PIA (product improvement analysis), has been used since then for all food studies in Unilever, and has proved to be a very flexible and powerful LCA tool. In 1991 an effort was made to apply the LCA-experience in order to assess the environmental impact of the various fat blends used for margarine and identify potential areas for improvement. An environmental score was calculated for 15 different oils and fats. These scores were used in the same linear programming approach that is usually applied to establish the least cost formulations (Bloemhof-Ruwaard et al. 1995). It was found that environmental improvements could be achieved although, in most cases, a cost penalty has to be incurred. Surprising, however, was the observation that the environmental improvement was mainly obtained by shifting from one-year crops (soya, rape) to multi-year crops (palm, coconut). In doing so the health aspects of the fat-blend deteriorates, as the fatty-acid composition changes substantially. This observation underlines that *in product improvement there are more than just environmental considerations.*

4. FUNCTIONAL UNIT

The above observation emphasises the need to accurately specify the function for which an environmental improvement is being sought. It also becomes clear that *functional units cannot always be unambiguously defined*, as is usually suggested in the LCA-literature. *Figure 36* shows that a functional unit is composed of two dimensions, viz. a technical dimension (e.g. nutritional value) and an emotional dimension (e.g. taste, pleasure).

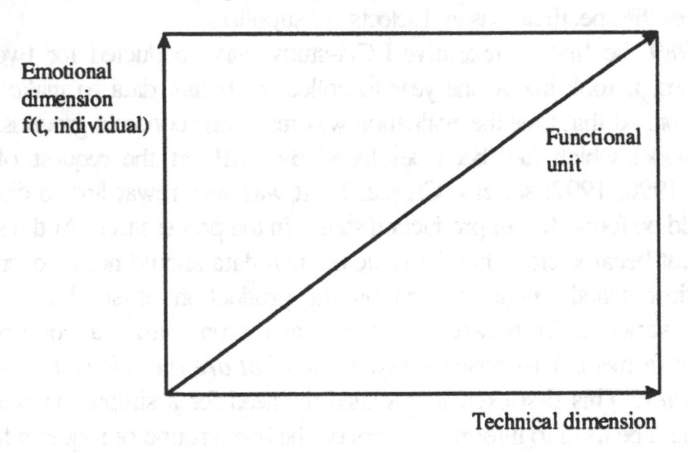

Figure 36. Schematic representation of the composition of a functional unit

The first dimension can be objectively measured, and compared with other products. The emotional dimension, however, is highly subjective and time dependant. The resulting functional unit, just like a vector, has both a specific length and a direction. For instance an ice cream consumed on impulse will have a large emotional dimension and a (very) small technical dimension. For astronaut-food the opposite will be true. It is almost impossible to compare products with greatly different emotional dimensions, particularly when the differences are also dependant on the time of use. Hence, if there is a substantial contribution from the emotional dimension to the functional unit, *it may be more relevant to define the range of products competing for the same function, than formulate the functional unit itself.*

5. WHO SHOULD CONDUCT LCA-STUDIES

There are great advantages if LCA-studies are conducted by the industries involved in the life-cycle of a product. That was demonstrated in 1992/3 when Unilever participated in comparative studies between one-way and multiple-way packaging systems. Those studies were conducted in the Netherlands as part of a political decision-making process within the frame of the Packaging Covenant. The key issue was which packaging material and design would lead to optimised recovery and recycling. The analysis covered not only the environmental impact, but also the market economic consequences of a change-over from one system to the other. The studies were politically sensitive, and therefore received a lot of attention from NGO's, politicians and scientists (SVM, 1994). Under these circumstances there were important benefits to conducting LCA-studies in-house, since this led to (a) good understanding of all processes involved, and therefore simpler identification of the most critical steps; (b) rapid access to data required. There were fewer problems regarding secrecy arrangements because of existing commercial relationships; and (c) all commercial and technical expertise which is required for conducting such studies being available within the same organisation. Therefore the studies could be carried out simultaneously, which greatly improved the quality of the outcome for both.

The transparent nature of the LCA-approach enabled third-party experts to conduct a peer-review. In this instance, however, the peer-reviewers not only reviewed the study itself, but they made additional calculations for new scenario's. This exercise clearly demonstrated the quality of the study, as it provided further transparency into the technical analysis, and allowed others to develop their own conclusions based on commonly-derived results. However, it also raised the question: *how far should a peer review go?*

6. LCA AS GUIDANCE FOR IMPROVEMENT

In 1994 a screening LCA-study which compared the environmental impacts of several products, including oil seeds and dairy raw materials, from different sources was conducted. Substantial differences were observed, but at the same time it became clear that *sometimes there is only a limited choice of alternatives.* The main reason is that for several major raw materials (e.g. soybean oil) there is a world-wide commodity market. In such cases it is almost impossible to select specific producers/suppliers unless one is prepared to accept substantial cost consequences. That clearly reduces the scope for improvement and highlights the fact that *an impact on the environmental profile of a product is most effective in a direct interaction between producer and customer.*

In 1995 the food-related environmental impact in households was investigated. That study revealed that about 5% of the total energy and water consumption and of the solid waste generation in the Netherlands can be directly attributed to food consumption in households (Dutilh et al., 1996). That impact is only partly related to the nature of food products. Two other factors also play a role, namely the type of domestic equipment used (e.g. microwave or conventional cooking) and, significantly, consumer behaviour. The last factor has the largest scope for improvement, but leaves the question for a company like Unilever: *how far does the influence of a food producer reach?*

In 1996, Unilever published it's first corporate environmental report. The company is determined to report it's overall environmental performance, coupled with the additional aim of attempting to assess the sustainability of a business such as Unilever's. In that process the question of how to estimate the total environmental impact related to Unilever's global operations arose. A new approach, applying LCA methodology, was developed called Overall Business Impact Assessment (OBIA) (Taylor, 1996), which estimates the overall impact of Unilever's business operations and compares them to global targets for various environmental themes. The methodology and it's application have been presented recently (Anderson, 1997), but the question remains: *are these the most relevant insights to support the improvement process?*

7. TOOLS FOR ENVIRONMENTAL STUDIES

In fact, the evaluation of environmental profiles may require different approaches, depending on the type of questions that need to be answered. Together with researchers from three Dutch institutes (IVAM-ER Interfaculty Department of Environmental Science - Environmental Research at the University of Amsterdam, CLM - Centre for Agriculture and Environment, Utrecht and DLO Agricultural and Economic Research Institute in The Hague) a scheme has been developed in which the various elements of environmental analysis, as described in this paper, are summarised (*Table 32*). In this scheme the information generated with one tool is used as an input for the conversion process on the following line.

Table 32. Interdependence between tools and information flows in environmental studies

Actor/role	Questions	Information input	Tool	Information output
producer	--	process description	data selection tool	environment related process information
LCA-technician	impact per process	environment related process information	LCA- approach	environmental profiles
environmental manager	how to set priorities	environmental profiles	evaluation tool	risks and opportunities
general manager	which direction	risks and opportunities		

It is recognised that the LCA-approach is not the panacea for all environmental questions, originating from different actors and requiring special responses. Therefore separate tools are distinguished. A "data-selection tool" is required to convert process data from different origins and nature into standardised environmentally related process information. That information can be used as input for an "LCA-approach", which generates environmental profiles for products, processes or indeed entire operations. A separate "evaluation tool" is required in order to translate those profiles into answers for strategic decision taking. That evaluation tool may contain elements of normalisation (either on geographical or on activity basis), cost efficiency and prioritisation or weighing.

The scope and application potential of the various tools is currently the subject of a study in which Unilever is involved together with twelve other major industries in the food-chain in the Netherlands. Unilever took the initiative for that study, which should lead to a common understanding on how to improve sustainability in the food-industry. The first reports on the outcome of this study will be published in the second half of 1997.

8. WHAT HAPPENED IN UNILEVER

Supply chain thinking is one of the basic elements in the Unilever culture. Hence, the environmental understanding generated over the last eight years has not led to dramatic changes. Sometimes the studies have confirmed earlier assumptions, sometimes new insights were obtained. In any case, environmental conditions have become of growing importance in supply contracts. In case of commodity markets Unilever makes an effort to raise general standards. A good example of that is the establishment of the Marine Stewardship Council in 1996 which should lead to sustainable fishing practices.

Within Unilever environmental tools are used to help the company in achieving its policy objectives. At present LCA- and OBIA-type of studies are the most appropriate as they provide a useful metrics on which to benchmark and judge improvements. Absolute outcomes of LCA studies are considered to be less relevant than their deviation over on time, space or parties involved.

REFERENCES

Anderson J.I.W. (1997). *Sustainability - Unilever's Approach.* World Conference on Environmental Challenges in Oilseeds Processing, Surfactants and Detergents and Oleochemicals. Organised by AOCS, Brussels, March.

Bloemhof-Ruwaard, J.M., H.G. Koudijs and J.C. Vis (1995). Environmental impact of fat blends, *Environmental and Resource Economics*, Vol 6, 371-387.

Dutilh, C., S. Velthuizen, H. Blonk (1996). Milieubelasting bij gebruik van voedingsmiddelen; besparingen mogelijk bij consument, *Voeding*, Vol. 57, no. 10, 12-15.

Krozer, J. (1990). Integraal ketenbeheer vertaald naar bedrijfssituatie, *PT Procestechniek* 1990, no.1, 27-30.

Krozer, J. (1992). Decision Model for Environmental Strategies of Corporations, in L. Preisner (ed.) *Conf. proceedings of 3rd annual meeting of European Ass. of Environmental and Resource Economists*, held in Krakau, 16-16 June 1992, 53-71.

SVM (1994). *Environmental care for our packaging, from cradle to grave*, Report issued by Stichting Verpakking en Milieu, Den Haag, The Netherlands.

Taylor , A.P. (1996). *Overall Business Impact Assessment (OBIA).* 4th Symposium for Case Studies, SETAC-Europe.

Unilever (1996). *Environmental Report*, Unilever Corporate Relations Dept., London.

Van den Berg, N.W., C.E. Dutilh, G. Huppes (1995). *Beginning LCA; a guide to environmental Life Cycle Assessment*, Report from National Reuse of Waste Research Programme NOH, no. 9453, ISBN 90-5191-088-6, Leiden, The Netherlands.

Chapter 17

AT&T

Materials, Industry, and Industrial Ecology

BRADEN R. ALLENBY
Vice-President, Environment, Health and Safety, AT & T, Warren, New Jersey[1]

Key words: industrial ecology, design for the environment, responsible material use

Abstract: This chapter states that the complexity of industrial development is such that human beings have only incomplete knowledge about appropriate material use and industrial processes. This makes it difficult to even ask the right questions about how the goal of sustainable development is to be achieved. This ignorance does not imply that existing research tools for industrial transformation should not be further developed. The author recommends that responsible material use is a key step towards industrial transformation.

1. INTRODUCTION: IGNORANCE IS PROFOUND

The question of environmentally appropriate material selection and use is a complex one. It is likely that we are not yet even asking the right questions, much less informed enough to make the right choices. This grave ignorance on our part, however, must not be seen as an excuse to halt progress on methodologies such as Design for Environment (DFE) and Life Cycle Assessment (LCA) which, as we practice, will begin to provide the increasing sophistication necessary if we are to hope to move towards a sustainable global economy. This effort is one in which every economic actor - the private firm, the consumer, non-governmental organizations (NGOs), and the government at all levels - has an important role to play.

An intellectually rigorous approach demands that we try to understand the rapidly shifting context of current environmental activity. Although environmental awareness - and associated levels of regulation, expenditure, and public concern - have been rising around the world for thirty years, environmental issues are generally still treated as overhead, not strategic, for consumers, firms, and society as a whole. Environmental issues are considered only after people and firms have done what they want to do, rather than integrated into all activities from the beginning. The new focus on "sustainability" and "sustainable development" is, in part, an implicit recognition of the limitations of the overhead approach (*Table 33*).

Table 33. Technology and Environment chararacteristics of principle approaches.

Primary activity	Remediation	Compliance	Industrial ecology/Design for Environment
Time frame	Past	Present/past focus	Present/future focus
Focus of activity	Individual site, media or substance	Individual site, media or substance	Materials, products, services and operations over life cycle
Endpoint	Reduce local anthropocentric risk	Reduce local anthropocentric risk	Global sustainability
Relation of environment to economic activity	Overhead	Overhead	Strategic and integral
Underlying conceptual model	Command-and-control intervention in simple systems	Command-and-control intervention in simple systems	Guided evolution of complex system
Disciplinary approach	Toxicology and environmental science; reductionist	Toxicology and environmental science; reductionist	Physical sciences; biological sciences; social sciences; law and economics; technology and engineering; highly integrative

As currently practiced, however, it is still inadequate, principally because of the failure to understand the many and fundamental scientific and technological issues which for the most part have yet to be recognized, much less addressed. Our ignorance is profound.

The developing field of industrial ecology (IE) is an attempt to begin to establish the scientific and technological base for progress towards an environmentally and economically efficient, sustainable global economy. IE may be thought of as the objective, multidisciplinary study of industrial and economic systems, and their linkages with fundamental natural systems - oversimplifying somewhat, as "the science of sustainability." The concept requires that human (artifactual) systems be viewed not in isolation from surrounding natural systems, but in concert with them. It is a comprehensive, systems-based approach which, among other things, seeks to optimize the total materials cycle involved in the provision of any goods or services, from virgin material, to finished material, to product, to obsolete product, to horizontal or vertical recycling of products and materials, to ultimate disposal.

Environmentally responsible material choice and selection is linked to several important concepts. These include the idea of a "functionality economy," where consumers purchase functionality rather than physical artifacts, and the responsibility for environmentally appropriate management of products and their constituent materials remains with industry. It is today certainly possible, for example, to identify opportunities to provide equal or greater functionality to consumers while using less energy and material per unit function. We may define the provision of equal units of function (or, more broadly, quality of life) using less material as "dematerialization," a process that over time offers an obvious path to achieve greater environmental and economic efficiency. Note that both "dematerialization" and "environmental and economic efficiency," unlike "sustainability," are terms that can, in fact, be quantified at various levels of economic activity, and thus can be integrated into design, manufacturing, service development and provision, and operating activities. Such strict definition, and, if possible, quantification are a key factor in encouraging environmentally responsible material choice and use by firms.

While dematerialization is frequently applied to consumer end products, it is obviously equally applicable to the production process - eliminating unnecessary cleaning steps in manufacturing electronics products, for example, which reduces solvent consumption - and in non-manufacturing fields such as agriculture. Reduction of pesticide use per unit crop, for example, dematerializes food production. It is also worth recognizing that the similar concept of providing equal units of energy while releasing less carbon as a result of fossil fuel combustion, commonly referred to as "decarbonization," is an important sub-class of dematerialization. Although we will not deal specifically with energy further here, the general concepts of dematerialization can and should be applied to energy systems as well. It is important to note that dematerialization, in itself, is not the primary goal. Rather, the primary goal is to achieve a state of "sustainable material use" as part of a sustainable economy. Dematerialization, material substitution (both by environmentally preferable materials and by services of various kinds), materials cycling systems, identification and use of environmentally preferable materials throughout the economy - all have a role to play.

2. SUSTAINABILITY: GLOBAL OR LOCAL

A major problem with many discussions of sustainability is that they confuse very different levels of activity. Thus, for example, a process or product designer may be asked to design a "sustainable process" or a "sustainable product", and a firm or municipality may call itself "sustainable." Because sustainability as a property only emerges at a global level, such claims are, of course, technically improper, although they may be valuable as a general indication of a benign attitude towards the environment, and as shorthand for "an environmentally and economically efficient global economy and culture, which can be maintained for a relatively long time". Because achievement of sustainable material use, and dematerialization, is in large part accomplished through scientific understanding and technological evolution, however, a slightly more structured framework is appropriate. *Figure 37* presents such a framework.

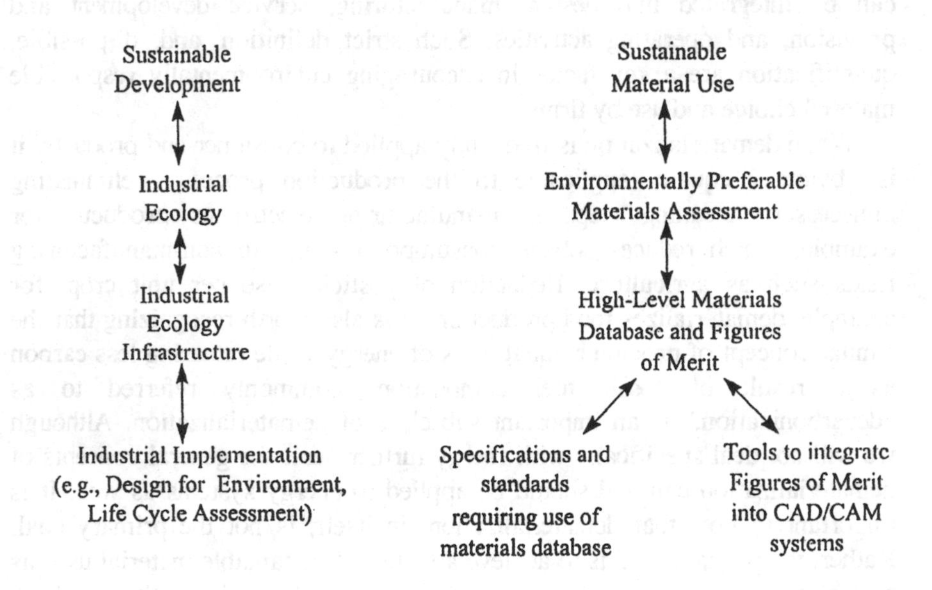

Figure 37. DFE Materials Example

"Sustainable development," defined by the Brundtland Commission that originated the term as "development that meets the needs of the present without compromising the ability of future generations to meet their own needs," is a worthy vision, but inherently ambiguous. For this reason, and because it may well be an emergent characteristic of a properly self-organized complex system (the global economy), and thus not definable on lesser scales, it is impossible to operationalize. Standing alone, it cannot provide guidance for either technology development or policy formulation. "Industrial ecology," defined above, is the field of study intended to provide the scientific and economic basis for the achievement of greatly increased environmental and economic efficiency: it is worth noting that it is an objective field of study, not a form of industrial policy or planning system. The "industrial ecology infrastructure" may be thought of as the set of policies, actions and activities which society and its components should undertake to support the implementation of industrial ecology. "Design for Environment" (DFE) occurs at the level of product, process, service and function design and implementation, and represents the integration of environmental considerations into the entire fabric of economic activity. At this stage, by the way, many environmental considerations appear not as separate issues, but simply as aspects of "good design:" energy efficiency in portable electronic goods, for example, is a design objective not just for "environmental" reasons, but because, by extending battery charge life, it makes a more desirable product for consumers. Environment succeeds best when it is invisible.

Consider material selection and use within this framework. At the level of "sustainable development," it is difficult to evaluate environmentally responsible material use and dematerialization, except for general principles. It is, for example, unlikely that current levels of material consumption extended to a global population of some 8 to 10 billion would be in any sense stable over time, and the concomitant implied energy expenditure, even if realistically achievable, would be highly problematic with any combination of proven technologies. At this level, dematerialization implies a decoupling of material acquisition from perceived quality of life, a cultural change of which there is no sign. Accordingly, we beg the question somewhat by noting that, at this level, society should attain a level of "sustainable material use."

And what is sustainable material use? It is at this point, in response to this query, that the vast and fundamental nature of our ignorance must become apparent. Not only do we not know, but we have not done the science, developed the methodologies, or even asked the right questions to be in a position to know. What materials are environmentally preferable? What substitutions are desirable? When does recycling of materials make sense, and when does it not, and what are the critical determinants? It is at the level of industrial ecology that we must begin to research such problems, and attempt to understand what materials are preferable in what applications.

Obviously, a systematic study of many dimensions of materials is necessary to provide such an understanding. Moreover, these data must become knowledge, and this knowledge transferred to those agents who make choices of materials as part of their economic activities, if the environmental efficiency of the global economy as a whole is to improve. Supporting the development of this capability is part of building the industrial ecology infrastructure. For example, it would be useful to build easily accessible databases, and perhaps generate figures of merit, which can inject environmental dimensions of materials directly into design software. Data - which in many cases are unreviewed, sporadic, or simply unavailable anyway - are not enough.

And then there is the final level, the level of actual design. Here the generic information on materials is fed into the specific systems which support the activity, be it designing a car, maintaining fleet networks for package delivery service, or growing corn. This is the province of the private firm which, in most developed economies, is the center of technological evolution and certainly the level at which material choice occurs in most cases. At this level, the designer should not be aware of any "environmental markers" at all; rather, she or he simply uses the tools available to select the most preferable material for the job. The environmental information is built into the system.

Despite the laudable efforts of the life cycle assessment community, and substantial toxicological data which have accumulated over centuries, it remains the case that our ignorance in this area is virtually complete. There are several reasons for this, but the most important is that our study of materials and their impacts has occurred within the context of reductionist science, and our ability to understand the systemic effects of different materials in different applications is primitive at best. Indeed, even now many life cycle assessments consider only one material in one application, and the substitution and systemic effects are frequently overlooked.

Thus, for example, substituting bismuth solders for lead solders in electronics applications would appear to be an obvious choice, given the (probable) lower toxicity of the former. A systematic comparison of the two options, however - and life cycle assessments should always be done in light of the full suite of options - indicates that, as a result of scarcity and low concentrations in ores, bismuth has greater environmental impacts during the mining and initial processing stages. Ironically, in fact, most bismuth is produced as a byproduct of lead mining, leading to the possibility that higher demand for bismuth might actually trigger more, not less, processing of lead ores. Even beyond that, however, bismuth solder in electronics applications is problematic because of the impacts on related technologies: such solders may require more aggressive fluxes to clean the parts to be soldered, which may, in turn, require chlorinated solvents rather than water-based ones. Moreover, bismuth may "poison" copper recovered from recycled electronics items, creating a need to use virgin, rather than recycled, copper. And changing these technologies, in turn, may change technological systems in other parts of the manufacturing or material management processes, or require product redesign. The point is not that such changes cannot, and, where appropriate, should not be made. Rather, it is that they must be thought about in systems terms, and individual data points, such as inherent toxicity, or even limited life cycle assessments, are seldom adequate to support reasoned decision making.

In an important sense, materials are not "things;" they are elements of a system and pattern, and certainly must be understood in that guise if the goal is sustainable materials use.

3. IMPLEMENTING RESPONSIBLE MATERIAL USE

Despite the nascent state of industrial ecology, it is possible to make several points concerning the implementation of dematerialization and the facilitation of responsible material use which can result in immediate improvement in material use even as further research lays the necessary basis for future, more fundamental, progress.

1. Environmentally responsible material use cannot be done by private firms alone. They lack the knowledge, the incentives, and the scale to adequately address all but the most trivial issues of material choice and use. Moreover, there are no mechanisms, explicit or implicit, that assure that even material choices and uses which appear environmentally reasonable at the level of the firm are sustainable when viewed from the perspective of the global commons and economy.

2. Dematerialization at its most basic is the reduction of material use per unit quality of life. It has a cultural dimension which, especially in the longer term, is not only increasingly important, but dominant. It can be achieved not only through the obvious means - lightweighting products, for example - but through reducing the velocity of materials through the economy (e.g., product life extension). It is also generally (although perhaps not necessarily) a byproduct of the evolution of a "functionality economy," where consumers purchase function, rather than physical product, from service providers.

3. In all these cases, it is necessary to understand when dematerialization is appropriate, and when it is less important. To do this, one must not only evaluate the direct environmental implications of materials in particular applications, which is a hard enough task, but also the indirect. These would include the environmental impacts which are already embedded in the materials before use.

4. Given the lack of systemic data, dematerialization and proper material selection will have to rely primarily on heuristics, especially at first. Thus, for example, it is desirable to use less toxic substitutes, all else equal; where that is not possible, product and material takeback can both extend the life of products and materials in the economy, and ensure proper management of materials to minimize environmental impact over their lifecycles.

5. The rub in "3.", of course, is that all else is not equal. Progress beyond heuristics, therefore, depends on supporting comprehensive research programs which generate the requisite knowledge to improve material use across the

economy. We need to learn how to evaluate the many tradeoffs which practitioners commonly see (toxicity versus energy efficiency, for example, is one which seems to appear frequently).

6. The economic impact which can result as knowledge about the environmental preferability of various materials is developed must be recognized. If shifts within material classes (from PVC to ABS, for example), and among classes (from wood to plastics, for example) occur as a result, significant transition costs, such as stranded manufacturing capacity, may be generated. Moreover, some firms will inevitably lose business during the process, which will undoubtedly give rise to considerable resistance to change. In one case , for example, a trade group several years ago produced a consensus document for its members discussing environmentally preferable packaging options, but its release was delayed considerably by a threat to sue (on antitrust grounds) from a few small packaging producers whose business would possibly have been adversely affected.

7. An interesting and defensible hypothesis is that a more sustainable (and more complex) economy will substitute information and intellectual capital for inputs of materials and energy. Thus, the evolution of the service economy should foster dematerialization. Proper costing of materials - albeit a politically unrealistic goal - would facilitate this process.

8. In the larger sense, however, developing knowledge about environmentally preferable materials, and increasing dematerialization, offer the potential for enhanced environmental performance throughout the economy at little cost and no loss of functionality to consumers (who presumably have the old options available as well in a competitive economy). It may not be a completely free lunch, but it is a very cheap one.

4. CONCLUSION

Environmentally responsible and efficient material use, or "sustainable material use," is a goal which will require considerable research. Initially, it requires us to recognize the depth of our current ignorance. That we cannot now determine the integrated real environmental impacts of materials in virtually any application is a clear signal that we have far to go before we develop the knowledge which must underlay an environmentally and economically efficient global economy andd, sustainability itself. Private firms, consumers, NGOs and governments must all contribute towards the dialog, and the development of fundamental science and technology, which are necessary if we are to have any hope of achieving that goal.

---

[1] The view points expressed by the author are his alone, and not necessarily those of AT&T.

Chapter 18

Dow Europe

Six Simple Sustainability Rules for A Complex World

CLAUDE FUSSLER
Vice-President, Environment, Health and Safety, New Businesss and Public Affairs, Dow Europe

Key words: eco-efficiency, material intensity, empowerment, factor four.

Abstract: The author argues that although modern life has become increasingly complex, there are six simple rules that could be adopted to enable the transformation towards sustainable development. These rules are summed up as increasing eco-efficiency, providing added value for customers, adopting clear objectives, empowering employees, citizens and communities, caring about people and the environment and getting 'out of today's box' and thinking beyond existing constraints. At Dow Europe a Six-Point Compass has been developed to stimulate ecological innovation.

1. COMPLEXITY IS A CONSTANT

Science and mathematics show us that complex systems are often influenced by a few variables and governed by a few simple 'rules'. One of the difficulties with the ideas of sustainability at present is that they are often dauntingly complex. It can seem that any individual action is futile because an entire system needs to be changed.

This complexity is real and cannot be wished away. It is, however, possible to offer a few simple rules to consider when approaching the topic from a business perspective. These rules will work best when they are supported by policy initiatives. For example, ones which internalise environmental costs or set long-term emission or energy targets for products and processes. But even in their absence the rules build awareness. And can move an organisation towards sustainability and the breakthrough innovations which are necessary over coming decades.

2. THE RULES OF ECO-EFFICIENCY

There are six rules, whose one word summaries form into the economic ECO/ECO:

- Eco-efficiency
- Customers
- Objectives
- Empowerment
- Care
- Out of the Box

ECO/ECO reminds us that sustainable development, eco-efficiency and cleaner production are about integrating *eco*logy and *eco*nomics. The economic imperative is to provide goods and services that people value because they bring them more quality of life. The ecological imperative is to reduce the environmental impacts and resource requirements of providing them. Both are equally important.

2.1 Eco-efficiency

Wastes, emissions and products which deliver poor environmental performance to customers who want better aren't just in conflict with sustainable development. They are business inefficiency and failure to convert expensive inputs into useful products.

The remedy to this is eco-efficiency. As the World Business Council for Sustainable Development (WBCSD) puts it: "the vision of eco-efficiency is simply to 'produce more from less'. Reducing waste and pollution, and using fewer energy and raw material resources is obviously good for the environment. It is also self-evidently good for business because it cuts companies' costs. Resource productivity is fundamental in the eco-efficiency approach. The potential for step-by-step improvements in resource productivity, to match the increases in labour productivity in recent years, is greater than often perceived."

And this doesn't just apply to your own operations. You need a life cycle perspective to identify opportunities for improvements in downstream or upstream sections of the product chain. The first rule is: *You must achieve more from less.* Business can use the eco-compass and eco-innovation principles to improve its efforts in this area.

2.2 Customers

WBCSD's concept of eco-efficiency also links environmental concerns with adding value for customers. The Council writes: "The goal is to create value for society, and for the company, by doing more with less over a life-cycle... To achieve long-term success, companies must create value for their shareholders and customers. Over time, customers will increasingly demand more than a product that simply fulfils a function." This implies that the second rule is: *Consider your customer's long-term needs even before they do.*

2.3 Objectives

Becoming sustainable is difficult. It encompasses a diversity of impacts - from long-term global problems to local nuisances. It requires co-operation between a wide range of internal and external stakeholders. It needs clear objectives.

To be successful, your objectives must be:

- *comprehensive* - covering not only processes but the life cycle of products, and issues of socio-economic security as well as ecology and resources
- translatable into *targets*, so that you can measure progress
- *clear*, so that they are easily understood and communicated
- capable of *implementation*, either immediately or by stretching your capability.

A quality approach is essential to meet all of these. The third rule is thus: *Set ambitious objectives and targets for sustainability.*

2.4 Empowerment

A quality approach requires the breakdown of organisational boundaries. The operational levels of an organisation share the best knowledge. Quality programmes try to unleash that knowledge by providing tools to analyse and shape it, and opportunities to act independently to implement solutions. They also reduce barriers between functions, and between an organisation and its customers and suppliers, in order to move knowledge and create solutions. The fourth rule is: ***Sustainability means empowered employees, citizens and communities.***

2.5 Care

Sustainable development is a complex issue. To understand it fully, and respond appropriately to its demands, you must extend your horizons. You have a duty and a moral responsibility to care for a wider range of people, organisms and natural systems. This caring has an ethical and spiritual as well as a utilitarian dimension.

The President's Council on Sustainable Development uses the term stewardship to describe this. In its definition the word: "calls upon everyone in society to assume responsibility for protecting the integrity of natural resources and their underlying ecosystems and, in so doing, safeguarding the interests of future generations. Without personal and collective commitment, without an ethic based on the acceptance of responsibility, efforts to sustain natural resources protection and environmental quality cannot succeed. With them, the bountiful yet fragile foundation of natural resources can be protected and replenished to sustain the needs of today and tomorrow" (The President's Council on Sustainable Development, 1996).

But the economic sphere also cares for human needs. The duty of care gives a deeper purpose to your business and a sense to your individual career. It's your lifetime's chance to make a difference. For yourself. For the billions who don't yet have their basic needs for food, water, shelter and medicine for the billions. And for future generations. The fifth rule is: *Sustainability is about ethics and socio-economic security as well as environment.*

2.6 Out-of-the-Box

Cleaner production, eco-efficiency, life cycle assessment, the eco-compass are all thinking tools. They are designed to challenge mindsets. To create discomfort with today's solutions. To provoke curiosity about the future. To excite with its opportunities. To introduce new, holistic, ways of thinking about the world. To encourage dialogue and - sometimes uncomfortable - partnerships with the forces which are reshaping it.

You must operate in the present. But don't be constrained by it. Remember that we're in a period of innovation lethargy, merely running to stand still. Get out of today's box and think of the coming decades. Think in the 30 year timescales over which inter-generational obligations, long-term trends and the innovative pathways to the future become clear and actionable. Think in terms of the minimum 'factor four' and ideal 'factor ten' improvements in performance which sustainability requires. The sixth rule is thus: Sustainable development is imperative and possible. Someone will make it happen. The choice is yours.

3. ECO-INNOVATION COMPASS POINTS THE WAY FORWARD

The nature of today's business dictates that many market and product development professionals focus their attention almost exclusively on short-term goals. It's fine to focus on the next quarter, but it's also fundamentally important to remain focused on the next quarter century.

Long-term business success will result from a lucid and creative approach to the fundamental shifts in societal values and human needs, the signs of which surround us even today. Demographers tell us we are about to experience an unprecedented explosion in world population in the not-too-distant horizon. We see the population of the Western world growing older. And polls tell us that we see a world growing even more concerned about environmental issues.

It *is* possible for an organisation to be focused on short-term success while at the same time positioning itself to respond to the market opportunities created by these fundamental changes.

3.1 Eco-Efficiency

The World Business Council for Sustainable Development (WBCSD) coined and defined the phrase "eco-efficiency" to define a new way of approaching product development and improvement. Eco-efficiency can be defined as a broad performance ratio to guide the entrepreneur in his role in society. It measures his simultaneous effort to minimise the ecological burden while he maximises the economical value he can afford to produce and his customers can afford to pay.

Eco-efficiency is not an absolute. The notion will evolve as a function of innovation, customer values and economical policy instruments. It represents the direction of an effort. The WBCSD has proposed the following definition: "*Eco-efficiency is reached by the delivery of competitively priced goods and services that satisfy human needs and bring quality of life while progressively reducing ecological impacts and resource intensity, through the life cycle to a level at least in line with the Earth's estimated carrying capacity*".

At Dow, we've built upon the WBCSD's work and developed a tool to help us become a more eco-efficient company - and in doing so to position ourselves to respond to the emerging social and demographic trends of the next 25 years. We call the tool we've developed the Eco-Innovation Compass.

3.2 The Six-Point Compass

Dow's compass has six points, which are intended to address all significant environmental issues. Two are largely environmental:

- Health and environmental potential risk; and
- Resource conservation.

Four are of business as well as environmental significance:

- Energy intensity;
- Materials intensity;
- Revalorization (remanufacturing, reuse and recycling); and
- Service extension.

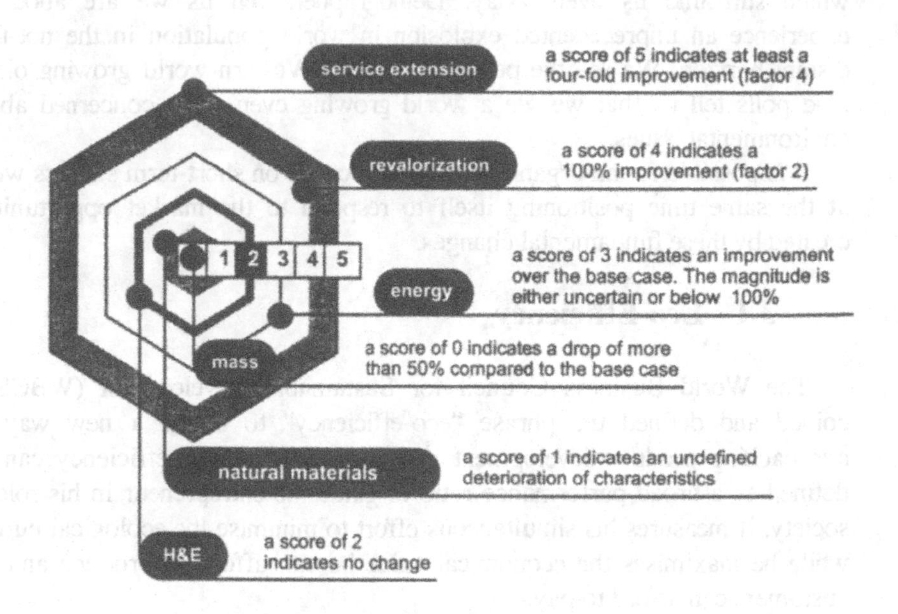

Figure 38. The Eco-Compass

The compass points to the concerns of significant actors and theorists in the environmental debate. The environmental movement, for example, places great emphasis on health and environmental risks, resource conservation and closing substance loops through revalorization (remanufacturing, reuse and recycling). None of the dimensions are independent of the others. They overlap. Energy intensity often correlates with material intensity. Revalorization often reduces both these parameters.

The Eco-Innovation compass isn't for stand alone use. It's meant for comparing an existing product with another or a current product versus with new development options. Our goal is to develop a product with the best profile on the Eco-Innovation compass. In doing so, we maximise the business potential of the product while minimising its environmental impact.

Before evaluating any product, a company must first change its perspective and look beyond the end-use product. One must consider the total design-make-supply-use system and the total life cycle of the product from initial raw materials to final waste products after use. This implies that one must look at the function fulfilled by the product rather than the product itself.

The compass can be read as follows:

- **Material Intensity:** The first point on the compass is material intensity, or mass. This means the total of raw materials, fuels and utilities consumed in the system during the life cycle to deliver the desired function. It is easy to overlook some large amounts of materials like the overburdens displaced and processed in mining a mineral or the translocations of water and salt in processing and cooling many of the basic chemicals. The opportunity is to significantly reduce the mass burdens and de-materialise the way the system provides quality of life and benefits to the market chain. Examples of companies that have made reducing material intensity a standard - and profitable - practice abound. P&G's compact powders and Sony's "Green TV" are just two.

- **Energy:** The second compass point is energy. Here we address not only the energy to assemble and react raw materials but the energy consumed or saved during the consumption and disposal phase. The opportunity is to spot the parts of the system and the life cycle which have the highest energy intensity and redesign the product or its use to provide significant energy savings.
- **Health and Environmental Risk:** The third point includes all the aspects of health and environmental risk. In nature nothing gets lost, all systems conserve mass element by element. As carbon, sulfur, chlorine or any element get involved at one end of a system they will continue to move through in various forms. Negative health impacts will threaten in those places where they reach or accumulate to a level beyond a critical dose for a link of the food chain. A good example of a company that has benefited from looking closely at this point is Dow's agricultural joint-venture, DowElanco. The company has developed a termite control system that uses about 1/10,000 of the active chemicals present in standard competing devices.
- **Recycling or Revalorization:** Material utilisation is improved by re-use in the same or another system. Designing for recyclability is important, recycling effectively and efficiently is even more important. Another opportunity is to design the system as part of a larger natural cycle. Materials are borrowed and returned to nature without negatively affecting the balance of the cycle. Plants are particularly efficient at converting solar energy, carbon dioxide and water into a variety of chemical intermediates and composite materials. Examples of companies benefiting from examining this point on the compass abound. Consider for example, Xerox, which has tapped into the large, thriving market for "recycled" copiers.
- **Resource Conservation** is the fifth compass point. The source of the material used in production is, of course, an important factor in improving the eco-efficiency of a product or service. Here we ask questions related to the nature and renewability of the energy and material used for a product. Is this a product where natural raw materials can be used? We've found a number of companies that have opted to use natural

materials for products ranging from washing liquids to oil well drilling fluids.

- **Service Extension:** The final point on our compass compels us to look at extending the life and functionality of products. Extending the durability and service life of any part of a system, especially in the usage phase, can improve eco-efficiency. Can we design a car that runs for 200,000 miles or more? Improving the functionality of products also increases their eco-efficiency. Best practice here seem to come from companies that make their products easy to upgrade and repair, or that simply last much longer.

Getting eco-efficient is a matter of re-designing a system in every possible respect. One must consider reductions in mass and energy utilisation, in reduction of toxic chemical and search for improvements in recycling or usage of renewable resources and innovation in service life and functionality extensions. The six dimensions are not independent. They overlap and inter-relate significantly. We're convinced that this compass can lead to Eco-Innovation and help business play an even more proactive role in the sustainability revolution.

REFERENCES

Fussler, C. and Peter Jansen (1996). *Driving Eco Innovation*, Pitman Publishing, London.

The President's Council on Sustainable Development (1996). *Sustainable America*, US Government Printing Office, Washington D.C. p. 110.

Chapter 19

The control of waste materials in Germany

ULF D. JAECKEL
Deputy head, Prevention and Utilisation of Waste Division, Federal Ministry for the Environment, Nature Conservation and Nuclear Safety, Bonn.

Key words: packaging, closed substance cycles, waste management hierarchy.

Abstract: This article presents a brief summary of how the German government has translated the concept of closing substance cycles into its policy. The government has adopted a policy which uses the waste management hierarchy, focuses on modifying the laws on packaging and on producer responsibility.

1. POLITICAL SITUATION

I would like to start with one really simple question: "Why don´t we raise taxes to fix the price of one litre of fuel at 5 German Marks and put an equivalent tax on the consumption of other energies and primary materials?" This would help to solve a lot of problems in the field of environmental policy. Many regulations that we already have - a lot of people say we have too many - would no longer be necessary. This would be a good general solution which would help to internalise many external costs for the use of the environment.

If Germany were an isolated island with no trade or external relations with other countries or if all other countries in the world were to apply this measure, this would be the right strategy. Unfortunately that is not the situation we have today.

While we should go on trying to influence the consumption of materials in the long-term, the political situation in most countries is not amenable to such a solution. We have to find different solutions which can work in the short-term. That is the reason why nowadays we do not try to reach out environmental goal by influencing material consumption at the beginning of the production process. Instead, we look at the end of this process, the waste. Waste management has become a central field of action of environmental policy in Germany and in the EU as a whole.

2. CLOSED SUBSTANCE CYCLE AND WASTE MANAGEMENT ACT

On 7 October 1996 the new Closed Substance Cycle and Waste Management Act (*Kreislaufwirtschafts- und Abfallgesetz*) entered into force. This Act means that both producers and consumers will have to undertake a radical re-think in the field of waste management. This Act contains two main parts and ideas. On the one hand, it builds upon the old Waste Management Act of 1986. Therefore it has to ensure the protection of the environment from a range of wastes, including hazardous waste. On the other hand it contains a frame for a new waste management policy. It tries to build a `closed loop economy'. This idea has been much discussed in Germany. It is very widely agreed that there is a need for closing materials cycles and waste prevention. The differences lie in the measures to be chosen to reach this goal.

From the general point of view, a "closed loop economy" will play an important role in the execution of the concept of sustainable development. Materials should be kept within the economy as long as possible to avoid waste and conserve natural resources. Regulations like the Closed Substance Cycle and Waste Management Act are therefore necessary. These acts follow the principle of "producer responsibility" first established in the Packaging Ordinance of 1991. It is a goal to implement this principles in most (or all) of the (important) product sectors. Therefore the Closed Substance Cycle and Waste Management Act contains the basic requirement of producer responsibility (Article 7, 22) and the possibility for the Federal Government to translate this principle into concrete legal acts (Articles 7. 23, 24).

The general industry point of view is that the goal should be to get a "closed loop economy," but that this can be achieved without regulation, and often without any assessment of environmental effects. There should be no legally-binding commitments, only voluntary commitments. Otherwise, it is argued, the competitiveness of the national industry would be weakened.

From the environmental point of view, it has to be said that a "closed loop economy" should not be a goal in itself. The environmental goals should lie on a higher level and include, the protection and conservation of natural resources, the reduction of emissions to the air, water and soil, and so on. Closing loops should be justified by the environmental benefits from an integrated life-cycle perspective. Many measures exist to further both goals. Each is a step towards a "closed loop economy" and helps to improve the environmental situation as a whole. But some measures like the recycling of "last little bits and pieces" are seen from the environmental point of view negative because of high environmental burdens, such as those arising from the use of energy. This should be kept in mind when following the goal of a closed loop economy.

2.1 Waste management hierarchy

The Closed Substance Cycle and Waste Management Act implements two new ideas. One is the waste management hierarchy and one is about producer responsibility. In order to promote an economy based on closed substance cycles and geared towards taking care of natural resources, the following hierarchy is introduced:

1. Avoidance: The legally binding obligation to avoid waste relates, on the one hand, to the introduction of low-waste industrial production processes and, on the other hand, to the promotion of the manufacture and use of low-waste products.

2. Recovery: Where avoidance is not possible, waste is to be recovered as its component substances or as energy. This process has to be environmentally friendly. The Act does not lay down any one-sided priority for the substance recycling of waste. Recycling and energy recovery are in principle of equal value. In principle in each individual case priority is given to the more environmentally sound form of recovery. But there is the possibility for the government to give priority to a certain form of recovery by an ordinance in special fields where general determinations can be made. The most important example in Germany is the Packaging Ordinance where priority is given to the recycling of substances.

Despite this, there are some prerequisites which shall ensure that only waste with high energy potential is subjected to energy recovery and that the combustion process is good enough to recover most of the energy of the waste. Energy recovery has to meet the following standards:

- the thermal value of the energy recovered must be at least 11,000kj/kg,
- the combustion efficiency of the recovery facility must be at least 75%, and
- the heat generated by energy recovery must be used either by the operator himself or by third parties.

These standards also help to make a clear distinction between energy recovery and "traditional" thermal waste treatment, which is merely a waste disposal measure.

3. Disposal: Where waste prevention and recovery is not possible or not permitted, the only option left is to dispose of it. The disposal has to meet some additional provisions concerning emissions into the air, ground and water meaning that a high level of safety must be reached.

2.2 Producer Responsibility

In general producer responsibility aims at integrating environmental protection into economic processes from the very outset. Whoever is responsible for the production, distribution and consumption of goods shall be responsible for the avoidance, recovery and environmentally-sound disposal of the wastes occurring. This means consistent the implementation of the polluter-pays principle in the field of waste.

This principle means that industry, trade and consumers should be encouraged to consider waste. Decisions on production, distribution (and consumption) of goods should also take into account what will happen to the product when it has reached the end of its life cycle. One shall be aware to take back and recycle products brought onto the market. Reflections shall be made - as early as during design and development - on how products can be made as environmentally friendly as possible.

The aim of environmental policy in Germany is to introduce producer responsibility in all of important product fields. Producer responsibility shall include a better internalisation of external environmental costs. Take-back obligations, recycling obligations and in some special cases even financial instruments like taxes will be the main measures used. This internalisation of external costs should lead to a reduction of material intensity and greater cyclicity. The reduction of packaging material and the increase of recycling are results of the internalisation of the obligations by the Packaging Ordinance (see .3.2).

3. THE GERMAN PACKAGING ORDINANCE

3.1 The Principle

The starting point, the prototype and the model example for the "new producer responsibility" and the start of an economy based on product recycling was the Packaging Ordinance of 12 June 1991. This Ordinance is the most important measure in German waste management policy.

This Ordinance aims at the following goals: (a) clearly reducing the packaging mountain by avoidance, reuse and recycling, (b) placing responsibility for product disposal on those who manufacture these products and put them on the market, (c)relieving local communities of the burden of waste disposal, and (d) clearly promoting substance recycling.

Comprehensive obligations were introduced on the acceptance of the return of packaging by producers, and on the levying of deposits have been introduced. The Ordinance contains the main following individual stipulations: (a) manufacturers and distributors have to take back packaging and arrange for their recovery, reuse or recycling; (b) consumers are able to leave secondary packaging behind in the shops. Distributors have to arrange for this so-called secondary packaging to be reused or recycled.

Industry was given the option of organising collection systems, independently of municipal waste disposal, which operate in the direct vicinity of the consumers themselves. This possibility was provided by the option of a so-called dual system. It was called *dual* because it is a scheme which operates side-by-side with traditional waste disposal facilities provided by the local authorities.

Industry seized this opportunity and established the Duales System Deutschland (DSD) GmbH in September 1990. The Packaging Ordinance provides specific quotas for collecting, sorting and recycling for a scheme such as this. These quotas were initially quite low but they have increased significantly since 1 July 1995. From mid-1995, the Packaging Ordinance as it stands at the moment stipulates that at least 64% of all sales packaging put on the market has to be recycled in the field of paper and cardboard, plastics and compounded materials; and at least 72% in the fields of glass, tin plate and aluminium. If these requirements are not met, the licence for this private enterprise collection system is revoked.

The costs which are met by DSD play an important role in the control of material flows. These costs are divided amongst the participants of the dual system. Therefore, licence fees are charged by DSD dependent on the kind of material and on weight (with an additional fee per item). The licence fee ranges from 0.15 DM/kg for glass packaging to 2.95 DM/kg for plastics. The fees will be equivalent to the actual costs for collecting, sorting, recycling or disposal. With these licence fees some external costs can be internalised.

3.2 Positive Experience with the Packaging Ordinance

Nearly six years after the entry into force of the Ordinance has proved successful in several fields. First, manufacturers have changed their packaging habits. Environmentally friendly disposal of packaging is a factor which is taken into account during the production process and is also increasingly used in advertising and marketing. For instance, a recent survey showed that two-thirds of packaging firms had abandoned composite packaging systems. Second, due to the differences in the licence fees for different materials and the fees themselves, changes in the packaging market can be seen. Packaging has become lighter and smaller. Packaging with higher proportional licence fees (i.e. plastics, glass) have been replaced by packaging with lower fees (i.e. cardboard). Useless packaging has disappeared. As a result, the use of packaging has been considerably reduced in Germany. In 1995, 1.3 million tonnes less packaging was used than in 1991, the year the Packaging Ordinance entered into force (see *Figure 39*).

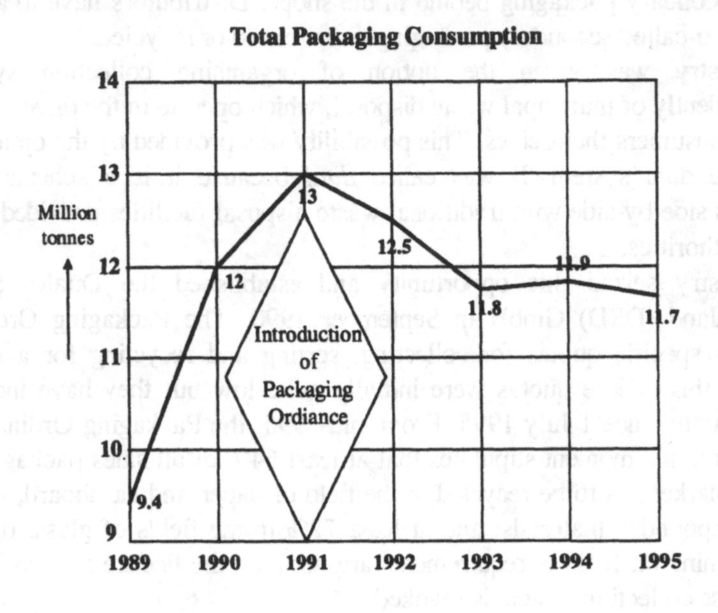

Figure 39. Developments of packaging consumption in Germany.

Third, in the field of transport packaging there is a trend towards reusable packaging especially in relation to the packaging of furniture, food, pharmaceutical products and bicycles. Fourth, industry has set up a nation-wide collection system for disposable packaging and has increased its recycling capacities for all packaging material. In 1995 the recycling quotas were quite high (see *Figure 40*).

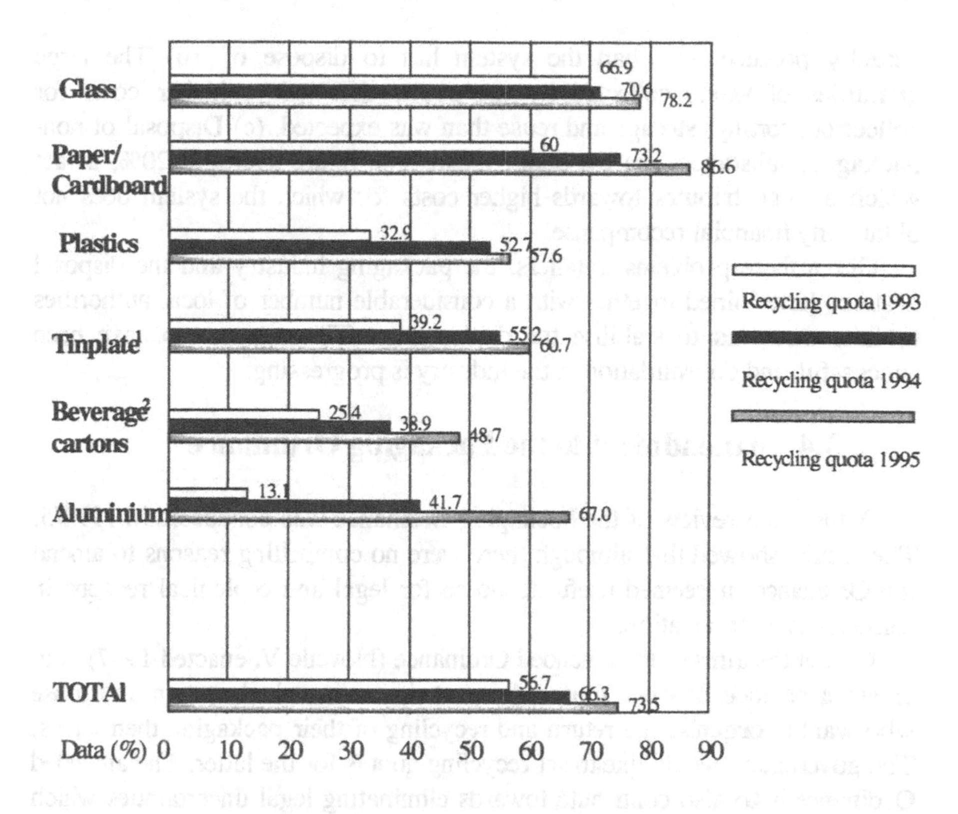

Figure 40. Recycling quotas for used packaging

3.3 Negative Experience with the Packaging Ordinance

A number of initial problems had to be overcome at the beginning of the scheme. Initially, a critical situation arose in the field of substance recycling of plastic packaging. Due to the zeal of the public in collecting packaging wastes (which was very much welcomed) led to the quantities initially collected being greater than the recycling capacities available. This situation has now changed. New technologies and processing capacity have emerged in areas where deficits were observed. In 1990 recycling capacity reached 20,000 tonnes, but this had increased to over 500,000 tonnes by 1996. Due to the Packaging Directive most of the packaging material collected will be recycled in Germany. Much criticised exports to European and non-European countries will be ended.

The Duales System Deutschland has also faced a number of problems. In the initial phase some serious financial problems emerged which had to be solved. The causes of the financial difficulties include: (a) "Free riders", that is firms which, although they imprint the green dot on their packaging as a sign that they are members of the system, pay for far less packaging than they actually produce and than the system has to dispose of. (b) The large quantities of waste collected by the public that led to higher costs for collection, sorting, storage and reuse than was expected. (c) Disposal of non-packaging substances via the dual system. The figure averages 20%, a fact which also contributes towards higher costs for which the system does not obtain any financial recompense.

Given these problems, retailers, the packaging industry and the disposal systems have joined together with a considerable number of local authorities to take measures to stabilise the dual system. This appears to have been successful, and consolidation in the industry is progressing.

3.4 Amendment to the Packaging Ordinance

A thorough review of the Packaging Ordinance was conducted in 1994/5. The results showed that although there were no compelling reasons to amend the Ordinance, it seemed useful to do so for legal and ecological reasons in some areas of application.

One of the aims of the amended Ordinance (Novelle V, enacted 1997) is to create a balance between those who participate in a dual system and those who want to organise the return and recycling of their packaging themselves. The government would like to set recycling quotas for the latter. The amended Ordinance is to also contribute towards eliminating legal uncertainties which have come to light during practical implementation. The amended version therefore does not contain any substantive changes, in particular with regard to the recycling requirement. It merely contains small modifications to the quota.

4. EXPERIENCE IN OTHER PRODUCT FIELDS

At this moment the government is trying to introduce producer responsibility in other product fields, i.e. in relation to cars, electronic appliances, graphic paper, construction wastes, textiles and furniture. This is not an easy task in a time of high unemployment, while priority is given to the economy. In this situation we are trying to proceed with Voluntary Commitments negotiated with industry. Up to 1997, voluntary commitments have been made on graphic papers, construction waste and end-of-life vehicles. The voluntary commitment for graphic papers undertook to increase the quota of secondary paper in newspapers, magazines and brochures to 70% by the year 2000. This quota is already fulfilled. In 1996 approximately 72% was reached. For construction waste the building industry has promised to reduce the waste for disposal from 1995 to 2005 to 50% of current waste arisings. For end-of-life vehicles, the automobile industry (German industry and importers) have committed themselves to reduce the waste for disposal to a maximum of 15% by 2002 and to a maximum of 5% by 2015. In the same voluntary commitment the automobile industry declared that it would take back scrapped cars that were up to 12 years old. That fact in particular was a compromise. Apart from this, the introduction of producer responsibility has not yet started.

5. CONCLUSION

Many things have been set in motion so far in the field of waste management in Germany. But it is obvious that efforts to control material flows are limited to the area of waste management. The opportunities for influencing the use of primary materials seems to be very small. Nevertheless efforts will be made to internalise costs for the use of environmental goods. Therefore it seems useful to stop concentrating only on certain products. It would be better to look at materials to optimise the use of scarce sorting and recycling plants, and to recycle where there are the best possibilities. It seems to be more useful to recycle products with higher volume than to recycle packaging of all sizes.

Chapter 20

Dematerialisation and innovation policy

Opportunities and barriers in the Netherlands

J.L.A. JANSEN
Professor, University of Delft, Delft.

Key words: **ecoefficiency, dematerealisation, material intensity per service, technological innovation, sustainable technology.**

Abstract: **This article argues that de-materialisation is a process that involves all the major societal stakeholders. Change is influenced by the degree of efficiency improvement, the complexity of the product system and the interactions between culture, institutional structure and technology. On the basis of examples from the policies on technological innovation in the Netherlands, the author argues for policies that are tailor-made to the solutions sought.**

1. THE CHALLENGE: GROWING NEEDS AND LIMITED RESOURCES

From the work of the Brundtland Commission (1987), "Our Common Future", and other analyses (RIVM 1992; Carnoules Declaration 1995) it is evident that the environmental capacity, the resources and the capacity of the environment to absorb and recover is limited and is already being exceeded far beyond acceptable boundaries. Although an exact identification of the boundaries now and in the future is subject to scientific and normative uncertainties, the situation is very serious and requires action (CLTM 1991; 1994). This was reaffirmed as a major political priority in the United Nations Conference on Environment and Development held at Rio de Janeiro in 1992, and again at the 1997 United Nations General Assembly in 1997. The needs of people to be fulfilled in the future will grow with the growth of the world population and the welfare per head. These factors determine future demands on the environment. Population growth and environmentally-unsound technological change may lead to severe tensions on global, regional and local scales (Homer-Dixon 1994). Sooner or later, the scarcity of essential environmental services will affect everyone, unless the challenge of a shift in our current modes of development onto a sustainable path is achieved.

2. DEMATERIALISATION: A PROCESS OF CHANGE

How can a shift to sustainable development be operationalised and specified in actions for responsible actors in society? Dematerialisation can be looked upon as a strategy in a process of concurrent changes for sustainability in which many stakeholders (consumers, producers, scientists, NGOs and government), each with its own concerns and responsibilities, are involved. Each stakeholder has its own role and responsibility to begin and manage this process of change.

The directions of change imply moving towards less material intensity, less energy intensity, reduction of emissions, a shift to renewables and so on. Knowing the general direction of change is not enough. We also need to understand the required degree of change and the appropriate level of targets to be set. Quantitative indicators of change over time, whether rough and global, or specific and local, are essential. Sustainable development is a continuous process of balancing the use of environmental resources for fulfilling present societal needs and the protection and conservation of the ecosystem for future generations. This requires a stable dynamic equilibrium. The specific objectives of sustainable development are determined by the degree of imbalance that exists between current consumption of environmental services and a sustainable level of consumption.

Depending on the assumptions, the related uncertainties and the specific environmental factors, an acceptable level of pressure on the environment will tend to be exceeded in the next half century by a factor of 5 to 50 (Speth, Febr. 1989 and more recently Weterings and Opschoor, 1992). Goodland and others (1993) and Schmidt-Bleek (1994) arrive at similar conclusions. Examples illustrating exhaustion, contamination and degradation, for different examples of environmental pressure have been worked out by Weterings and Opschoor (1992).

3. MATERIAL INTENSITY AS AN INDICATOR FOR ECOEFFICIENCY

Matter, energy and space are the physical inputs from the environment to the economic process. Solid wastes and emissions are the outputs to the environment. From a cradle-to-grave perspective, the use matter, energy and space and expulsion of wastes and emissions are continuous and parallel processes. The reduction of waste and emissions by end of pipe measures in environmental effect oriented approaches (as discussed in Chapter 19) usually require extra inputs of matter, energy and space. On the other hand, increasingly efficient use of inputs result in reduced outputs to the environment as well. Input factors therefore are the most effective indicators of environmental efficiency. Among those, matter (including energy carriers) is the primary carrier of the means to fulfil needs. Energy and space usually function to transform or transport matter. Renewable energy sources demand space. At the back-end removal of solid wastes demands space. The use of energy and space can roughly be related to the use of matter. This leaves matter or material intensity as a handy indicator to set targets by, and to judge the effects of measures to increase the environmental efficiency of products and services.

One approach has been to state that the Material Intensity Per Service (MIPS; see Chapter 4; Schmidt-Bleek 1994; Schmidt-Bleek and Tischner 1996)' must be reduced by a factor 10 or more over the next few decades. Factor 10 improvements should be achieved across a range of activities, ranging from simple products, up to complex technological. Many needs in contemporary societies are met by complex technological systems operating at a variety of levels, as shown in *Table 34*.

Table 34. Applying factor 10 to needs, functions and products in relation to marine transportation

Mobility (transportation across water)			
simple product	complex product	technological systems	interlinked systems
propeller	engine	ship	shipping system

The achievement of a target to de-materialise society by a factor 10 or more requires intensive changes in culture, economic structure and technology: culture legitimates the nature and volume of societal needs to be fulfilled, expressed in consumption patterns (sufficiency) (Thoenes 1991); economic and institutional structures fulfil legitimate needs (effectiveness); and technology provides the technical means by which needs can be fulfilled (efficiency).

These three closely interacting elements characterise the development of society's approach to sustainable development. The acceptability and viability of environmentally-efficient technologies is directly connected to economic and institutional conditions (structure) and to the demands of society (culture). In this context it should be understood that these conditions and demands are not static and may change radically as a result of environmental developments and policies. Against this background, it becomes clear that the development of sustainable technology cannot be realised without taking into account: (a) the cultural and structural limiting conditions under which the technology must function, determining the possibility, the desirability and acceptability of an innovation (Schwarz 1992; Schwarz and Thomson 1990); and (b) the cultural and structural requirements which must be met for technologies and systems to work effectively.

4. THE CHALLENGE OF 'SUSTAINABILITY': TIMELY AND APPROPRIATE ACTION

The dimensions of change have to be understood to design policies for change taking the roles, responsibilities and abilities of stakeholders (consumers, producers, scientists, NGOs and government) into account. So far three dimensions of change for de-materialisation and technology development for sustainability have been discussed: (a) the degree of targeted efficiency improvement expressed in factors such as MIPS; (b) the complexity of a product system in the sequence from single product to inter-linked technological systems; and (c) the interlocking nature of culture, structure and technology in the de-materialisation processes. These dimensions are dependent of each other. The higher the degree of efficiency improvement or the more complex the product system, the stronger the relationship between culture, structure and technology will be. Together they determine other dimensions of change: its intensity and speed (See *Table 35*).

Table 35. The interaction of the factors influencing change

Dimensions of change				
Degree of efficiency improvement	▸	intensity	▸	time
Complexity of product system	▸	of	▸	for
Culture/structure/technology interaction	▸	change	▸	change

A successful de-materialisation strategy calls for the participation of industry. A phased policy process that aims to enhance opportunities for de-materialisation should be matched with the capabilities and planning horizons business is accustomed to for implementing decisions. In designing an appropriate transition strategy it is important to work along concurrent tracks with various timescales at the same time. These tracks can be characterised by:

1. **Care**, good housekeeping, in industry corresponding to operational processes like quality management, maintenance, auditing, efficiency drives etc. with time scales up to 5 years.

2. **Adaptation and improvement**, leaving fundamental structures and technologies unchanged but implementing incremental changes with time scales from 5 to 20 years.
3. **Renewal**, by fundamental jump-like changes, resulting from long term research and affecting structure, culture and technology fundamentally, with time scales of over 20 years. Renewal of technology means "redefining actual technology development trajectories and promoting new ones" (Jansen 1991).

Each of the tracks can be coupled with targets as a function of time on the route to sustainable technology. In a continuous attempt to lower the environmental burden each of the tracks is essential: the first two to implement known techniques and incremental improvements which result in a primary efficiency increase and which help to gain time in the preparation for the renewal along the third track (see *Figure 41*).

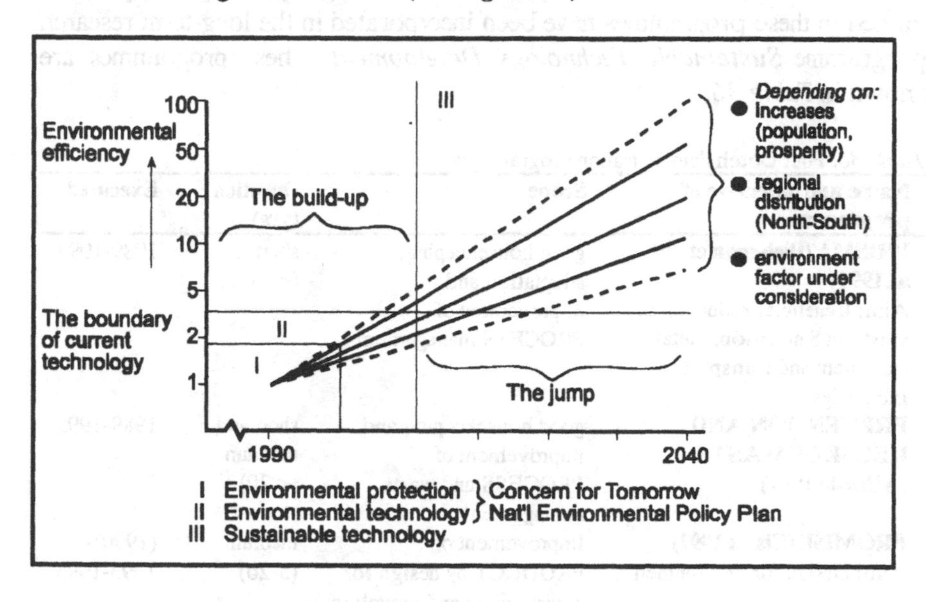

Figure 41. Technology in sustainable development: targets

5. OPPORTUNITIES, BARRIERS AND CHALLENGES IN THE NETHERLANDS

Stakeholders in the process of development and application of technology on the three levels of sustainable development are citizens and consumers, producers, distributors, researchers and government. In the process towards sustainable development these stakeholders interact with each other on the basis of their interests, opportunities, barriers and mutual drives. A first step to influence key actors in their behaviour, and to develop effective policies, is to recognise these interests, opportunities and barriers. With regard to the first two levels of change, much experience has been gained in the last two decades (Schmidt Bleek 1994, Graedel and Allenby 1995, Dieleman et al. 1991, Clark 1997) and in governmental programmes on environmental care (VROM 1996) and technology development programs.

Dutch programmes designed to promote more environmentally sound practices in process design and management not only resulted in new technological options, but also highlighted a number of questions in relation to cultural and structural factors. In the framework of the Dutch environmental target-group policies (VROM 1989, VROM 1993), these programmes encourage technology development to support short- and medium-term environmental policies. The objective is 'learning by doing'. Experience gained in these programmes have been incorporated in the long-term research programme *Sustainable Technology Development.* These programmes are shown in *Table 36.*

Table 36. Four Dutch demonstration programmes

Name and objective of programme	**Scope**	**Duration (yrs)**	**Executed**
PRISMA (Dieleman et al.1991) Aim: Beneficial reduction of waste in 8 nutrition, metal treatment and transport industries	good housekeeping, adaptation and improvement of PROCESS management	short (< 5)	1989-1993
PREVENTION AND REUSE OF WASTE (Allessie 1989)	good housekeeping and improvement of PROCESS and waste management	short and medium (< 20)	1989-1995
PROMISE (Clark 1997) Aim: Design for environment	Improvement of PRODUCT by design for environment and recycling	medium (5-20)	(1990)-1993-1995
SUSTAINABLE TECHNOLOGY DEVELOPMENT (Jansen 1993)	renewal by sustainable design of PRODUCT AND PROCESS for FUNCTIONS AND NEEDS.	long (>20)	1993-(1998)

The first three programs: PRISMA, PREVENTION of WASTE, and PROMISE signalled opportunities, demonstrated economic advantages of environmental approaches, recognised barriers and experienced ways to overcome them. The PROMISE project assessed the drivers for action in firms: (a) *internal drivers* including increased personnel motivation, innovative power, reduced costs, better product and company image, increased product quality and managers' sense of responsibility; and (b) *external drivers* like industrial customer and end use demands, social responsibility, competitive position, trade organisations and government policy.

In most industrialised countries care and good housekeeping is well under way. Product and process change oriented towards environmental goals has begun in many firms, but a lot more can be done. Only few initiatives have however been undertaken which seek to redefine needs and functions over the long term. Based on the technology strategy of the first National Environmental Policy Plan (VROM 1989) a research programme was developed.

6. THE DUTCH "SUSTAINABLE TECHNOLOGY DEVELOPMENT" PROGRAMME

6.1 Starting points

The objective of the Dutch inter-ministerial research programme has been to investigate and illustrate how innovation processes could enable the development of credible options for renewal of technology in the next decades. The programme was based on the following considerations.

A common interest: In the long run the different stakeholders tend to have a common interest in an orderly economy which meets the needs of peoples at an acceptable price and quality. The inability of economic and institutional providers to meet these needs leads to spasmodic adjustments. The challenge is to anticipate in a timely manner the effects of changing conditions and changing demands. Future changes in the availability of production factors and their direct and indirect costs will fundamentally alter production structures and infrastructures, including transportation. The uncertainty is not "whether or not" but "when and how" (Factor Ten Club, 1995; Schwarz, 1992; Schwarz and Thomson, 1990; Jansen, 1991). Given that the time required for fundamental change is in the order of a few decades, it is urgent that preparatory measures are adopted sooner, rather than later. This calls for a proactive interpretation of the precautionary principle (UNCED 1992).

A common barrier: A common barrier for all stakeholders is the uncertainty of future developments. Uncertainty in the development of sustainable technology is expressed in the practical question of uncertain returns on investment for technology which could be profitable only under different economic conditions. There is uncertainty in how conditions will change, in the expected time of its occurrence and in its nature. This shortcoming of the market could be managed through governmental intervention which, for example, leads to a change in the taxation basis with respect to production factors. However, for the long-term the credibility of administrative measures is doubtful, unless a deeply-rooted social consensus emerges.

Markets do not effectively signal the steadily increasing costs of energy and land which may be expect over the next three to four decades. Moreover, business is only concerned with planning for limited periods of time (up to about ten years). This dilemma has to be solved in the interest of society and of business itself. Hence, to push technology towards sustainability at least three conditions should be fulfilled (Bakker et al., 1991). There should be: (a) consensus on the goal, route and starting point of development; (b) agreement among relevant actors with respect to their roles and responsibilities in the process of technology development; and (c) agreement on incentives, policy instruments and institutional arrangements for the development of sustainable technology. In international decision-making on sustainability we lack almost everything: accepted common fundamental philosophies, principles and quantifiable targets, adequate institutions, and enforceable legislation and instruments (Vonkeman and Maxson 1993).

6.2 Core elements and results

There are two core elements of the research programme. First the programme uses the method of backcasting (Goldemberg et al. 1985) on two dimensions: formulating development paths back from needs to products and services, and from the future (2 generations ahead, around 2040) back to the present. Such a methods should indicate the barriers to be overcome in meeting a redefined need, and prospective combinations of technologies resulting into research and development programmes. Second the programme uses an approach based on interactive and iterative development processes (Grootveld 1989), with participation from the relevant stakeholders incorporating their interactions in the 'culture, structure and technology-complex' (see Jansen 1993 for details).

Although the programme has not yet been completed the results of the experiments so far completed indicate the following:

- Scoping of technology for a sustainable long-term future can be initiated and managed;
- Backcasting from needs to products and from future to present is a powerful tool for creative approaches of the innovation process;
- Shared future visions, even if they are rough and ready, are essential elements in the innovation process;

- It appears possible to bridge the tension between the industrial need for economic prospects in the medium-term and the necessary orientation on long-term targets for sustainability. Targets can be constructed "ex ante" towards which technological development trajectories can be oriented which have intermediate benefits and spin-offs;
- The essential interaction of culture, structure and technology can be operationalised at different systems levels from 'future visions' to 'product viability';
- Networks can be set up, maintained and operationalised at the interfaces of different technological and non-technological disciplines;
- Ways have been developed to approach and mobilise stakeholders in specific technology development projects; and
- Ways to maintain an interactive and iterative process in innovation and decisionmaking have proved to be successful.

7. CONCLUSION

Different characteristics of the short-term, middle-long term and long-term components of policy justify different approaches (see *Table 37*). Policy must take account of what can be done in the near-term, but place this in the context of deeper long-term changes which are being sought. The STD programme has provided a framework for putting into practice a sustainable technology policy.

So far social-science research in the environment in the Netherlands has been concentrated on the determinants of the behaviour of individuals, on environmental care systems and on legal and communicative instruments (Mandele et al. 1995). In other countries like Sweden the situation does not seem to be essentially different (Linden 1996). This leaves questions about institutional behaviour and changes with respect to renewal over the longer term. Cramer (1997) underlines the need for research on the internal mechanisms of organisational structure and culture, and on the role of external stakeholders in the response to environmental challenges.

Table 37. Characteristics of technology for sustainability

Innovation trajectory	Optimise production and products (good housekeeping)		Process and product improvement	Function directed innovation
Period of implementation	up to 5 years	up to 10 years	up to 20 years	20 to 50 years
Environment characteristic	End of pipe / End of product	Integrated in Process / Product		
Technology characteristic	Application	Adaptation and Improvement		Renewal
Innovation characteristic	Incremental	Incremental	Radical	New systems, paradigm shift
Environmental efficiency factor	1,2 - 1,5	1,5 - 5		5 - 50
Environmental burden reduction %	20 - 30%	30 - 80%		80 - 98%
Nature of challenge	Operational	Implementation		Conceptual
Operational Level	Factory	Division		Board
Driving forces	Real costs, savings, public image	▶ ▶ ▶ ▶ ▶ ▶ ▶ ▶ ▶ ▶		Shared vision and values
Policy measures	direct regulation	direct regulation, development support	standards, taxation, sectoral agreements	strategic development, cooperation agreements
Impacts on infrastructure	-	-	-/+	+
Change in company attitude	moderate	involve employees	internal and external communication	change of culture and structure
Degree of uncertainty	low	low	moderate	high
Scope of action	micro	micro	micro/macro, intrasectoral	macro, intersectoral
Specification level	process	process / materials	materials / product / function	function / need
Relevant actors in technology development	Private enterprise	Private enterprise	Private enterprise / Research organisations	Research organisations / Private enterprise

Source: Cramer and Jansen (1995)

REFERENCES

Allessie, M.M.J., (1989 Jan/Febr/March). An approach to the prevention and recycling of waste, UNEP Industry and Environment, p 25-29.

Bakker, H.J., J.Dronkers and P.Vellinga (1991). Veranderend Wereldklimaat, reacties en commentaren, Delwel, Den Haag, p 54.

Clarke,R. (1997). Ecodesign: A promising approach to sustainable production and consumption, UNEP Industry and Environment, ISBN 92-807-1631-X

CLTM (Dutch Committee for Long-term Environmental Policy) (1991): Highlights from THE ENVIRONMENT: ideas for the 21st century, Kerckebosch bv Zeist, pp13-17, pp35-43.

CLTM (1994). The Environment: Towards a sustainable future, Kluwer Academic Publishers, Dordrecht / Boston / London, 608 pages.

Cramer, J. and J.L.A. Jansen, Milieuforum March 1995 no 3, p 10. Samson and Tjeenk Willink, Alphen aan de Rijn, Netherlands

Cramer, J. (April 11, 1997). Milieumanagement van 'fit' naar 'stretch' (Environmental management from 'fit' to stretch'). inaugural speech, Katholieke Universiteit Brabant.

Dieleman, H., and others (1991). Kiezen voor preventie is winnen, NOTA (now Rathenau Institute). Den Haag, 224 pages.

Factor-10 Club, (1995). Carnoules Declaration, Wuppertal Institute for Climate, Environment and Energy, Wuppertal, Germany.

Freeman, Chr. and C.Perez (1988). Structural crises of adjustment, business cycles and investment behaviour, (in G.Dosi et al, Technical Change and Economic Theory, Pinter, London and New York). pp 38-66.

Goldemberg J., Johansson T.B., Reddy A.K.N. and Williams R.H.(1985). An End-Use Oriented Global Energy Strategy, Annual Review of Energy, 10:613-88.

Goodland, R. and H.Daly (1993). Why Northern income growth is not the solution to Southern poverty, Ecological Economics, 8, Elsevier Science Publishers B.V., Amsterdam, pp 85-101. Goodland, R., Daly H., El Seraphy S. (October 1993). The urgent need for rapid transition to Global Environmental Sustainability, Environmental Department, The World Bank, Washington DC 20433, to be published.

Graedel, T.E., Allenby B.R. (1995). Industrial Ecology, Prentice Hall, Englewood Cliffs, New Jersey 07632, ISBN 0-13-125238-0.

Grootveld, G. van (1989). Interactieve beleidsbepaling, in Instrumenten van Milieubeleid, under redaction of Besemer, de Groot and G. Huppes, Samson HD Tjeenk Willink, Alphen aan de Rijn.

Homer-Dixon, Thomas F. (summer 1994). Environmental scarcities and violent conflict, International security, Vol 19, no 1, pp 5-40.

Jansen, J.L.A. (May 22 1991). Duurzaam denken, duurzaam doen, inaugural speech Technical University Delft.

Jansen J.L.A. (1993). Towards a sustainable future: En route with technology!, in: The Environment: London, Towards a sustainable future., p 497-523, Kluwer Academic Publishers, Dordrecht / Boston . London.

Lindén, A.L. (March 1996). Environmental sociology in Sweden, Environment and Society, Newsletter Research Comittee International Sociological Association, no 9.

Mandele, H.D. van der, Y.M van Sark, S.A.Wink (July 1995). Duurzame ontwikkeling, onze tweede natuur, Advice to the (Dutch) government by the Advisory Council for Research on Nature and Environment, Rijswijk.

Perelet, R.A. (1993). The environment as a security issue' in 'The Environment: Towards a sustainable future, Kluwer Academic Publishers, Dordrecht / Boston / London p 147-173.

RIVM (National Institute of Public Health and Environmental Protection) (1992). The Environment in Europe: a Global Perspective, Bilthoven, 120 pages.

Schmidt-Bleek, F. (1994). Wieviel Umwelt braucht der Mensch?, Birkhäuser Verlag, Berlin, Basel, Boston, ISBN 3-7643-2959-9, pp 13-30.

Schmidt-Bleek, F. U.Tischner (1996). Produktentwicklung, Natur gestalten - Natur schonen, WIFI Österreich.

Schwarz, M. (February 1992). Report to the program direction Environment & Technology, Ministry VROM-DGM, The Netherlands, De Balie, Amsterdam, 19 pages.

Schwarz, M., M.Thomson (1990). Divided we stand, redefining politics, technology, and social choice, Harvester Wheatsheaf, New York, London, Toronto, Sydney, Tokyo, pp 103-152.

Speth, J.G. (1989). Can the world be saved?, *Ecological Economics*, vol 1, pp 289-302.

Thoenes, CLTM (Dutch Committee for Long-term Environmental Policy) (February 1991). Highlights from THE ENVIRONMENT: ideas for the 21st century, Kerckebosch bv Zeist p 25.

UNCED (1992). The Rio Declaration, United Nations Conference on Environment and Development.

Vonkeman, G.H. and P.Maxson (1993). International views on long term environmental policy, in The Environment: Towards a sustainable future, p 219-245, Kluwer Academic Publishers, Dordrecht / Boston / London.

VROM (1996). Milieu & Bedrijven, informatiebulletin, Ministry of Housing, Spatial Planning and Environment

VROM (1989). National Environmental Policy Plan: To Choose or to Lose, Ministry of Housing, Spatial Planning and Environment, SDU, The Hague.

VROM (1993). National Environmental Policy Plan II (1993). Ministry of Housing, Spatial Planning and Environment SDU, The Hague.

Weterings, R.A.P.M., J.B.Opschoor (April 1992). The Ecocapacity as a challenge to technological development, (advise on request of the program direction for Environment and Technology, Ministry for Housing, Spatial Planning and Environment). Advisory Council for Research on Nature and Environment, Rijswijk, pp 5-30.

WRR (Scientific Council for Government Policies) (1994). Duurzame Risico's: een blijvend gegeven, Sdu Uitgeverij, Den Haag, ISBN 90-399-0179-6, pp 27-49.

Chapter 21

The substance flow approach

Implications for European Union policy

MARIUS ENTHOVEN

Director General, Directorate General XI, European Commission.

Key words: eco-labelling, eco-efficiency, integrated approach, best available techniques

Abstract: This chapter provides a brief overview of the policy of the European Union aimed at dematerialisation of the European economy. The Community policy aims at sustainable production and consumption, and supply and demand-oriented environmental policy.

1. THE MAIN CONCEPTS

1.1 Introduction: Industrial metabolism

The concept of *industrial metabolism* was originally developed in the 1980s in the broader context of *industrial ecology*. As shaped by Robert Ayres and Robert Frosch, the idea of industrial metabolism is based on an analogy between the biosphere and the industrial economy, with both considered as systems for the transformation of materials. It covers production and consumption, energy and the process itself of economic development (cf. Robert Ayres, *Industrial Metabolism*, 1989; see also chapter 1). Following this holistic approach, the main elements of industrial metabolism are the inputs and the outputs of the system. They account for the mass flows of key industrial materials of environmental significance and for the waste emissions associated with them and finally for the necessary actions to carry on proper recycling.

How valid today are the concepts of industrial ecology/substance flow analysis? What lessons can be drawn from this approach for the Community policy and for industrial management? This paper outlines the implications that these ideas could have for an overall strategy on Sustainable Production and Sustainable Consumption within the general context of the relevant Community policy. Understanding the concept of 'industrial metabolism' can be very useful for reaching a more unified, continuous and comprehensive consideration of production and consumption processes from an environmental point of view.

A first consequence of considering the industrial metabolism perspective is the recognition that, in many activities, the major sources of environmental pollution have been shifting from production to consumption processes. Several industrial sectors are now increasingly successful in controlling the material flows in their production processes. Therefore, from a public policy point of view, there is a need to follow the consumption side and to develop new concepts for monitoring the flows.

A second consideration is the recognition that a large number of materials used are, at the end of the day, lost to the environment. Many materials are degraded, dispersed and lost to the economy in the course of a single normal use. The main result of such a dispersion and dissipation of materials in the environment is to make problems global. We face a situation in which, on the one hand, problems of production tend to be industrial and local, on the other hand, problems of consumption will tend to be problems for everyone and global.

A practical application of the industrial metabolism viewpoint implies a detailed accounting and monitoring of the flow of materials and energy through human activities. Of course it is important not to refer to industrial metabolism as a dogmatic model. It would certainly be more profitable to use it as a reference background. It will make us more aware of the complex interaction of man-made activities and the need for a comprehensive examination of materials and energy use, emissions sources, transport, and of environmental impacts of different pollutants. This systematic look at materials and energy flows can lead to earlier identification of problems, taking into consideration a broader range of monitoring, including technological and socio-economic trends, as well as traditional environmental indicators.

The concept of de-materialisation was also developed within the framework of industrial metabolism. It is used to stress the decline, within a specified time, of the weight of materials used in industrial end products, or in the energy content of the products. Obviously a positive de-materialisation trend would be extremely important for the environment, because the principle of using less material could result in the generation of smaller quantities of waste in both production and consumption. The trend is definitely visible in many industrial sectors, however it is difficult to generalise it. From the very well known paradox of the "paperless office" to the certainly not diminishing generation of municipal solid waste, we can see examples of persistence of the old approach to production and consumption.

Specific reference to the de-materialisation concept could be useful to show the relative unimportance of considering production processes per se.

De-materialisation of products in order to minimise environmental effects all through the life-cycle and using resources in the most optimal way is also one of the characteristics of *eco-efficiency*. The World Business Council on Sustainable Development has defined eco-efficiency as follows: "Eco-efficiency is reached by the delivery of competitively priced goods and services that satisfy human needs and bring quality of life, while progressively reducing ecological impacts and resource intensity throughout the life-cycle, to a level at least in line with the earth's estimated carrying capacity". The underlying idea is "to produce more from less" by (a) reducing material and energy intensity; (b) reducing toxic dispersion; (c)enhancing (material) recyclability; (d) maximising the use of renewable resources; and (e) extending product durability.

1.2 Life-cycle assessment and material flow analysis

The accent that we need to put on material management must be supported by proper measurement and data collection. It is certainly a truism that what gets measured gets managed. As the growing importance of environmental issues to business requires a management response, there is increasing emphasis on appropriate measures to track performance and guide organisational actions.

Material flow analysis and life cycle assessment are becoming quite well known environmental policy tools. They are certainly based on well established industry practices. In fact, measures of energy or material utilisation efficiency have long been a feature of the manufacturing industry. As high efficiencies mean diminished use of resources, they now have environmental significance as well.

Proactive environmental industries have been using the control of inputs and outputs of their production processes to measure their environmental performance. A growing number of companies are making corporate environmental reports and presenting the results of this exercise to the public. Environmental reporting through the use of environmental performance indicators improves the credibility of industry in the eyes of government authorities and public opinion.

According to the definition developed by the Society of Environmental Toxicology and Chemistry (SETAC) "Life cycle assessment is a process to evaluate the environmental burdens associated with a product, process, or activity by identifying and quantifying energy and materials used and wastes released to the environment; to assess the impact of those energy and material uses and releases to the environment; and to identify and evaluate opportunities to affect environmental improvements. The assessment includes the entire life cycle of the product, process, or activity, encompassing extracting and processing raw materials; manufacturing, transportation and distribution; use, re-use, maintenance; recycling, and final disposal" (cf. SETAC, 1991). A full LCA study should include a description of the product system and its mass and energy balances (inventory phase), an accounting of the environmental and resource impacts of the system (classification and characterisation) and finally some valuation of the different impacts (normalisation and valuation; see for details Chapter 5).

In reality, most of the life cycle studies cover only a limited set of life cycle components. More limited life cycle inventory studies (LCI) in which no formal impact assessment is used are usually performed in order to provide mass and energy balances and environmental emissions inventories. In these cases, downstream assembly, use and waste management steps are not considered. Most life cycle studies have been comparative assessments of substitutable products delivering similar functions, but there has been a recent trend towards a more complex use of life cycle assessment in comparing material flows and services provision at a macro level (e.g. life cycle assessments of waste management options and transport systems).

Industry is now considering a more integrated perspective of the environmental impacts of its products. This approach has been driven by both market and regulatory considerations. The primary motivation for the adoption of the life-cycle approach appears to be the desire to capitalise on policy-driven strategic opportunities and to provide customer satisfaction through product improvements.

Practical adoption of life cycle approaches has resulted in the setting up of specific instruments such as eco-profiles of products and eco-indicators aimed at communicating with the public. According to the work currently going on in ISO with regard to environmental labelling, eco-profiles could be defined as "a quantified declaration of a product's environmental performance under a group of pre-set environmental indicators, based on the findings of a life-cycle assessment (LCA)". They constitute an environmental map of the product comprising the different impacts of all the materials and the processes involved.

One application of LCA is the development of the Eco Indicator method, which is rapidly gaining popularity in those countries where there is more attention to environmental product policy. With this mathematical procedure it is possible to compile all environmental impacts of a product into a single figure, the Eco Indicator. The magnitude of this Eco Indicator is a simple indicative measure of the product's environmental load. The usage of the Eco Indicator method for environmental evaluation of product design is gaining popularity within Dutch industry and it is officially supported by the government.

Companies are facing more and more product-related environmental pressures from government policy, environmental organisations, consumers, and from the market in general. LCA has been used by business as an important analytical tool to tackle these environmental pressures. Global and local problems create a political need for action and again LCA seems to be an effective tool for addressing these issues.

Consumers have begun demanding more information about the environmental impact of products. Following this request, we are now experiencing a proliferation of environmental labelling initiatives and other forms of industrial environmental reporting at national level. The Community response was the establishment of the Eco-label scheme at European level.

The EU Eco-label programme, as laid down in Council Regulation (EEC) 880/92 (now under revision), is one element of a wide strategy aimed at promoting sustainable production and consumption. With this aim in mind, the strategy embraces the promotion of environmentally aware behaviour patterns, in particular by identifying and promoting "green" products. The main purpose of the Community Eco-label is to influence the market directly by guiding consumers towards products with a reduced environmental impact. Indirectly the measure encourages producers to bring cleaner products to the market.

First of all, the scheme is selective. The label is awarded to products with the lowest environmental impact in a product range. Product categories are carefully defined so that all products that have direct equivalence of use as seen through the eyes of the consumer are included in the same product group. The European Commission develops ecological criteria for product groups in close collaboration with Member States. Indeed, the Commission cannot adopt criteria before environmental experts have given their opinion by voting in the Committee of representatives of Member States.

Transparency and widespread participation are further enhanced by the considerable input of representatives of industry, commerce, environmental and consumer organisations. The consumer, therefore, should recognise that the eco-label logo represents ecological criteria which have been established according to scientific and technical guidelines with widespread participation from independent and neutral bodies (see Chapter 22).

2. THE RELEVANT COMMUNITY POLICY

2.1 Industry and Environment

The challenge of today's environmental policy is to integrate environmental requirements into industrial activity from both a supply- production, and a demand-consumption, point of view. It is imperative to reconcile the objectives of improving the environmental performance of industry and strengthening its competitiveness. Consumption patterns will play a key role in meeting this challenge. Key players in the process are public authorities, industry, NGOs, consumer groups and trade unions.

The overall objective of the Community policy on Environment and Industry is to contribute to sustainable development. As mentioned in the proposed Decision on the Review of the Fifth Environmental Action Programme, this move will be possible by achieving sustainable production and consumption patterns. The following few years will be extremely important in completing the framework within which these patterns are to develop. Completion of the framework requires the setting of a number of targets. These include (a) scrutiny of the environmental performance of industrial activities; (b) identification of the most environmentally efficient production processes; (c) identification of guidelines to reduce industrial risk and waste; (d) identification and promotion of "*green products*"; (e) identification of the most environmentally friendly consumer behaviour; and (f) development and transfer of knowledge required for environmental policy-making.

2.2 Sustainable Production and Consumption

Sustainable production and sustainable consumption can be achieved in the context of a "*framework for an integrated life cycle oriented product policy*", as indicated in the proposed Decision on the Review of the Fifth Environmental Action Programme. In order to achieve sustainable production and sustainable consumption the main issues are risk, product and waste management .

The objective of a product-oriented environment policy based on a life cycle approach is to *identify the environmental aspects of a product through its entire life cycle* ("cradle-to-grave") so that policies can be developed which seek to reduce the overall negative effects of products and their production and thus avoid shifting environmental problems from one stage of the life of a product to another and from one medium to another. Identifying these environmental aspects allows for choices to be made, both by policymakers and regulators, as well as by all actors in the product chain. As we have already seen, material flow analysis and life cycle assessment can be effective tools for the establishment of a product-oriented environmental policy.

In essence, the "product" is the key to both sustainable production and consumption. By focusing on a product we are forced to confront all aspects of the way it is designed, produced, used and discarded, including not only emissions from installations during the production process, but also the pollution caused before and after the production process. In other words, the environmental effects throughout the whole product life-cycle.

Focusing on the life-cycle of a product has the immediate result of involving all the actors throughout the whole chain of a product: producers (suppliers, manufacturers, retailers) and consumers. Taking the life-cycle of a product as a basis for environmental policy measures provides an *integrated approach* to pollution. In the same way that it is essential to avoid shifting environmental problems between different media during the production process, it is essential to avoid shifting environmental problems between the various stages within the life-cycle of a product. This is possible by ensuring that those at the start of a product chain (designers, providers of raw materials, manufacturers) are aware of what happens to the product at a later stage (use, waste phase) and take that into account. For example, a product must be designed so that less energy is consumed during its use and it must be designed so that it can be repaired more easily or recycled once it becomes waste.

In order to minimise the environmental effects of a product throughout its entire life-cycle, *all actors* in the chain must be aware of the environmental issues at stake: information must be gathered and made available to the others in the chain.

2.3 Supply-oriented environment policy

When we come to supply-oriented environment policy at Community level, the successful implementation of the Integrated Pollution Prevention and Control (IPPC) Directive must be the central element in the strategy for promoting sustainable production. To fit in with the IPPC framework, all activities concerning sustainable production must take an integrated approach to the control and reduction of emissions to the different environmental media and to the efficient use of energy and raw materials, and must also be guided by the concept of Best Available Techniques (BAT). Promoting sustainable production will also require that special attention is paid to small installations because of their crucial importance in the creation of employment as well as their significance in terms of pollution.

Continuous improvement of the environmental performance of industry will be further strengthened by the continued implementation of measures such as the Environmental Management and Audit Scheme (EMAS). Further development of the EMAS approach will take into account the ongoing examination of the interaction between the existing Regulation and the IPPC Directive.

In the context of an integrated, life-cycle-oriented product policy producers will be encouraged to favour the continuous improvement of products and production processes, for example by developing concepts of *extended producer responsibility*. In this perspective, the EU Eco-label scheme can be seen as the necessary complement to sustainable production and the link towards sustainable consumption. In fact, it provides the framework for producers to identify, on the basis of the life-cycle approach, more environmentally friendly products, and then to communicate to consumers the environmental added value of their products. The Eco-label makes a link with the demand-side of Community environmental policy.

2.4 Demand-oriented environmental policy

Information is going to be the main character of a Community demand-oriented environmental policy. The Review of the Fifth Environmental Action Programme stresses the importance of creating awareness amongst all sectors of society in order to encourage environmentally friendly behaviour patterns all through the product chain, from product designers to producers to consumers.

Diffusing information about the environmental effects of a product during its whole life-cycle will be essential in order to support sustainable consumption. It is difficult to introduce policy measures which encourage the various actors in the chain to change their behaviour if it is not known what the environmental effects of their current behaviour are and what the best behaviour would be. On the basis of life-cycle approaches, more environmentally friendly products can be identified. The easiest way for consumers to recognise "green" products is by labelling them, as is happening in the EU Eco-label scheme. Consumers should be encouraged to buy these products, so as to put pressure on producers.

In this way new consumption patterns will have an impact on production. Companies should redefine customer satisfaction through information. Usually both marketing and economic theory have concentrated on the consumer's satisfaction with a product or service, only taking account of the particular moment of consumption. This attitude is today clearly insufficient. The Community framework of sustainable production and consumption aims to encourage a more holistic and better-informed concern with the impacts of consumption throughout the life cycle of products and services.

Proactive companies have to redefine customer satisfaction because they are concerned not only with the performance of the product at the time of consumption, but also with what happens to it afterwards. This 'cradle to grave to cradle again' approach, perfectly in line with the concept of industrial metabolism, involves the whole of a company's activities and the whole life history of a product in reinforcing customer satisfaction and building loyalty. Companies have to realise that many of the demands which may emerge as a result of environmental pressures could actually help to develop a stronger, longer-term relationship between the company and the consumer.

3. CONCLUSION: COMMUNITY ACTION

Finally, it is clear that the European Commission is changing its emphasis with respect to environmental policy. In the move towards sustainable development, both the role of producers (retailers as well as manufacturers) and consumers represents a key element. A shift is taking place from detailed, medium-oriented legislative measures to framework legislation and additional policy instruments based on an integrated and life cycle oriented approach to pollution abatement. The aim is to develop a general framework for a product-related environmental policy based on a life cycle approach. This framework will take the form of a Communication on Integrated Product Policy, addressing all actors in the chain. Both producers and consumers will be encouraged to change their behaviour patterns in a way which is less damaging to the environment. In this context the lessons learnt in defining and applying concepts such as industrial metabolism, material flow analysis, LCA are once again valid and can still play an important role within more integrated environmental policies which take into account patterns of consumption and waste generation.

REFERENCES

Robert Ayres (1989). *Industrial Metabolism*, in J.H.Ausubel and H.E. Sladovich (eds.), Technology and Environment, National Academy of Engineering, National Academy Press, Washington D.C., pp. 23-49.

SETAC (1991). *Guidelines for Life-Cycle Assessment: A 'Code of Practice'*, Brussels/Pensacola.

Council Regulation 880/92, *Regulation on a Community ecolabel scheme*, Brussels , 11.4.1992.

Chapter 22

The EU Eco-label

Supporting democratic decisions

MARCO LOPRIENO

The author works at the Directorate General XI of the European Commission, in Brussels.

Key words: LCA, eco-labelling, EU policy

Abstract: The chapter focuses on the policy and methodological implications of the use of LCA in the EU Eco-labelling scheme. The fundamental contribution of LCA methodology to the scheme is the definition of a tool to support the decision making process. The proposed Revision of the Regulation 880/92 clarifies the original generic "cradle to grave approach". The formalisation of the Procedural Principles for establishing Eco-label criteria provides a specific framework for the involvement of the interested parties and therefore a "democratic control" of LCA.

1. INTRODUCTION: THE EU ECO-LABEL IN THE BROADER CONTEXT OF A PRODUCT-ORIENTED ENVIRONMENT POLICY

The EU Eco-label programme, as laid down in Council Regulation (EEC) No. 880/92,[1] is one element of a wide strategy aimed at promoting sustainable production and consumption. With this aim in mind, the strategy embraces the promotion of environmentally aware behaviour patterns, in particular by identifying and promoting "green" products. The purpose of the Community Eco-label is to influence the market by guiding consumers towards products with a reduced environmental impact. The consumer, therefore, should recognise that the Eco-label logo represents ecological criteria which have been established according to scientific and technical guidelines with widespread participation of independent and neutral bodies.

Being a so-called "market based instrument", the primary function of the EU Eco-label is to stimulate the dynamics of supply and demand of products with a reduced environmental impact. This economic approach is also reflected in the measure to "provide guidance to consumers" which has considerable implications with regard to economic efficiency. In fact: "the eco-label could reduce costs for consumers and retailers by lowering the time and effort needed to obtain and provide information."[2]

The identification and promotion of eco-labelled products is based on a policy oriented use of the Life Cycle Assessment methodology. In this framework the EU Eco-label is probably one of the most important tools of the European environmental policy to manage material flows in industrial products.

First of all, the EU Eco-label award scheme is selective. The label is awarded to products with the lowest environmental impact in a product range. Product categories are carefully defined so that all products that have direct equivalence of use, as seen through the eyes of the consumer, are included in the same product group.

The Scheme is transparent. The European Commission develops ecological criteria for product groups in close collaboration with the Eco-label national competent bodies and the Member States. Indeed, the Commission cannot adopt criteria before environmental experts have given their opinion by voting in the Committee of representatives of Member States, the Eco-label Regulatory Committee.

Transparency and widespread participation are further enhanced by the considerable input of representatives of different interest groups (industry, commerce, trade unions, environmental and consumer organisations) represented at the level of the Eco-label Consultation Forum.

Other national and supranational eco-label schemes started with environmental evaluations made on a rather qualitative basis, most of the times developing a simplistic methodological approach. The EU Eco-label adopted a more quantitative basis that, together with the use of a multi-criteria approach, set from the beginning the distinctive characteristic of the EU Scheme.

The overall objective of the Community policy on Environment and Industry is to contribute to sustainable development. The EU Eco-label scheme is well in line with the principles, goals and priorities selected by the Fifth Community Environmental Action Programme and its Review, as well as of Agenda 21.

Sustainable production and sustainable consumption can be achieved in the context of a "*framework for an integrated life-cycle oriented product policy*", as indicated in the proposed Decision on the Review of the Fifth Environmental Action Programme.[3]

The objective of a product-oriented environment policy based on a life-cycle approach is to *identify the environmental aspects of a product through its entire life-cycle* (cradle-to-grave) so that policies can be developed which seek to reduce the overall negative effects of products and their production and thus avoid shifting environmental problems from one stage of the life of a product to another and from one medium to another. Identifying these environmental aspects allows for choices to be made, both for policy makers and regulators, as well as for all actors in the product chain.

In essence, the "product" is the key to both sustainable production and consumption. By focusing on a product, the way it is designed, produced, used and discarded, not only emissions from installations during the production process are addressed, but also pollution caused before and after the production process. In other words, the environmental effects throughout the whole product life-cycle.

Focusing on the life-cycle of a product has the immediate result of involving all the actors throughout the whole chain of a product: producers (suppliers, manufacturers, retailers) and consumers.

Taking the life-cycle of a product as a basis for environmental policy measures provides an *integrated approach* to pollution. In the same way that it is essential to avoid shifting environmental problems between different media during the production process, it is essential to avoid shifting environmental problems between the various stages within the life-cycle of a product. This is possible by ensuring that those at the start of a product chain (designers, providers of raw materials, manufacturers) are aware of what happens to the product at a later stage (use, waste phase) and take that into account. For example, a product must be designed so that less energy is consumed during its use and it must be designed so that it can be repaired more easily or recycled once it becomes waste.

In order to minimise the environmental effects of a product throughout its entire life-cycle, *all actors* in the chain must be aware of the environmental issues at stake: information must be gathered and made available to the others in the chain.

Information is going to be the main character of a Community demand-oriented environmental policy. We have already seen how the Review of the Fifth Environmental Action Programme stresses the importance of creating awareness amongst all sectors of society in order to encourage environmentally friendly behaviour patterns all through the product chain, from product designers to producers to consumers.

Diffusing information about the environmental effects of a product during its whole life-cycle will be essential in order to support sustainable consumption. It is difficult to introduce policy measures which encourage the various actors in the chain to change their behaviour if it is not known what the environmental effects of their current behaviour are and what the best behaviour would be.[4]

2. ANALYSIS OF THE ROLE OF LCA IN REGULATION 880/92

According to Articles 1 and 5 (4) of Regulation 880/92, the Community Eco-label scheme is based on a "cradle-to-grave approach." The criteria to be set for each specific product group (e.g. copying paper, indoor paints and varnishes, etc.) should be based on the consideration of the *environmental impacts* in all the life stages (pre-production, production, distribution, use, disposal) of the products considered. Moreover, the objective of Regulation 880/92 is, in fact, to promote "products that have a reduced environmental impact during their entire life-cycle" (Article 1).[5]

The use of the LCA methodology to study the environmental impact of product in the EU Eco-label Scheme has fuelled many criticisms. From many parts it has been questioned the legitimacy and the competence of public authorities intervention to improve the environmental performance of products. "Experts who are charged with the task of developing an ecolabelling scheme (ideal or otherwise) are likely to encounter a number of more-or-less intractable problems: they will not be able rationally to select product categories; they will not be able rationally to set product categories boundaries; they will not be able to take into considerations all the physical effects which a product has on the environment during its life-cycle; they will not be able accurately to estimate the impacts of these effects and they will not be able continuously to update the ecolabel product selection criteria. It should be stressed that these are not merely 'technical' problems: they cannot be resolved or avoided. So the claim that ecolabels might guide consumers to more environmentally sound purchases is untenable."[6]

Obviously this point of view is the result of a limited understanding of the role of Life Cycle Assessment within the actual functioning of the Programme. In particular, it must be clear that LCA is only a decision support tool and that the EU Eco-label is an environmental marketing instrument. Both are not the panacea of the European environmental policy but, nevertheless, they contribute in a pragmatic way to shape a sustainable consumption policy. In this framework the definition of a clear and accountable set of procedures is much more important than just aiming at the definition of a very sophisticated theoretical instrument with very little impact on the real world.

Moreover any eco-label scheme which has criteria as its basis is by nature discriminatory since the purpose of the scheme is to allow consumers to make a choice between products based on their environmental impact. In other words Eco-label is not a "win-win solution" because, by definition, it is an instrument aimed at foster competition on environmental grounds. As Giandomenico Majone stressed more than twenty years ago talking about choices among policy instruments for pollution control, "The search for a system that would resolve most of the political conflict over the environment (...) is bound to lead to disappointment."[7]

The Regulation just sets a framework for the Community Eco-label. The scheme can only be applied to products for which ecological criteria have been established by the Commission in accordance with the principles and procedures of the Regulation. The entry into force of the Regulation was therefore only a starting point for the preparatory work for the actual launch of the scheme. The period since the entry into force of the Regulation has mostly been devoted to establishing product groups and the corresponding ecological criteria.

The following Commission Decisions establishing ecological criteria have so far been adopted and published: Washing machines, Dishwashers, Soil Improvers, Toilet paper, Paper Kitchen Rolls, Laundry Detergents, Single-ended Light Bulbs, Indoor Paints & Varnishes, Bed-linen and T-shirts, Double-ended Light Bulbs, Copying Paper and Refrigerators. The considerable amount of work behind the definition of the different product groups allowed a constant refining and improvement of the LCA methodology and a better understanding of its role with regard to the policy framework. As Prof. Udo de Haes notices about the recent situation of the EU Eco-label: "Although many methodological points are still unresolved, the standardising of LCA methodology has helped remove roadblocks in the process. This is reflected in the number of product groups that are currently involved in the criteria-setting process for European ecolabels (...). In some cases ecolabels have significantly increased the market share of the products at stake."[8]

As a practical example of product development caused by the EU Eco-label based on LCA we can take the case of Hoover the appliance manufacturer company that first applied to the Eco-label for washing machines in 1993.

In particular regarding the "New Wave" range of machines: "the LCA for the eco-label confirmed Hoover's focus on reducing the water, energy and detergent consumption of New Wave, but also indicated other areas of environmental impact that should be considered before the launch of the range; if it was to meet the eco-label criteria. Hoover thus began to consider the environmental impacts arising from production, distribution and disposal of the New Wave."[9]

The use of the EU Eco-label has been perceived by Hoover as an important success factor "especially in the environmentally-aware German market in which the company doubled its market share in 1994. It also enabled the company to enter other environmentally-conscious markets such as Denmark and Austria."[10]

Especially in the field of appliances industrial designers from all the major companies are usually taking more and more into account the ecological criteria developed for washing machines, dishwashers, refrigerators with regard to energy and water consumption, refrigerants, etc.[11]

One of the main characteristics of the relationship between LCA and the EU Eco-label scheme is certainly the "work in progress" and "learning by doing" approach. As the *Groupe des Sages* defined it: LCA "... is a methodology still in the process of development, requiring additional research and systematic data collection. Therefore policy makers, competent bodies and practitioners must remain aware of the current capabilities and limitations of LCA and should support its continuous development."[12]

In order to ensure consistency, effectiveness and sufficient quality of the criteria setting process, it has been necessary to define a six-phase procedure, based on the "*Procedural Guidelines for the Establishment of Product Groups and Ecological Criteria*" which was agreed between the Commission services and the Competent Bodies of the Member States.[13] The different phases are: feasibility study; market study; inventory; environmental impact assessment; setting of criteria; presentation of a draft proposal for a Commission Decision.

The preparatory work for the setting up of the *Procedural Guidelines* started in December 1993 when the Commission established the so-called *Groupe des Sages* to advise on the methodological implications of using LCA in the general framework of the Eco-label. The *Groupe des Sages* produced three reports. The main themes of their research were summarised in the key recommendations of the first report, where the relationship between LCA and analysis of the market conditions and innovation was already clear. "Life-Cycle Assessment can make a significant contribution in providing a scientific, unifying and transparent basis for the EU Eco-labelling programme. It is central to this programme because it compares different products on the basis of their common function. It relates environmental impacts, at all stages from cradle to grave, to both market changes and technology improvements."[14]

The *Groupe* played an important role in fitting the LCA methodology already established at international level by SETAC, the Society for Environmental Toxicology and Chemistry, and ISO, the International Standards Organisation, within the context of the EU Eco-label.

Therefore, the objectives of the guidelines set by the *Groupe des Sages* have been extremely useful in supporting the Member States and the Commission in establishing Eco-labelling criteria based on a scientifically sound and practicable methodology. Moreover, the methodological guidelines have improved the uniformity in the methods applied in different Member States.

The work of the *Groupe* has been instrumental in clarifying the distinction between the decision-making phases and the proper LCA study. In other words, between the policy aspects, the proposal of the key ecological criteria, and the scientific analysis on which the LCA is based. "LCA is only a tool for decision-making; which cannot replace the actual decision-making itself"[15]

Such an approach is fundamental in clearing all the possible misunderstandings regarding LCA. In fact, when it is clear that LCA is only a decision support system, which cannot replace the actual decision making, it is possible to make a clear distinction between the Eco-labelling procedure, under the responsibility of the Eco-label Competent Bodies and the Commission, and the LCA study under responsibility of the LCA-practitioners.

The policy aspects were stressed by internal guidelines agreed between the Commission and the Eco-label Competent Bodies which have constituted an indispensable reference point for the development of principles to be complied with in the establishment of product groups and ecological criteria.[16]

In practice, the functioning of the Eco-label scheme consists of three general phases: the establishment of criteria, the award of the label to products and the revision of the criteria. Whereas responsibility for establishing and revising the criteria lies mainly with the Commission, the award of the label to products is a matter for the National Competent Bodies. These Competent Bodies, which are independent and neutral, have been designated by the Member States to implement the Community Eco-label scheme at national level.

The initiative for selecting a group of products is taken either by the Commission, or by the Competent Bodies. In the initial stage of the operation of the Community scheme, priority was given to the latter possibility. More recently, the Commission has assumed sole responsibility for selecting groups of products. This is in line with the wishes of the Member States and the interest groups for greater consistency in the application of the scheme. The interest groups, i.e. industry, commerce, consumer organisations, environmental protection organisations and trade unions, are consulted on the choice of product groups.

A feasibility study is carried out to collate data on the following aspects: the market structure, the interests of the parties concerned, the relevance and potential benefits of the label for the environment, the risks of distortion between the various national segments of the internal market and finally international aspects. An ad hoc workshop composed of experts from the Member States and representatives of all the parties concerned evaluates the feasibility study. On the basis of these results, a complete analysis of the life-cycle of the group of products is made. This study comprises a market study, an inventory and evaluation of the environmental impact of the group of products, and a proposal for criteria.

The proposal for ecological criteria is officially presented to a Forum provided for in the Regulation for consultations with interest groups. The proposal is discussed and voted upon in a Regulatory Committee. A formal Decision by the Commission concludes the adoption procedure.

Given the nature of the Eco-label, which involves a range of responsibilities and the internal procedural rules of the Commission's departments, those departments collaborate closely in the various stages of the process of drawing up the criteria. In particular, the draft Decision to be presented to the Regulatory Committee is the subject of prior interdepartmental consultation.

Under the Regulation, the Competent Bodies are responsible for awarding the label. A summary of each application is circulated to all the Competent Bodies, whereas the complete dossier on the evaluation of the product is sent only on request.

3. LCA AND THE PROPOSED COMMISSION REVISION - COM (96) 603

The operation of the Community Eco-label scheme has recently made substantial progress (November 1997). Eco-label criteria have been now published for 12 product groups and the Eco-label has been awarded to 23 producers for a total of more than 182 products.

Since the launching of the EU Eco-label, certain difficulties have been encountered in the implementation of the Regulation. In particular, there is a need for improving and streamlining the approach, methodologies and working procedures of the scheme, in order to increase its effectiveness, efficiency and transparency. Therefore, on 10 December 1996, the Commission approved a proposal to amend Regulation 880/92, in accordance with its Article 18, which provides for the revision of the Community Eco-label scheme within five years of its entry into force.[17]

Several concepts and requirements in the present Eco-label Regulation 880/92 have appeared difficult to interpret and implement in the absence of sufficient operational indications. In particular, the concept of "reduced environmental impact" during the entire life cycle of a product.[18] No methodology exists to determine the total environmental impact of a product in absolute terms. Strictly interpreted, this concept cannot be properly implemented and it has fuelled all sorts of criticisms regarding an objective use of LCA. Therefore, the methodological principles for establishing the ecological criteria are clearly stated in the proposed Revision.

It should be clear that the methodological approach includes a life-cycle analysis applied to the product group concerned, on the basis of which a limited number of key environmental aspects are selected and the improvement and substitution options are identified. The criteria should refer to these aspects and also take into account the practical possibilities for improvement of the product in a life-cycle perspective.

The provision for excluding products classified as dangerous in accordance with Directives 67/548/EEC and 88/379/EEC,[19] if applied in a rigid way, leads to the exclusion of entire categories of products such as compact detergents (classified as irritant) and solvent based paints (flammable).

Another difficult measure is the mandatory consideration of all the life stages of a product. Without qualifications, this requirement might imply the development of criteria for all the raw materials used in manufacturing a given product. In most cases such an extensive application of the "cradle-to-grave" approach is not practically feasible.

The proposed Revision clarifies even more the nature of the Community Eco-label scheme by stating that the programme is intended to guide consumers towards products which represent more environment-friendly alternatives compared to other products in the same product group. It should also be clear that the Eco-label is nothing more than an indication of the potential for reducing certain impacts. Eco-label criteria are, in fact, based on a generic assessment of such impacts, not on a study of the actual environmental effects related to the life cycle of each specific product.

We have already seen how Regulation 880/92 stresses the objective of providing consumers with better information on the environmental impact of products. In the proposed Revision the shape of the label itself should correctly reflect this objective by including more information on the key aspects which motivate the awarding of a label to a given product. Again, a correct implementation of the LCA methodology will be fundamental in providing the basis for the necessary policy decisions.

The present approach is based on a "pass-fail" system. In the case of the EU scheme, which is multi-criteria based and applies to a wide variety of conditions throughout the Community and internationally, this approach has proven to be insufficiently flexible and to involve substantial difficulties in setting suitable hurdles for the parameters under consideration. Therefore, it is proposed to introduce a rating for each of the quantitative criteria considered. The hurdles corresponding to the first level (one "flower") would represent the base-line for a product in order to be awarded an Eco-label.

Further improvements on one or more of the parameters would be recognised by attributing two or three "flowers". This would provide an incentive to and recognition of producers for such further improvements and information to consumers on the specific characteristics of each product labelled.

At present, Regulation 880/92 does not include criteria for selecting product groups for the Community scheme. Only food, drinks and pharmaceuticals are excluded *a priori*. It should be clarified that the Community scheme should not apply to products which are of minor interest at Community level in terms of the internal market and environmental policy. The lack of criteria for the selection of product groups, together with the opportunity which is presently offered to a Competent Body to require the opening of the procedure for setting Eco-label criteria, involve a risk of dispersion and waste of resources in the operation of the Community scheme.

The selection criteria should also take account of the suitability of the Eco-label as a policy tool for the promotion of improving a specific product sector. No guidance is given at present in Regulation 880/92 on how to establish the selectivity level of Eco-label criteria. This has resulted sometimes in the need to reconcile divergent points of view between Member States where product technology and market structures are substantially different. In fact, certain Competent Bodies tend to interpret the Eco-label as a sign of excellence, whereas others are more interested in broader participation in the scheme and its overall potential for promoting improvements.

The necessary dialectics among different national approaches have never resulted in practices of "horse-trading" between product groups and key ecological criteria in order to defend some hidden economic interests, as some commentators are trying to suggest.[20] On the contrary, the harmonisation process among the Eco-label Competent Bodies is definitely heading towards higher environmental protection standards as the latest Decisions on copying paper and refrigerators show.

Without generalising we can very well understand Eco-labelling in the context of the European environmental policy making. As Alberta Sbragia notices "The policy process in the environmental arena illustrates how a minority of states can use the EU to force an upgrading of standards in the rest of the Union. Rather than 'environmental dumping' the Union's policy-making process has led to 'up-market environmentalism'."[21]

Setting the selectivity level of the Eco-label will be easier under the proposed graded approach which introduces more flexibility for adaptation to the specific circumstances of the various Member States.

4. CONCLUSIONS: THE "DEMOCRATIC CONTROL" OF LCA

As we have already seen a distinctive character of the EU Eco-label scheme has always been the involvement of the parties concerned and the balanced participation of all the relevant interest groups. The ecological criteria setting process for each product group is developed within the framework of a specific Ad Hoc Working Group involving all the interested parties. This practice has been used more and more within the scheme. In 1996-97, for instance, the Commission organised more than 40 Ad Hoc Working Groups to discuss different phases of technical projects.

The attention given to the procedural aspects of the EU Eco-label Scheme, in order to make it more efficient and workable, starts to pay off. In fact, if we compare the awards situation in the first half of 1997 with the same period of 1996 the number of products Eco-labelled multiplied by four spreading out in nine different countries: Denmark, Finland, France, Ireland, the Netherlands, Portugal, Spain, Sweden and the United Kingdom. "It is certainly too early to assess the consumers' response and the users' perceptions and reactions. However, the very encouraging reactions from manufacturers in recent months show that the dynamics of supply and demand in a free market are working to the advantage of firms applying for the EU Eco-label."[22]

The formalisation of the Eco-label Policy Principles developed within the present Regulation will set out, in the proposed Revision, the procedures for the operation of the scheme, aimed in particular at ensuring the efficiency and transparency of the Eco-label criteria setting process.

According to Annex IV of the proposed Revision, the widespread dialogue among the different stakeholders is foreseen as a fundamental element for the development of Eco-label criteria. "The involvement of the parties directly or indirectly concerned by the mandate and a balanced participation of all the relevant interest groups, such as industry, including SMEs and hand crafts through their business organisations, trade-unions, retailers, importers, environmental protection groups, consumer organisations, shall be actively pursued. A specific ad hoc working group involving the interested parties mentioned above shall be established for the development of Eco-label criteria for each product group." "A specific work programme and a corresponding time table shall be established including, in particular, the following phases: Market study; Life-Cycle Assessment; Proposal of the criteria. Each phase and step shall be concluded by at least a meeting of the ad hoc working group in order to consider the results and indicate further orientations."[23]

The open consultation and transparency is even more stressed by the following provisions:

"A final report containing the main results shall be issued and published. Interim documents reflecting the results of the different stages of work shall be made available to those interested and comments on them shall be considered.

A draft version of the report including also the draft Eco-label criteria shall be published. An open consultation on the content of this draft report shall be carried out. A period of at least 60 days for the submission of comments on the draft criteria will be allowed before adoption of the criteria. Any observations shall be considered. On request, information on the follow-up to the comments will be provided."[24]

Again this practice of democratic confrontation and dialogue among different stakeholders has been perceived by some companies and commentators as "potentially wrong and deceptive" or as a source of "political manipulation."[25] We believe that this is not the case. In particular because the EU Eco-label allows full participation of all possible interest groups to discuss LCA methodology in a complete transparent way. Trade associations from different industrial sectors (appliances, detergents, paper, textiles, etc.) have participated to the various Eco-label working groups. This procedure can be considered as a workshop for testing the applicability of new environmental policy measures based on life-cycle approach.

By definition, in the move towards sustainable development, both the role of producers and consumers represents a key element In order to support sustainable consumption, more environmentally friendly products can be identified. The easiest way for consumers to recognise "green" products is by labelling them, as is happening in the EU Eco-label scheme. The most logical way for producers to avoid inaccuracies and distortions is to participate in the criteria-setting process sharing information. The necessary co-operation that follows is at the hearth of the democratic process. It can be analysed, using the theory of games. In fact, according to the framework developed by Robert Axelrod, the "reputation" of a seller is the main reason for making truthful statements about the product which he or she is selling. Moreover, the higher the probability of future interaction with the buyer (the consumer) the less incentive the seller (the producer) has to lie, since the producer is expecting the buyer to continue to buy the same products.[26]

The example could be extended to an "Economic Theory of Democracy" but it can be used also to exemplify the interactions of the different stakeholders in the framework of the EU Eco-label. However, it is always important to ensure the transparency of the "rules of the game" by a widespread participation and a general accountability. In other words, there must be a clear distinction between the decision support tool (LCA) and the decision-making process. This particular approach is the essence of what we could refer to as the "democratic control" of the LCA methodology.

---

[1] Council Regulation (EEC) 880/92 of 23 March 1992 on a Community Eco-Label award scheme, Official Journal of the European Communities, L 99, 11.04.1992.

[2] Frans Oosterhuis, Freder Rubik, Gerd Scholl (1996), Product Policy in Europe: New Environmental Perspectives, 'Kluwer Academic Publishers', Dordrecht, p. 145.

[3] Proposal for a European Parliament and Council Decision on the review of the European Community Programme of policy and action in relation to the environment and sustainable development "Towards sustainability", COM (95)647 final, p. 9.

[4] For an analysis of the need for a product-oriented approach in environmental policy see: Frans Oosterhuis, Freder Rubik, Gerd Scholl, op. cit., pp. 33-5.

[5] Council Regulation (EEC) 880/92, Article 1, p. 2.

[6] Julian Morris (1997), Green Goods? Consumers, product labels and the environment, Institute of Economic Affairs, London, p. 49. The recent book of Mr Morris is a rare concentrate of misleading information regarding the EU Eco-Label scheme. As an example the main rationale behind a possible confrontation between a national approach to Eco-Label and a "more centralised European system" is the following: "The primary reason for these differences of opinion seems clear: the greater the control any particular person has over the decisions made by the ecolabelling authority, the more likely that person is to benefit (either electorally or financially) from the decisions made - so national politicians and civil servants favoured an ecolabelling authority on home turf, whilst Commissioners and Eurocrats favoured an ecolabelling authority in Brussels" (ibid., pp. 58-9).

[7] Giandomenico Majone (1976), "Choice Among Policy Instruments for Pollution Control", Policy Analysis, 2,

[8] Udo de Haes (1997), "Slow Progress in Ecolabelling. Technical or Institutional Impediments?", Journal of Industrial Ecology, 1 (1), pp. 4-5.

[9] Robin Roy (1997), "Design for environment in practice - development of the Hoover 'New Wave' washing machine range", The Journal of Sustainable Product Design, April, p. 39.

[10] Robin Roy, op.cit., p. 40.

[11] Personal communication to the author by Prof. Han Brezet, Delft University of Technology, September 1997.

[12] Groupe des Sages (1997), Guidelines for the Application of Life-Cycle Assessment in the EU Eco-label Award Scheme, European Commission, p. 16.

[13] See the special issue of the Eco-label newsletter, European Commission, DGXI, Procedural Guidelines for the Establishment of Product Groups and Ecological Criteria, Commission Information on Eco-labelling, no. 6, Brussels, June 1994.

[14] Groupe des Sages, op.cit. p. 16.

[15] Groupe des Sages, op.cit. p. 7.

[16] European Commission, DGXI, A Framework of Policy Principles for the Establishment of Product Groups and Ecological Criteria, Brussels, 1994.

[17] Proposal for a Council Regulation (EC) establishing a revised Community Eco-label Award Scheme, COM (96)603 final.

[18] Council Regulation (EEC) 880/92, Article 1.

[19] Council Regulation (EEC) 880/92, Article 4 (2).

[20] See again on this point: Julian Morris, op. cit., p. 67. According to the analysis carried on by Mr Morris we could easily identify the EU Eco-label with a general conspiracy theory against the Western economy.

[21] Alberta Sbragia (1996), "Evironmental Policy the 'Push-Pull' of Policy-Making" in Helen Wallace, William Wallace (eds), Policy-Making in the European Union, Oxford University Press, Oxford, p. 253.

[22] Marco Loprieno (1997), "The European Union Eco-label scheme: an environmental policy marketing tool", Industry and Environment, no. 1-2, p. 38.

[23] COM (96) 603 final, p. 37.

[24] op. cit., p. 38.

[25] In reality behind most of the attacks to the EU Eco-label there is a concerted campaign against eco-label schemes certified by third parties. See: "Procter & Gamble steps up attack on eco-labelling", ENDS Report 265, February 1997, pp. 26-7. Where the move is led by a coalition of US companies and industry associations ideologically more in favour of informative labels.

[26] See on this point: Robert Axelrod (1984), The Evolution of Cooperation, Basic Books, New York, pp. 126-32.

Chapter 23

Towards Industrial Transformation

The way ahead

PIER VELLINGA, JOYEETA GUPTA AND FRANS BERKHOUT

Pier Vellinga is Director at the Institute for Environmental Studies, Vrije Universiteit, Amsterdam, Joyeeta Gupta is a researcher at the Institute for Environmental Studies, Vrije Universiteit, Amsterdam, and Frans Berkhout is a Senior Fellow at SPRU, University of Sussex, UK.

Key words: industrial ecology, industrial transformation, dematerialisation, incentive structure, consumers, production and consumption patterns, evolutionary economics.

Abstract: This article highlights the key arguments and paradoxes that have been presented by the authors in this book. It illustrates that analysis promoted under the emerging research paradigm of industrial ecology provides a basis for the transformation of production and consumption processes in more sustainable directions. Research into the effects of the macro incentive structure such as the fiscal and trade incentives and research into the preferences and behaviour of consumers in their role as voter, worker, shareholder and consumer should complement existing approaches. The article promotes a transformation approach that includes evolutionary economics, political sciences, consumer and market sciences thus expanding the present industrial ecology approach. Such an approach can provide a clearer picture of what drives the system of production and consumption and it can identify and clarify more sustainable pathways. The article concludes with the presentation of an international research agenda that builds on existing activities of the International Human Dimensions of Global Change Program on Industrial Transformation.

1. EVOLUTION OF RESOURCE MANAGEMENT

In the 1990s a number of concerns have come together providing the basis for international efforts and new approaches to safeguard the global environment. One is the concern about depletion of resources and the need to consider resources in economic analysis and decision making, originally focusing on limits to the availability of resources on the input side of the economy. Over time the need to safeguard environmental life support systems such as biodiversity and soils, water and climate have also come to be seen as resource questions (as in the 'ecospace' concept). Scarcity of resources can now be seen in the round, including inputs, processes and waste outputs from economic activities.

Second is the pollution agenda, starting as an end-of-pipe response to clearly visible disasters evolving into an industrial ecology approach with a focus on technological and organisational structure, diversity and efficiency. This includes a changing role for technology moving from abatement technology towards a focus on core processes, such as renewable energy production and the recovery and re-use of materials.

The third concern was about fulfilling the needs of a growing world population. In the event that the developing world follows a development path similar to that of the industrial world, the volume of goods and services will have to grow by many orders of magnitude. Technological analysis on the basis of extrapolating existing trends in resource efficiency indicates that the global resource base may be able to support such demands, but major concerns remain. For example, it is not obvious that the technological systems required for the requisite efficiency increases will be available and will be introduced in time. There are enormous political and institutional hurdles to be overcome. Moreover, meeting the priority needs of shelter, food and energy is likely to go at the expense of the quality of the environment in terms of biodiversity and resilience, and may have an impact on quality of life. Fears of global scarcity and global interdependency have triggered many initiatives for international co-operation in the field of environmental policy and research.

A concern with resource scarcity, pollution prevention and economic and social trends has provided the basis for this book about substance and material flows in the economy and the need for a transformation of production and consumption processes. We believe that society, as it enters the 21st century, faces the challenge of facilitating a process of industrial transformation that meets both environmental and societal concerns.

At this stage research can play an important role in identifying and analysing a range of transformation scenarios. Such a research agenda includes industrial ecology, but it is more than that. It includes a re-analysis of human needs and preferences, and a re-analysis of incentive structures at a macro and micro level.

This chapter analyses and sums up the key concepts, research tools, emerging paradoxes, and implementation issues that arise in this book. It presents a research agenda for the 21st century and draws conclusions on how such an agenda can be promoted internationally.

2. FROM INTUITION AND BELIEFS TO OPERATIONAL PRINCIPLES AND CONCEPTS

The debates on environmental protection have taken place against the background of our philosophical insights about how the world is constructed. When a system is to be improved, it is necessary to distil the kinds of principles that will guide such improvements. Such principles can be derived from concepts of morality prevalent in a society, from nature, or from scientific rationality. For pioneers on the issue such as Robert Ayres, it is important to understand the metabolism of the industrial system, the processes of material cycles and energy flows within the industrial system as a subsystem of the comprehensive biogeochemical cycles. An understanding of the underlying scientific principles would provide the answers to the problem of resource scarcity and environmental pollution. In Chapter 1 Ayres makes a case for using quantitative approaches to simplify the complexity of production and consumption systems, in order to make predictions and to inform policy. He argues that the simplifications introduced by neo-classical economics may not prove to be reliable assumptions for predicting the future. Ayres and others have sought to apply thermodynamic principles to understanding the relationship between resource consumption and organisation in economies. Economies are seen as 'dissipative structures' which consume materials extracted from concentrated resources, process them relatively rapidly, and then dissipate them back into the environment. Primarily non-renewable chemical energy is used to drive this 'linear' use of resources. Through combustion to provide work and to provide energy services such as heat and light, a parallel process of dissipation of the potential energy locked up in fossil energy resources therefore occurs. Many of these insights have been borrowed from the thermodynamic view of ecosystems which has developed over the past 40 years. The main conclusions are that industrial systems should become less materials intensive, and should aim to be based on renewable resources.

However, there are limitations to the application of metaphors taken from nature to understand economic processes. Chadwick argues in Chapter 2 that many unfounded ideas persist about the relationship between resource use and organisation in natural biogeochemical cycles. He cautions that these cycles are not always balanced, or functioning in equilibrium. Moreover, nature is not necessarily "efficient". There are many examples of accumulations of "wastes" that appear to be resources in a later era. While it is useful to be inspired by nature, it does not always provide a reliable model for economic or industrial development.

Chadwick's caution carries a salutory message for the concept of industrial ecology which has grown out of the notion that the industrial system should adopt the best features of natural systems in which residual products from one set of organisms are used as energy and mineral resources by other organisms. The industrial ecology metaphor recognises that the industrial system is part of the larger biosphere and needs to become compatible with it (Patel 1992, Socolow et al. 1994). In applying the concept to wastes, Frosch argues in Chapter 3 that wastes should be considered as raw materials for production, now or in the future. Wastes are only a re-sorting of the elements and energy in a system. The design of the production process should include a design of the generation and disposal of related wastes in a manner similar to the interconnected food web. Frosch argues that even after a process of resource use minimisation, there will inevitably be some waste left over. The waste that cannot be reused, or recycled in the current manufacturing processes should be separated and stored in `filing cabinets' for future use, or they may be used as sources of energy. Thus Frosch argues that concentrating wastes may be more useful than diluting them - the opposite of the truism that the solution to pollution is dilution. Typically, the recovery and reuse of resources will depend on the emergence of new markets and economic agents. Frosch also points to the trade-offs which are necessary in encouraging the development of networks of materials exchange. One trade-off is between the energy used and the environmental impact of extracting virgin resources and the impacts associated with waste recovery and reprocessing. Another is between the impacts associated with the dissipation of wastes and those of recovery. There will be many different ways of reorganising industrial materials stocks and flows, the key will be to make this change in an economically and environmentally effective way.

The slow response of the market place and consumers to the perceived need for industrial transformation, led proactive thinkers from the scientific, non-governmental and governmental sphere to argue in favour of strategic policy intervention. Weizsäcker argues in Chapter 4 of the need for vision and urges others to adopt a Factor Four approach; i.e. reducing material and energy intensity by a factor of four and thereby halving environmental damage while doubling wealth. The Factor Four objective is to be seen as a medium-term first step towards a long-term, Factor Ten objective. While such goals do not as yet have broad societal support, they have been applied to developing a far-sighted innovation policy (see Jansen in Chapter 20).

3. FROM CONCEPTS TO RESEARCH TOOLS

Research tools are needed to translate the principles of industrial ecology into useful analysis, theories and knowledge. Whether at the level of the product, the process, the firm, or a whole national economy, systems analytical tools are necessary which contain data on physical stocks and flows. This data can then be used to study the structure and dynamics of stocks and flows in the system (substance flow analysis), to develop efficiency measures (for example, eco-efficiency or material intensity for product service, MIPS), to match an analysis of the physical economy to the monetary economy, or to match physical stocks and flows to an evaluation of their impacts in a systematic way (life cycle analysis). Udo de Haes et al. (Chapter 5) present a review of what Verbruggen terms 'environment cum economy' models.

Typically these tools are data hungry. Wernick discusses the key choices that need to be made in creating material flow accounts (Chapter 6). He calls for the focus to be placed on material flows directly associated with financial transactions, although the boundary between "nature" and the "economy" is not always clear-cut. Trace flows, as well as commercial flows of materials should be accounted for, and where possible materials flows should be coupled to energy flows. Finally, Wernick asserts that the greatest value will be derived when physical accounts can be linked to indicators which capture the heterogeneity of materials and their impacts. A key challenge for the future is to map data on social and economic variables onto materials accounts.

These modelling approaches all share a characteristic that they aim to integrate different (apparently incommensurable) perspectives and measures of the economic and ecological system. Monetary and physical indicators are linked, and processes across diverse spatial and time scales are analysed together. The problem of valuation (is a tonne of carbon dioxide worth more than a gramme of cadmium in the environment?) is fundamental to all of these assessments, and in this way industrial ecology is no different from other branches of environmental studies. This is the central theme of the chapter on economic valuation of environmental impacts in LCA studies by Powell, Craighill and Pearce (Chapter 9). Powell et al. argue that there are theoretical and practical advantages to attaching monetary values to all the environmental impacts associated with an economic system.

A trade-off between simplicity and complexity is also apparent in all these tools. They may need to be simple and transparent to be useful to the policymakers and entrepreneurs (Udo de Haes et al. in chapter 5). The risk is that simplicity may come at the cost of relevance or veracity. Even in large systems, local details may be important. At the same time, limits in knowledge, data and computer software mean that the opportunities for sophisticated analysis are today constrained. Most existing models are partial and static. Dynamic modelling using the concepts of pertri nets and expert systems is only just becoming possible in a few cases. At the same time, these models limit the flexibility of the researcher because of their in-built biases (Boelens and Olsthoorn in Chapter 8).

Verbruggen (Chapter 7) discusses the relationship between physical and economic models. He argues that disciplinary differences of perspective still limit the potential for truly integrated assessments of the relationship between the economy and the environment. Fundamental differences remain in natural and social science approaches on questions such as the definition of sustainable development and the objective of analysis (a cause-effect-response chain in natural sciences, a concern with welfare and cost-benefit in economics).

Finally, Udo de Haes et al. warn that results from analysis can rarely be translated into a specific decision. The tools can never in themselves provide the only reference point for transparent and democratic decision-making. They are appropriate for policy processes only when they are embedded in clear decision-making processes. Social inputs are needed to define the scope of the study, to identify alternatives and to evaluate impacts.

4. TESTING HYPOTHESES

Industrial ecology studies have already spawned a number of debates. One objective of this book is to present a number of discussions in the form of hypotheses. The intention is not to test these hypothesis in formal terms, but to allow alternative views to be stated.

4.1 De-materialisation and re-materialisation

The first hypothesis discussed is: *With economic development and a corresponding shift from an 'industrial' economy to a 'service' economy, resource use and related environmental pressure will decrease.*

The 'dematerialisation hypothesis' is used both as an empirical reality and as a normative objective for environmental management and policy. A number of studies have purported to show that the materials and pollution intensity of economies tends to grow in early phases of development, while in later phases shifts in the sectoral composition of the economy, technological advances or environmental policies lead to a break in the link between materials use and economic growth (Jänicke et al, 1989). The picture of energy demand in industrialised countries since the late 1970s indeed suggests such a phenomenon.

While there is some evidence that the inverted `U' curve (the Environmental Kuznets Curve (EKC)) holds for some materials (and for some local pollutants), it is argued by de Bruyn in Chapter 10 that a more comprehensive measure of throughput reveals that de-materialisation is not a permanent process and that periods of de-materialisation are followed by periods of re-materialisation. Using a different set of data, Berkhout shows in Chapter 11, that the materials intensity of industrialised countries is convergent, and falls within the surprisingly narrow range of 16 to 21 tonnes of materials per capita per year. He also shows that despite structural change in advanced economies and the growth of services, materials flows in all economies are still growing in absolute terms. Relative improvements predicted by the de-materialisation thesis therefore appear to be over-compensated for by a real growth in demand. These findings appear to show that achieving radical improvements in materials efficiency at the level of the whole economy, as promoted by the advocates of Factor Four, may prove very difficult.

However, this does not imply that technological change and the process of de-materialisation cannot be fostered. De Bruyn argues that resource intensity does not follow a smooth process along a path of equilibrium, but goes through rapid step-jump changes followed by periods of stasis. He interprets this pattern using evolutionary economic theory. During periods of stability in technological paradigms and institutions, the path of learning and selection moves around an attractor point, but as markets and institutional structures change, new opportunities to save energy and materials are adopted, and the path may shift towards a new attractor point. This analysis suggests that technology alone will not achieve de-materialisation. Processes of learning, new incentive structures and institutional changes will also play a vital and interconnected role.

4.2 De-materialisation - a guiding principle ?

The second hypothesis discussed is: *Dematerialisation should be a guiding principle in policies aiming at sustainable development.*

Commentators like von Weizsäcker contend that since economies are basically linear (and, by implication, wasteful users of resources) a good way of reducing the impact of anthropogenic resource use would be to reduce materials throughput. One of the great appeals of this 'more from less' approach is its simplicity. It is argued that an understanding of the relationships between economic and natural flows of materials, or a consensus over robust measures of 'natural' and 'man-made' capital is not currently feasible. In the absence of this knowledge or agreement, but given the need for clear objectives, dematerialisation is held up to be a key to guiding policy and business and consumption decisions.

But "fates matter", as Socolow and Thomas show in Chapter 12. Looking only at fluxes of materials, and disregarding *how* they are dissipated back into the environment and their consequent health and other impacts, can lead to sub-optimal choices. The discussion in this chapter concerns the relative health impacts of lead used as an additive to gasoline and lead used in batteries to power electric vehicles. Lave et al. (1996) have offered the provocative argument that a transport system based on battery-powered vehicles would lead to greater lead emissions into the environment than a system based on vehicles with leaded petrol. In mass terms, therefore, the latter alternative may be better. However, Socolow and Thomas show that not all uses of lead are equally harmful. Lead used in gasoline is a dissipative use of lead which disperses lead in the atmosphere and onto sediments where it can have toxic effects on humans. By contrast, lead used in batteries can be cleanly recycled and a potentially hazardous product can be effectively managed. The conclusion is that while materials efficiency is a useful goal, the differential impacts of dissipated materials residues and different uses of materials must be carefully considered in choosing between alternative options.

4.3 Closing materials loops globally or locally

The third hypothesis discussed is the critique of free trade on environmental grounds: *Global flows of resources make it more difficult to achieve radical improvements in the materials efficiency of economic activities.*

Global commodities and industrial markets are regarded with great suspicion by many environmentalist commentators. They see global flows of materials and manufactured goods as environmentally damaging in themselves, while also encouraging the wasteful use of virgin non-renewable resources. This is because global competition between virgin resources depresses prices. The critics also argue that the process of global economic development tends to push more resource- and environment-intensive industrial activities to less developed economies. Consumption in the north is therefore built upon exploitative resource relationships with the south. This critique is carried through to arguments about the management of “wastes”. International agreements like the Basel Convention have promoted the concept of “self-sufficiency” which prohibits the trade in wastes and requires national management schemes.

Van Beukering and Curlee (Chapter 14) challenge this latter argument. They show that recycling is increasing on a global scale and the trade in recyclables is also increasing. Trade in recyclables increases the opportunities for recycling in a cost-effective manner. The trade issue raises the question: On what scale are materials cycles to be closed? Should cycles be closed locally or globally? Van Beukering and Curlee point out that it would not be optimal for small countries and small facilities to close materials cycles autonomously. This might lead to incineration and the dissipation of resources, when an economic and environmental case for recycling exists. On the other hand, exports of waste should not lead to exploitation of another country's vulnerabilities. The export should not imply dumping of wastes in countries where disposal costs are lower due to less stringent environmental regulations. Apart from the environmental impacts of such dumping practices, the incentive to develop better recycling techniques may be lost. At the same time, the exported wastes might compete with the local recycling industry in the importing country. Jaeckel, (Chapter 19) in his discussion of the German packaging waste law, explains that the export of wastes had the effect of undermining government efforts to encourage recycling until this trade was managed.

Trade in minerals is also discussed by van der Veeren. Nutrients imported in feed stocks for intensive livestock production concentrate in soils, groundwater and surface waters in Europe, while leading to a loss in the source countries. For example, the cassava trade between Thailand and Europe leads to soil exhaustion and deforestation in Thailand and an excess of nutrients in the soil, in the Netherlands, where the cassava is used as animal feed. Excess nutrients are used as manure, but a part of the manure leaches through the soil into the ground water. The option of exporting excess animal manure in the Netherlands to meet a perceived nutrient deficit elsewhere in the world is now being discussed.

These assessments take a pragmatic course. Seeking to reverse the globalisation of the world economy is perhaps as foolhardy as trying to hold back the waves. Dematerialisation depends on two interrelated processes: the improvement of the materials efficiency of economies; and the greater cyclicity of economies. The first depends on full environmental costing of virgin resources and the adoption of efficient systems of production and consumption. The second depends on recycling carried out at the appropriate level. For some materials this will mean local recycling, for others cycles will be regional and global. Efficient markets for these materials to the extent that recyclates begin pervasively to be seen as substitutes for virgin materials can only be developed if trade barriers are removed. Of course, this does not diminish the need to safeguard the vulnerable against the abuses of the powerful.

4.4 Dissipative Flows: Manageable or Inevitable

The fourth hypothesis concerns the management of flows of resources out of the economic system which are inherently dissipative: *Dissipative flows cannot be managed.*

Van der Veeren argues in Chapter 13 that this view is over-simplified. He shows that nutrient emissions are basically inevitable in agricultural production, and that such emissions tend to dissipate. Harmful impacts on the environment (such as threatening biodiversity) may result if concentrations exceed threshold levels. However, if high agricultural productivity is to be promoted along with biodiversity, two mutually exclusive goals, nutrient emissions need to be managed. He argues that management options exist at different levels of the agricultural system, and that by integrating measures creatively dissipative flows can be minimised. Partly this is through the recovery of nutrient streams which can be recycled. These management approaches are knowledge-intensive, requiring high quality site-specific understanding of nutrient budgets and adaptive management of nutrients. At the farm-level, many strategies can be adopted to reduce the dissipative emissions of nitrogen and phosphates, but intensive farming techniques may still lead to an excess of nutrient emissions.

5. EMERGING APPLICATIONS IN INDUSTRY

The book presents a series of private sector experiences and experiments of the electronics industry (Philips), the food industry (Unilever), the chemical industry (Dow Europe) and the communications industry (AT&T). How are the ideas of industrial ecology being applied, and what can we learn about industrial transformation? Firms, as holders of technology and knowledge of materials flows are key agents in the change process.

Cramer argues in Chapter 15 that, although useful as an input to decisions on incremental changes to products, life cycle analysis (LCA) is impotent as a methodological tool, since it cannot support technology strategy. Technology strategies depend on the relationship between markets and technologies. The consideration of the resource implications is only one of many factors to be taken into account. Cramer proposes a formal method for gaining insights into the future developments of the market (STRETCH) which incorporates an iterative process of brainstorming. Her critical message is that environmental improvements must make sound business sense to industry if they are to be adopted.

Dutilh argues further in Chapter 16 that there are limits to what industry can do, even if it is committed in theory to dematerialisation and has been applying LCA to its products. These limits are defined by consumer preferences and choices; trade-offs between environmental benefits and nutrition value, international trade and prices associated, and consumer behaviour in relation to the disposal of the products. He explains that Unilever has adopted the Overall Business Impact Assessment methodology to evaluate the impact of Unilever's business operations in relation to their targets for various global environmental themes. Furthermore, within the company different tools are needed at different levels of management. The data selection tool is needed to generate environment related process information, the LCA approach is needed to develop environmental profiles and the evaluation tool is needed to study risks and opportunities. Dutilh emphasises the crucial role of the final consumer in shaping business attitudes, and in determining the environmental profile of products in use. Optimisations of process and product can quickly be overshadowed by the behaviour of consumers. According to this view, consumer sovereignty remains a substantial obstacle to the goal of reducing resource use intensity.

Allenby also argues in Chapter 17 that there are no easy answers for industry since the available information is limited and since there are limits to substitution. He gives the example of lead solders substituted for by bismuth solders in the electronics industry, since lead is a toxic substance. However, a life-cycle analysis indicates that bismuth has a higher mass rucksack. Moreover, since bismuth is produced as a by-product of lead mining, an increase in the demand for bismuth might ironically lead to a higher supply of lead. He argues further that if firms are to change their approach to production, they need clear economic incentives and good information.

Finally, Fussler, who works at Dow Europe and with the World Business Council for Sustainable Development confidently states in Chapter 18 that environmental improvements *do* make sound business sense. He argues that firms need both a set of principles which embed sustainability objectives into the strategic aims of the business, and simple tools which can be applied in the management of innovation and technology. Fussler proposes six sustainability principles which integrate best practice management theory with new ideas about the greening of business, seeing these developments as complementary. Eco-efficiency and ambitious sustainable development objectives are seen as tied to the empowerment of employees, citizens and communities, innovative far-sightedness, and customer satisfaction. At Dow Europe a tool called the "eco-compass" has been developed to guide innovation management. This uses a simple mapping technique for characterising the life cycle performance of new products, compared to the product it replaces.

The picture of the adoption by industry of the ideas of industrial ecology is patchy. Systematic assessment techniques like LCA have been applied in the defence of existing products, and in limited cases to the optimisation of individual products. In general, however, firms find full-scale LCAs too cumbersome to integrate into their innovation processes. There have also been attempts to include simple rules for strategic planning, sustainability criteria which aim to reduce the materials intensity, and improve the cyclicity of product systems. Finally, there are signs that the growing importance of services (energy services for example), and the extension of the producer responsibility concept, will encourage market changes in which the environmental performance comes to be seen as a central part of consideration of the function of goods and services. Environmentally-optimised services could in some markets replace products.

In the meanwhile industry is paralysed by inertia. Industrial inertia, a new metaphor in industrial studies, implies resistance to applying a decision (cf. Byé 1997; Mokyr 1992; DiMaggio and Powell 1983; den Hond 1997). The inertia can be inherent in the organisational framework of an industry, there may be innovation inertia or it can be the result of a strategic decision. Fussler and James (1995: 9) claim that the innovation inertia or industrial lethargy is the result, inter alia, of corporate anorexia (a term coined by Gary Hamel) which is the result of years of cost-cutting and re-engineering leading to job insecurity and the lack of creativity. Den Hond (1997) argues further on the basis of an analysis of the auto industry in Germany that sometimes the inertia is a strategic decision.

The discussion in Chapters 15-18 shows that the major constraints before industry appear to be three-fold. Industries lack the economic incentives that would justify the adoption of an industrial ecology perspective on their activities (resources and waste management are still too cheap); they lack information about how best to change their production process (innovation is always risky); and they fear that they will not be able to appropriate value in the market from changes which they do make (the consumer may not support industry in their efforts to improve). Clearly there is a role for governments, pushed by civil society, to correct these market failures.

6. EMERGING GOVERNMENT POLICIES

Evidence of an emerging framework of government policies is reported in the next set of chapters. For the most part these can be seen as first steps in a policy process. If the 1980s saw the acceptance of the need for integrated approaches to industrial environmental policy focused on the production site, the 1990s have seen the growing awareness of a need for a wider, more holistic, product and consumer oriented environmental policy. The final form of the new policy architecture is as yet unclear, and policymakers have become more cautious about their role. While individual producers can be regulated, chains of producers and consumers are much less promising as targets for regulation. The role of policy must then be to facilitate, encourage, educate and to modify incentive structures.

Jaeckel argues in Chapter 19 that despite the good intentions of the German government, policies related to materials flows seem to be limited to the back-end (waste management). The 1996 Closed Substance Cycle and Waste Management Act aims in principle to reshape materials fluxes in the whole German economy, and endorses the principle of producer responsibility, but the focus of its measures is on waste management. The Packaging Ordinance also aims at reducing packaging, promoting substance recycling and relieving the local communities of the burden of disposal, by placing responsibility for packaging waste disposal onto manufacturers. The experience thus far has been a qualified success according to Jaeckel. More plastic has been collected and this has stimulated domestic recycling, while the growth in packaging waste volumes has been halted. The costs of transition to a system of packaging waste recovery and reuse have been high.

Jansen reports in Chapter 20 on the application of the dematerialisation concept to a technology policy. As with de Bruyn and Dutilh, Jansen emphasises the need to integrate an analysis of culture and socio-economic structures with technology foresight. He argues in favour of setting goals for preferred future technological and economic systems, and then to backcast to the present day. Technology policy then aims to evaluate the different routes which can be taken to reach the final goal. The process of fostering companies along this track should be undertaken by Governments by *inter alia* designing demonstration programmes. The Dutch Government has developed a Sustainable Development Programme aimed at understanding the driving forces of industry, providing administrative incentives such as direct regulation, standards, economic taxation, removing environmental subsidies and strategic sectoral development agreements.

Marius Enthoven argues in chapter 21 that the European Commission has tried to take an integrated approach to environmental policy by widening its scope and by encouraging the integration of environmental considerations into all areas of Commission policy. In relation to supply, the approach attempts at preventing the shifting of environmental problems from one phase to another, from one medium to another and through risk, product and waste management. In relation to demand, the EU currently focuses on the eco-label, but Enthoven also discusses a more ambitious product-oriented environmental policy which will employ a range of measures. As Loprieno explains in his chapter on the EU Ecolabel (Chapter 22) there are great complexities in developing scientifically-based information instruments acceptable to industry, and useful to consumers.

The role of governments in environmental policy is being reassessed as voluntary approaches and market-based instruments look set to play a greater role. Traditional regulatory approaches are not matched to the new objectives of environmental management, partly because political systems are out of tune with the systemic and temporal nature of environmental problems. Decision-making systems have short-term reporting mechanisms and cannot easily "reconcile tensions among cross-purpose goals, conflicting jurisdictions, cultures and horizons by using methods of decisionmaking that depend on ideological and organisational considerations that are no longer valid" (Gerlach 1992: 70). Moreover, governments are not able to manage the global scope of markets, technologies and firms, while frequently being hampered by short-term political pressures. Many governments have chosen a disjointed, incremental approach to environmental policy which is hard to undo.

The legal and fiscal tools can steer the production processes to some extent. However, an analysis of the market also shows that there are limited chances of the market sending clear signals about the direction in which industries should strive for in the future. The record on subsidies and taxes also indicates that there are still many unresolved issues. "Many of today's subsidies encourage practices that are economically perverse or trade-distorting or ecologically destructive or socially inequitable" (Moor and Calamai 1997:1). The total amount of annual subsidies amounts to 490-615 billion dollars in OECD countries and 217-272 billion in non-OECD countries (Moor and Calamai 1997: 49). The economic system in OECD countries is intermeshed in a complex system of subsidies. Unilateral removal of these subsidies is likely to have an impact on the competitive position of some of the export oriented industries and as such presents certain major political and economic challenges; even though they are likely to lead to industrial transformation, positive for the environment.

7. CONSUMPTION OF RESOURCES

Governments claim that they lack societal support for strategic long-term decisions. Industry claims that the bulk of the pollution problems are caused by consumer choice and consumption behaviour. Here we are faced with another paradox. A common-sense perspective of the environmental debate over the past thirty years shows that consumer and environmental movements have been among the first to signal the need for change. To what extent are consumers and their non-governmental organisations interested in environmental issues? What does research into consumption reveal? These two questions are dealt with in some detail below.

A brief history indicates that the consumer and environmental movements have been among the first to signal for change. It was Ralph Nader's persuasive books and lobbying on air and water pollution and waste management that led to government action in the United States in the early

1970's. It was because of the campaigns of Greenpeace in the 1960's and 1980's that nuclear wastes and the transboundary movement of hazardous wastes came on to the international agenda. From fragmented and local environmental and consumer groups in the early part of this century, there has been a shift to national and international groups and networks that assert themselves in the policy process. But the increasing complexity of production and consumption processes and the embeddedness of lifestyles has had its impacts. Although consumers are not in the fore-front of the debate on industrial transformation, they may not be so far behind. The Consumers in Europe Group is now lobbying for changes in consumption patterns based on better information to be provided by governments and industry.

Consumers do not know how they can reduce their own environmental impact, nor do they have clear incentives to make environment-friendlier choices (Jenkinson 1997). In response to consumer pressure the Directorate General on Consumer Policy of the European Commission now has a mandate to promote sustainable consumption.

Consumers have not attracted much attention thus far from researchers. Stern et al. (1997) argue that there are four critical questions that need to be investigated: a) which human activities are the significant environmental disrupters? b) who are the key actors responsible? c) what forces cause or explain environmentally disruptive behaviour? and d) how can environmentally disruptive behaviour be changed? There is, therefore, a 'hard' and 'soft' side to studies on consumption.

Data can be collected on the per capita use of materials, energy, products and transportation and the per capita contribution to waste including emissions. The data can be aggregated and disaggregated in different ways to identify which individual economic choices in which countries or regions are the significant environmental disrupters. However, such aggregations and disaggregations do not often take into account the `social' as opposed to the `technical' aspects of consumption (Lutzenheiser 1997).

Faye Duchin (1997) argues that people live in households, and households make decisions about their future in a manner similar to that of firms. However, such households are rarely presented with the possibility to radically alter consumption patterns. She argues in favour of household research through e.g. social accounting matrices (SAM's) and household classifications to learn more about the factors that motivate household consumption patterns. Such research is being carried out by different universities in the Netherlands. Vringer, Gerlagh and Blok (1997) of Utrecht University have done research on the consumption of both direct and indirect energy by households and conclude that there is a more or less linear relationship between household expenditure and indirect energy consumption (the energy embodied in consumer products). Over the last five decades consumption patterns have changed significantly in the Netherlands leading to a threefold increase in the consumption of direct and indirect energy if changes in production efficiencies are neglected (Vringer and Blok 1995). Gatersleben and Vlek (1998) have shown that the psychological drivers of consumption behaviour typically dominate knowledge about environmental impacts of this behaviour. They use the Needs, Opportunities and Abilities (NOA) model to identify factors that have influenced consumer behaviour in the Netherlands. They conclude that the opportunity and ability to consume are the key psychological factors motivating consumptive behaviour and even though consumers are aware of the environmental impacts of their behaviour, they are not quite as willing to change it. The purpose of such projects is to understand consumer behaviour and to identify suitable policies to stimulate environment friendly consumer behaviour. At present the Household Metabolism Project, an initiative of the Universities of Groningen, Utrecht and Wageningen, is investigating the options for reducing greenhouse gas emissions by changes in household consumption patterns.

Does consumer behaviour change over time? Social surveys reveal that consumer preferences do change. And new preferences, stimulated always by new knowledge may lead to more environmentally friendly beliefs. Surveys reveal that once consumers are informed about environmental issues and are provided with services to facilitate change, then new patterns of behaviour can be established rapidly. A survey shows that three quarters of the Dutch population separates and delivers batteries, used paint, glass and paper. A similar proportion does not throw waste onto the street (a dissipative use of resources) and carries reusable shopping bags. Most of the population close the water taps while washing the dishes, only wash clothes when the machine is full, purchase unbleached coffee filters, do not purchase drinks in cans; and separate organic wastes (quoted in Hoevenagel et al; 1996: 25; although the authors note that there is a gap between what people say they do and what they do). Jaeckel also claims in his chapter that the consumer response to waste collection policy in Germany was much higher than expected. At the same time, consumption behaviour does not become less resource and pollution intensive autonomously. Consumers have to be provided the information and the justification for taking action. The government has a role to play in providing infrastructure and fostering organisational and cultural attitudes which are more environmentally friendly (Hoevenagel et al. 1996: 159). This process will become more and more important in the future for, as Enthoven puts it, the major cause of pollution is no longer the production sector but the consumption of products. Furthermore, only by involving the consumer actively in the process of research, as Udo de Haes et al. point out in their chapter, can the social support for the research activity and consequent policy decisions be generated.

8. TOWARDS A RESEARCH AGENDA

In line with the present initiatives of the International Human Dimensions Programme (IHDP) and in follow-up to the analysis in this book, we propose a research agenda called Industrial Transformation. Such an agenda consists of research into the (a) macro incentive structure, (b) industrial ecology and (c) consumer perspectives and behaviour. These are discussed below.

8.1 Research into the Macro-Incentive Structure

Macro incentives refer to the incentives provided by governments and the economic system that drive present day economic development. A first factor driving economic development is the world-wide liberalisation of markets and the related trend towards specialisation and up-scaling of international production and service companies. A second important factor is the role of labour in the tax system relative to natural capital. A third and maybe most important factor is the lack of adequate incentives to call for a internalisation of environmental costs in the pricing of resources. Research in this field should include an analysis of liability allocation i.e. investigate how responsibilities (and risks) are and can be shared between producers, consumers and government.

A research agenda into macro incentives would include research into the following types of questions:

- What are the different types of macro incentives, what are the major driving forces and how can they be classified?
- How does the existing macro incentive structure encourage or subsidise unsustainable development? This would call for an analysis and evaluation of fiscal instruments, trade and international investment related instruments, liability issues, land tenure systems, information instruments and property right systems, and tradable resource use instruments. It should also include empirical research into (international) material flows and theoretical studies about the relation between economic growth, resource use and environmental degradation.
- How can macro economic and macro ecological indicators be developed for monitoring resource use and international substance flows and the impacts of macro incentives on these?
- How can changes in incentives encourage a more environment friendly economic development process? This type of research could focus on small steps (such as incremental changes in the tax system) as well as on large steps (e.g. introduction of a system of tradable resource use permits); on the incentives at sectoral, regional and global level.
- How can the effectiveness of the proposed changes be measured and analysed? This would include research into the effects/effectiveness of concepts being

applied and/or considered in international agreements affecting the use of environmental resources.

The ultimate aim of this type of research is to identify and evaluate how changes in incentive structures could help producers and governments to meet human needs, wants and preferences and help consumers adopts sustainable patterns of behaviour.

8.2 Industrial Ecology

Research under the banner of industrial ecology has evolved from the earlier research into end-of-pipe and process efficiency research. Research presented under this heading mainly centres around efficiency and environmental impacts of the entire chain of resource use, including technological innovation, technology assessment and organisational issues.

Industrial Ecology research would examine, inter alia, the following questions:

- How is the production and innovation process organised and managed? What is the nature of the interaction between producers within and across sectors? How can the transition processes between sectors and firms be analysed?
- How does the corporate sector respond to legislation and incentives from the government? How does the private sector respond to civil society pressures and concerns. How do the different producers involved in different steps of the production process respond to command and control, fiscal instruments, liability instruments and corporate governments (shareholder issues)?
- What are the promising technological developments and development trajectories, products and promising configurations/arrangements of production? What are the most efficient ways of organising the production process across the different sectors? The research should identify and analyse various possible technological and organisational trajectories that production units and entire sectors could go through, moving from one dominant way of producing goods and services to another way of doing things.
- What tools and indicators can be developed to facilitate the assessment of the environmental performance of production units and/or products such as standardisation, benchmarking and the development of life cycle analysis type of tools?
- What is the potential for changing the way the production process is organised to make it more environmentally beneficial? Seen in the context of industrial transformation this type of research should include an analysis of changes in the sectoral division of production. Specific elements of the research agenda are the identification and evaluation of promising technology development trajectories, responsibility and liability schemes within and between firms, environmental auditing and corporate strategy analysis and evaluation that can improve the resource use efficiency.

The Industrial Ecology agenda includes both technology and organisational issues and its goal is to optimise resource use across the production and consumption process. As such it is almost as broad as the Industrial Transformation research agenda promoted by IHDP. The main difference is that the IT agenda is more explicit in introducing the need for research into the macro incentive structure, that should bring macro-economists and political scientists more clearly into the picture. Moreover the IHDP agenda stresses the need for research into consumer and citizen issues. This should bring sociologists, psychologists, and political scientists into the game.

8.3 Sustainable consumption

We are all both producers and consumers. In order to produce goods and services, or more broadly, to generate welfare, we need to consume inputs and we produce outputs. The new focus on consumption in environmental studies is therefore the other, until recently little explored, side of production. Although traditionally research into consumption has been concerned with *final* consumers in the household, a new generation of research into sustainable consumption is developing a wider definition which includes all actors in the economic system. Systems of consumption are conceived of as networks of actors involving producers and consumers. Research into these systems needs to clarify the information available to each of the actors, the incentive structures which determine their behaviour, and the degree of control which each has over his or her decisions. Normatively, this research is also concerned with elasticities of behaviour, with the degree and rate at which behaviour may change.

The short-term choices of consumers, broadly defined, are generally determined by prices, the available physical and institutional infrastructure, and by their knowledge. However, consumers also pay a role (directly or through intermediary organisations) in decisions regarding changing the infrastructure or the institutions, or in driving production through social and cultural change. Relatively little research has been carried out on consumption and the environment.

Research into the role and influence of the consumer would cover the following questions:

- What are the different roles of the consumer and how do they influence the process of economic development? An important element of research is the analysis of the role of consumers as consumers of products and services, citizens (voters), employees, shareholders and as supporters of environmental NGOs demanding the protection of the environment.
- How do human needs, wants and preferences evolve and how do these influence corporate decisions about product development and government decisions about changing the macro-incentive structure?

- How do consumers respond to product development, marketing and public relations strategies of producers of goods and services ? Other elements are consumer taxonomy, monitoring and analysis of developments in awareness and behaviour, and the feasibility of introducing potentially cleaner but controversial technologies such as biotechnology.
- What kinds of policy measures and instruments can be developed to effectively support the consumer's own concerns and to assist him or her in being an aware and conscious consumer?

The aim of such research is to understand the role of the consumer in influencing industrial transformation and how incentives can be provided to the consumers to change their behaviour in response to new information about how industrial processes influence the environment.

9. THE DEVELOPMENT OF AN INTERNATIONAL RESEARCH AGENDA

Addressing the environmental issues mentioned above requires international co-operation in the development of an incentive structure that promotes the introduction of more environmentally friendly products, services and consumption practices. The basis for international co-operation in the field of policy development can be strengthened by co-operation in research. Actually the development of an international research agenda in itself helps to explore and analyse the range of options available.

International co-operation in the natural sciences has a long history. Co-operation in technological research is very common too although commercial investments and related property rights issues do set certain limits. In the social sciences international co-operation has proceeded in the context of the international academic associations such as the International Social Sciences Council (ISSC), one of the founding organisations of the IHDP programme (see below). Other organisations with a record are the International Political Science Association, the International Law Association, the International Society for Ecological Economics, but the developments in these areas have been hampered by the lack of common paradigms and the subsequent lack of adequate financial resources. However, now that a number of environmental problems have been recognised as global problems there is a specific incentive to enhance international co-operation in social sciences research.

Although the need for international co-operation in research is recognised it is not yet clear what the best ways are to enhance such co-operation. A start has now been made through the International Human Dimensions Programme (IHDP). This programme builds on earlier successes in the field of natural science research co-operation through the International Geosphere Biosphere Programme (IGBP).

Now the two programmes are connected and sponsored both by ICSU (the International Council for Scientific Unions) and ISSC (the International Social Sciences Council). Industrial Transformation is one of the research areas that is presently being promoted and sponsored by the IHDP.

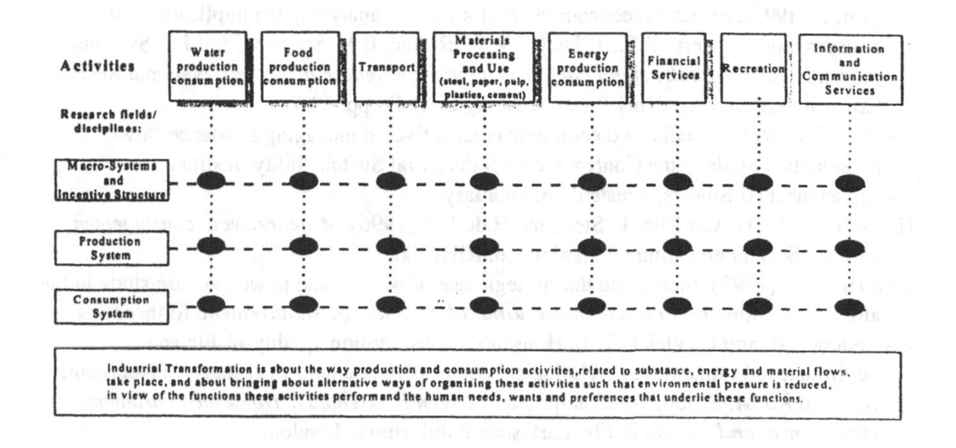

Figure 42. Research fields and activities undertaken by IHDP

This book provides a survey of concepts, tools and applications of holistic approaches to the environmental analysis of industrial activities. At the same time it has provided a framework in which to place future research into industrial transformation. We have highlighted three broad areas that we see as priorities for such a research programme: the macro-incentive structure; industrial economy; and sustainable consumption. This is reflected in *Figure 42* which shows more clearly the scope of research opportunities. There is much work to do, some of it at a national level, but also through richer collaboration at a regional and global level. Radical enabling technologies such as e-mail, conference calls and videoconferencing make true international collaboration in the social sciences, such as within the context of the IHDP, a reality. We invite researchers, policymakers and business to rise to the challenge of looking at environmental problems in the round; to develop further hypotheses and concepts to explain interaction of human activities and the environment; and to apply these concepts in changing policies and behaviours.

REFERENCES

Byé P. (1997). Productive inertia and technical change, in P. Byé, J.J. Chanaron and A. Richards (eds.) *Industrial history and technological development in Europe*, research papers of the West European Science and Technology Forum, March 21-22 , the Newcomen Society and the Science Museum, London.

DiMaggio, J. and W.W. Powell (1983). The iron cage revisited: institutional isomorphism and collective rationality in organisational fields, *American Sociological Review*, Vol. 48, pp. 147-160.

Duchin, F. (1997). Structural economics: A strategy for analyzing the implications of consumption, in Stern, P.C., T. Dietz,, V.W. Ruttan, R.H. Socolow, and J.L. Sweeney (eds.), *Environmentally significant consumption: research directions*, National Research Council, National Academy Press, Washington D.C., pp. 63-72.

Duchin, F. (1997). A social and economic perspective on managing substance flows, presentation made at the Conference on Substantial Sustainability, Institute for Environmental Studies, Amsterdam, February.

Hoevenagel, R., U. van Rijn, L. Steg, and H.de Wit (1996). *Milieurelevant consumenten gedrag*, Sociaal en Cultureel Planbureau, Rijswijk.

Den Hond, F. (1997). Inertia and the strategic use of politics and power: A case study in the automative industry, *International Journal of Technology Management*, forthcoming.

Gatersleben, B. and C. Vlek (1998). Household consumption, quality-of-life and environmental impacts: A psychological perspective and empirical study, in Noorman, K.J. and A.J.M. Schoot-Uiterkamp, (eds.) *Green Households? Domestic consumers, environment and sustainability*, Earthscan Publications, London.

Jenkinson, Kay (1997). The consumer interest in the environment, *European Environment*, Vol 8, pp. 85-91.

Kuik, O., R.S.J. Toll and H. Verbruggen (1997). The impacts of climate change policies of Annex I countries on the economies of developing countries - A critical review, Paper presented at the 2nd seminar on Environment and Resource Economics, Girona, May 1997.

Lutzenheiser, L (1997). Social structure, culture and technology: Modelling the driving forces of household energy consumption, in Stern, P.C., T. Dietz, V.W. Ruttan, R.H. Socolow, and J.L. Sweeney (eds.), *Environmentally significant consumption: research directions*, National Research Council, National Academy Press, Washington D.C., pp. 77-91.

Moor, A.M. and P. Calamai (1997). *Subsidizing unsustainable development: Undermining the earth with public funds*, Earth Council (San Jose) and Institute for Research on Public Expenditure (The Hague).

NRLO (1992): International trade and the environment, Nationale Raad voor Landbouwkundig Onderzoek, Report no. 92/19, The Hague

Rabe, B.G. (1991). Exporting hazardous waste in North America, *International Environmental Affairs*, No. 3, pp. 108-123.

Ruigrok, E. and F. Oosterhuis (1997). *Energy subsidies in Western Europe*, Greenpeace International.

Stern, P.C., T. Dietz, V.W. Ruttan, R.H. Socolow, and J.L. Sweeney (1997). Consumption as a problem for environmental science, in Stern, P.C., T. Dietz, V.W. Ruttan, R.H. Socolow, and J.L. Sweeney (eds.). *Environmentally significant consumption: research directions*, National Research Council, National Academy Press, Washington D.C., pp. 1-11.

Vellinga, P., S. de Bruyn et al. (1998). *Industrial Transformation: Research Directions*, IHDP-IT, no. 10.

Vringer, K., T. Gerlagh, K. Blok. (1997). *Het directe en indirecte energiebeslag van huishoudens in 1995 en een vergelijking met huishoudens in 1990*, Report 97071, NW&S-UU, Utrecht.

Vringer, K. and K. Blok, (1995). *Consumption and energy-requirement: a time series for households in the Netherlands from 1948 to 1992*. Report 95016, NW&S-UU, Utrecht.

Wilk, R.R. (1997). Emulation and global consumerism in Stern, P.C., T. Dietz, V.W. Ruttan, R.H. Socolow, and J.L. Sweeney (eds.). *Environmentally significant consumption: Research directions*, National Research Council, National Academy Press, Washington D.C., pp. 110-115.

Index

accounting, 1, 56, 94, 188
acid rain, 47
agriculture, 10, 12, 13, 51, 53, 75, 84, 85, 86, 217, 218, 220, 221, 222, 224, 227, 252, 261
assessment, 6, 11, 35, 50, 51, 59, 65, 66, 67, 76, 79, 81, 82, 83, 84, 98, 101, 109, 111, 126, 127, 128, 129, 136, 145, 166, 184, 207, 209, 237, 239, 245, 246, 247, 248, 249, 251, 264, 270, 276, 299, 300, 302, 312, 315, 325, 326, 329, 331, 338, 339
AT & T, 259
batteries, 75, 116, 184, 191, 192, 194, 195, 196, 200, 201, 202, 203, 204, 207, 208, 209, 210, 213, 214, 215, 328, 336
best available techniques, 297
Business Council for Sustainable Development, xi, 54, 244, 249, 268, 271, 331
cars, 50, 52, 156, 191, 194, 195, 212, 214, 215, 282
CFC, 65, 151, 183
closed substance cycles, 275, 277
company chains, 55
consumer, 8, 11, 45, 52, 55, 75, 77, 80, 82, 88, 91, 92, 107, 115, 119, 131, 150, 152, 155, 156, 194, 209, 210, 218, 240, 241, 242, 243, 247, 248, 251, 256, 259, 260, 261, 263, 265, 266, 276, 278, 279, 286, 288, 290, 300, 301, 302, 303, 304, 307, 308, 309, 310, 311, 313, 315, 316, 317, 318, 321, 324, 331, 332, 333, 334, 335, 336, 337, 338, 339, 340, 343
consumers, 241, 261, 305, 308, 318, 331, 335, 336, 339, 340
consumption, 3, 4, 5, 13, 15, 38, 52, 57, 60, 65, 71, 73, 82, 84, 88, 89, 90, 91, 122, 123, 137, 147, 148, 149, 150, 151, 152, 154, 156, 159, 160, 162, 163, 164, 165, 167, 168, 170, 171, 184, 189, 199, 200, 201, 204, 213, 233, 243, 245, 246, 256, 261, 263, 273, 275, 278, 280, 286, 287, 295, 297, 298, 301, 302, 303, 304, 305, 307, 308, 309, 310, 311, 312, 318, 321, 322, 323, 328, 331, 334, 335, 336, 337, 340, 342, 343
consumption pattern, 335, 336, 337
dematerialisation, 43, 46, 51, 53, 80, 147, 148, 151, 152, 153, 154, 156, 160, 165, 166, 169, 170, 171, 173, 176, 177, 178, 179, 181, 187, 297, 326, 328, 333
design, 31, 32, 37, 39, 59, 83, 85, 89, 117, 123, 125, 150, 162, 177, 185,

195, 196, 209, 237, 240, 241, 243, 244, 245, 247, 248, 251, 255, 259, 261, 262, 263, 264, 270, 273, 274, 278, 288, 290, 291, 295, 300, 302, 303, 304, 309, 312, 313, 324, 333
design for environment, 259, 260, 263
developing countries, 52, 150, 156, 162, 204, 205, 229, 231, 232, 233, 235, 337, 343
dynamic ecosystems, 21
eco-compass, 246
eco-design, 251
ecodesign, 295
eco-efficiency, 245, 246, 248, 267, 268, 270, 271, 274, 297, 299, 325
ecoefficiency, 285, 287
ecolabelling, 55, 56, 59, 82, 310, 319
ecological rucksack, 45, 48, 49, 50, 51, 71
ecology, 11, 31, 32, 40, 108, 167, 191, 192, 210, 211, 213, 215, 238, 248, 259, 260, 262, 263, 265, 268, 269, 297, 321, 322, 324, 325, 326, 330, 331, 332, 337
economic development, 123, 147, 148, 150, 156, 160, 165, 169, 171, 176, 177, 181, 297, 326, 328, 337, 338, 340
Economic valuation, 127, 132
EMAS, 57, 83, 303
employment, 7, 53, 100, 158, 229, 234, 282, 303
empowerment, 267, 331
energy analysis, 6, 55, 168
energy consumption, 15, 65, 147, 150, 151, 152, 159, 160, 162, 243, 245, 246, 336
energy efficiency, 52
energy intensity, 162, 181, 246, 273, 286, 299, 324
energy use, 28, 38, 90, 129, 142, 147, 162, 243, 298, 324, 339
environment-economy models, 97, 98, 99
European Commission, 59, 83, 139, 297, 301, 304, 307, 308, 319, 333, 335
evolutionary economics, 147, 321
factor 4, 29, 51, 52, 54, 165, 166, 168, 178, 180, 189
factor ten, 45, 270, 324
food products, 251, 252, 256
for externalities, 128
GDP, 52, 90, 91, 149, 159, 161, 171, 172, 176, 178, 179, 187, 188, 189
gasoline, 191, 199, 204, 209, 213, 328
incentive structure, 321, 322, 327, 333, 337, 338, 339, 340
incentives, 39, 171, 231, 237, 265, 292, 321, 331, 332, 333, 337, 338, 339, 340
incineration of waste, 131
industrial ecology, 11, 32, 40, 108, 167, 191, 192, 210, 211, 213, 215, 259, 260, 262, 263, 265, 297, 321, 322, 324, 325, 330, 331, 332, 337
industrial metabolism, 11, 108, 167, 297, 298, 304, 305
industrial transformation, 321, 337, 339, 342
industry, 5, 13, 14, 31, 32, 33, 34, 35, 36, 37, 38, 40, 41, 42, 43, 51, 75, 77, 82, 86, 89, 91, 92, 94, 112, 115, 125, 140, 162, 177, 179, 180, 187, 189, 194, 195, 201, 203, 206, 210, 215, 226, 231, 234, 239, 257, 261, 276, 278, 280, 282, 288, 289, 299, 300, 301, 303, 308, 313, 317, 320, 329, 330, 331, 332, 333, 335, 342
innovation, 20, 105, 155, 156, 157, 179, 186, 208, 237, 240, 248, 249, 267, 268, 270, 271, 274, 285, 288, 291, 293, 294, 312, 324, 331, 332
integrated approach, 183, 297, 302, 303, 309, 332, 333
integrated models, 97, 101
interdisciplinarity, 97
Intergovernmental Panel on Climate Change, 6
International Standards Organisation, 312
international trade, 229, 231, 234, 237
inventory analysis, 59, 61, 65, 83, 129, 136, 137
labelling program, 59, 312
landfills, 40, 127, 203, 229
LCA, 6, 55, 56, 57, 58, 59, 62, 65, 67, 68, 69, 70, 71, 76, 78, 79, 80, 81, 84, 111, 112, 113, 115, 116, 117, 118, 120, 121, 122, 125, 126, 127, 128,

129, 130, 131, 133, 134, 135, 136, 137, 140, 142, 143, 144, 145, 146, 236, 239, 249, 251, 253, 254, 255, 256, 257, 258, 259, 299, 300, 305, 307, 310, 311, 312, 313, 314, 315, 317, 318, 319, 325, 330, 331
lead acid batteries, 214
lead electric vehicles, 191
lead recycling, 191, 203, 214
legislation, 183, 230, 292, 304, 339
life cycle analysis, 6, 111, 126
life cycle analysis of future products, 239
life cycle assessment, 65, 82, 83, 84, 127, 128, 129, 245, 249, 251, 264, 270, 299, 300, 302
mass balance, 1, 12, 81, 114, 119, 123, 151, 161, 168, 170, 209
mass balances, 12
mass flows, 1, 7, 10, 12, 69, 297
material flow account, 85, 99, 114, 116, 117, 169, 181, 325
material flows, 11, 13, 19, 50, 52, 69, 70, 80, 85, 86, 87, 89, 91, 93, 94, 105, 111, 113, 114, 115, 119, 121, 122, 123, 125, 233, 237, 279, 283, 298, 300, 308, 325
material intensity, 50, 273, 287
material intensity per service, 285, see also MIPS
material productivity, 51, 155
materials consumption, 90, 147, 150, 151, 160, 163, 171, 243
materials cycling, 21, 22, 24, 27, 28, 261
metals, 7, 10, 12, 13, 31, 36, 40, 45, 46, 48, 49, 73, 89, 90, 93, 149, 169, 175, 185, 194, 209, 214, 238
MFC, 55
MIPS, 45, 50, 51, 120, 287, 288, 325
modelling, 6, 69, 70, 73, 75, 97, 98, 99, 101, 104, 107, 108, 111, 112, 114, 120, 121, 122, 123, 124, 125, 162, 226, 325, 326
multidisciplinarity, 97
NGOs, 82, 259, 266, 286, 288, 301, 340
north-south, 229
nutrients, 217, 329
OECD, 46, 52, 93, 152, 156, 162, 163, 189, 230, 237, 334
oxygen, 10, 20, 22, 33, 68, 87, 196
ozone, 66, 67, 68, 76
packaging, 36, 60, 136, 137, 138, 235, 251, 255, 258, 266, 275, 278, 279, 280, 281, 282, 283, 329, 333
packaging waste, 281, 329, 333
paper industry, 115
pesticides, 13, 37, 77, 217
plastics, 12, 33, 57, 143, 144, 231, 236, 266, 279, 280
pollution control, 181, 183, 185, 192, 205, 206, 215, 311
prices, 2, 8, 14, 36, 80, 100, 101, 102, 105, 107, 132, 133, 134, 135, 138, 150, 161, 170, 236, 328, 331
production pattern, 233
recycling, 2, 13, 27, 29, 31, 36, 39, 40, 46, 50, 51, 55, 63, 78, 90, 120, 127, 129, 136, 137, 138, 139, 140, 141, 142, 143, 144, 145, 149, 174, 181, 182, 185, 186, 191, 192, 194, 201, 203, 207, 209, 210, 212, 213, 214, 225, 229, 230, 231, 232, 233, 234, 235, 236, 237, 238, 244, 255, 261, 263, 272, 273, 274, 277, 278, 279, 280, 281, 282, 283, 290, 295, 297, 299, 329, 333
resource conservation, 55, 203, 230, 272, 274
resource efficiency, 86, 165, 166, 170, 173, 177, 178, 179, 180, 181, 322
responsible material use, 259, 263, 265
reuse, 31, 32, 33, 36, 38, 39, 40, 41, 173, 185, 186, 234, 272, 278, 279, 282, 324, 333
revalorization, 272, 273
risk, 19, 40, 56, 67, 77, 79, 81, 103, 132, 140, 187, 194, 207, 209, 221, 235, 236, 257, 260, 271, 272, 273, 302, 313, 316, 325, 331, 332, 333, 338
services, 1, 2, 3, 5, 7, 8, 10, 16, 28, 45, 50, 51, 56, 58, 103, 107, 114, 131, 136, 155, 156, 179, 241, 242, 260, 261, 268, 271, 286, 287, 292, 299, 300, 304, 312, 322, 323, 327, 332, 336, 339, 340
SETAC, 58, 79, 82, 83, 84, 126, 129, 136, 145, 146, 258, 299, 305, 312
SFA, 55, 56, 57, 70, 76, 80, 85, 86, 87, 93, 94, 111, 112

soil degradation, 218, 225
soil erosion, 90, 133
Solar energy, 38
stakeholders, 57, 61, 82, 245, 247, 269, 285, 286, 288, 290, 291, 292, 293, 294, 317, 318, 319
standards, 56, 65, 68, 78, 125, 126, 130, 183, 207, 210, 213, 235, 257, 277, 294, 316, 333
strategic environmental product planning, 239
subsidies, 14, 53, 56, 236, 333, 334, 335
substance flow analysis, 55, 85, 108, 297, 325
sustainable consumption, 302, 303, 304, 308, 309, 310, 318, 331, 335
sustainable development, 45, 102, 262, 269, 270, 286
sustainable environmental management, 217
sustainable technology, 285, 288, 289, 292, 293
system fatigue, 21, 28
tax, 2, 8, 14, 39, 53, 56, 186, 275, 278, 292, 294, 333, 334, 337, 338, 340
technological innovation, 105, 179, 240, 285
technology, 5, 35, 38, 39, 99, 107, 127, 135, 149, 155, 157, 163, 174, 183, 185, 192, 196, 213, 224, 233, 235, 241, 260, 262, 266, 285, 287, 288, 289, 290, 291, 292, 293, 294, 295, 296, 312, 316, 322, 327, 330, 331, 333, 338, 339, 340, 342
toxics, 13
trade, 2, 57, 73, 82, 91, 98, 100, 101, 102, 104, 105, 109, 132, 185, 208, 214, 225, 229, 230, 231, 232, 233, 234, 235, 236, 237, 238, 251, 265, 266, 275, 278, 291, 301, 308, 313, 317, 321, 324, 325, 328, 329, 331, 334, 338
traffic, 52, 131, 140, 141
transport, 13, 39, 40, 41, 50, 53, 59, 60, 62, 77, 85, 89, 94, 104, 140, 141, 142, 143, 144, 152, 170, 193, 196, 201, 209, 215, 218, 219, 225, 228, 231, 242, 280, 287, 290, 291, 298, 299, 300, 328, 335, 339
Unilever, 251, 253, 255, 256, 257, 258, 330, 331
waste, 4, 5, 6, 7, 8, 10, 12, 13, 14, 31, 32, 33, 34, 35, 36, 37, 38, 39, 40, 41, 43, 45, 46, 47, 50, 51, 53, 56, 57, 59, 62, 63, 65, 66, 73, 75, 77, 78, 86, 87, 88, 89, 90, 91, 92, 107, 115, 117, 119, 120, 127, 128, 131, 133, 136, 137, 138, 139, 140, 141, 142, 143, 144, 145, 154, 155, 161, 165, 170, 173, 174, 181, 182, 183, 184, 185, 189, 201, 203, 211, 214, 217, 220, 221, 225, 226, 227, 229, 230, 231, 233, 234, 235, 236, 238, 242, 243, 252, 256, 268, 273, 275, 276, 277, 278, 279, 281, 282, 283, 287, 290, 295, 297, 298, 299, 300, 302, 303, 305, 309, 316, 322, 323, 324, 327, 328, 329, 332, 333, 335, 336, 343
Waste management, 46, 238, 276, 277
water, 2, 3, 7, 8, 12, 13, 20, 22, 23, 24, 32, 47, 48, 49, 50, 51, 53, 61, 62, 65, 69, 71, 74, 86, 87, 90, 92, 105, 117, 129, 171, 196, 202, 203, 217, 218, 219, 220, 221, 222, 223, 225, 226, 227, 233, 252, 256, 264, 270, 273, 276, 278, 287, 311, 312, 322, 329, 335, 336, 342
water pollution, 75

ENVIRONMENT & POLICY

1. Dutch Committee for Long-Term Environmental Policy: *The Environment: Towards a Sustainable Future*. 1994 ISBN 0-7923-2655-5; Pb 0-7923-2656-3
2. O. Kuik, P. Peters and N. Schrijver (eds.): *Joint Implementation to Curb Climate Change*. Legal and Economic Aspects. 1994 ISBN 0-7923-2825-6
3. C.J. Jepma (ed.): *The Feasibility of Joint Implementation*. 1995 ISBN 0-7923-3426-4
4. F.J. Dietz, H.R.J. Vollebergh and J.L. de Vries (eds.): *Environment, Incentives and the Common Market*. 1995 ISBN 0-7923-3602-X
5. J.F.Th. Schoute, P.A. Finke, F.R. Veeneklaas and H.P. Wolfert (eds.): *Scenario Studies for the Rural Environment*. 1995 ISBN 0-7923-3748-4
6. R.E. Munn, J.W.M. la Rivière and N. van Lookeren Campagne: *Policy Making in an Era of Global Environmental Change*. 1996 ISBN 0-7923-3872-3
7. F. Oosterhuis, F. Rubik and G. Scholl: *Product Policy in Europe: New Environmental Perspectives*. 1996 ISBN 0-7923-4078-7
8. J. Gupta: *The Climate Change Convention and Developing Countries: From Conflict to Consensus?* 1997 ISBN 0-7923-4577-0
9. M. Rolén, H. Sjöberg and U. Svedin (eds.): *International Governance on Environmental Issues*. 1997 ISBN 0-7923-4701-3
10. M.A. Ridley: *Lowering the Cost of Emission Reduction: Joint Implementation in the Framework Convention on Climate Change*. 1998 ISBN 0-7923-4914-8
11. G.J.I. Schrama (ed.): *Drinking Water Supply and Agricultural Pollution*. Preventive Action by the Water Supply Sector in the European Union and the United States. 1998 ISBN 0-7923-5104-5
12. P. Glasbergen: *Co-operative Environmental Governance: Public-Private Agreements as a Policy Strategy*. 1998 ISBN 0-7923-5148-7; Pb 0-7923-5149-5
13. P. Vellinga, F. Berkhout and J. Gupta (eds.): *Managing a Material World*. Perspectives in Industrial Ecology. 1998 ISBN 0-7923-5153-3; Pb 0-7923-5206-8

KLUWER ACADEMIC PUBLISHERS – DORDRECHT / BOSTON / LONDON